2022 29th International Workshop on Electric Drives: Advances in Power Electronics for Electric Drives (IWED 2022)

Moscow, Russia
26 – 29 January 2022

IEEE Catalog Number: CFP22N62-POD
ISBN: 978-1-6654-6787-2

**Copyright © 2022 by the Institute of Electrical and Electronics Engineers, Inc.
All Rights Reserved**

Copyright and Reprint Permissions: Abstracting is permitted with credit to the source. Libraries are permitted to photocopy beyond the limit of U.S. copyright law for private use of patrons those articles in this volume that carry a code at the bottom of the first page, provided the per-copy fee indicated in the code is paid through Copyright Clearance Center, 222 Rosewood Drive, Danvers, MA 01923.

For other copying, reprint or republication permission, write to IEEE Copyrights Manager, IEEE Service Center, 445 Hoes Lane, Piscataway, NJ 08854. All rights reserved.

*** *This is a print representation of what appears in the IEEE Digital Library. Some format issues inherent in the e-media version may also appear in this print version.*

IEEE Catalog Number:	CFP22N62-POD
ISBN (Print-On-Demand):	978-1-6654-6787-2
ISBN (Online):	978-1-6654-6786-5
ISSN:	2767-7834

Additional Copies of This Publication Are Available From:

Curran Associates, Inc
57 Morehouse Lane
Red Hook, NY 12571 USA
Phone: (845) 758-0400
Fax: (845) 758-2633
E-mail: curran@proceedings.com
Web: www.proceedings.com

TABLE OF CONTENTS

KEYNOTE ADDRESS: GAN MHZ POWER CONVERSION: FASTER, SMALLER AND TOUGHER... 1
Miroslav Vasic

KEYNOTE ADDRESS: WBG-DEVICES ON MODERN POWER ELECTRONICS AND ELECTRIC DRIVES.. 2
Moumita Das

KEYNOTE ADDRESS: EXPLORING INVERTER TOPOLOGIES FOR MACHINE-TOOL DRIVES 3
Alecksey Anuchin

KEYNOTE ADDRESS: IMPEDANCE-SOURCE NETWORKS: FEATURES, BENEFITS AND CHALLENGES FOR INDUSTRIAL APPLICATION 4
Oleksandr Husev

KEYNOTE ADDRESS: BENEFITS OF SIC SEMICONDUCTOR DEVICES FOR TRACTION APPLICATIONS.. 5
Mikhail Chervinsky

VIRTUAL LABS: AN EFFECTIVE ENGINEERING EDUCATION TOOL FOR REMOTE LEARNING AND NOT ONLY 6
Lev Rassudov, Alina Korunets

NUMERICAL ANALYSIS OF AN ELECTRIC HIGH-SPEED RIM-DRIVEN ROTOR 10
Jhon Paul C. Roque, Robert Cameron Bolam, Yuriy Vagapov, Alecksey Anuchin

DEVELOPMENT OF DOUBLE-MOTOR ELECTRIC DRIVE FOR SCREEN WIPERS 14
Igor Polyuschenkov

INITIAL DESIGN ESTIMATION OF SYNCHRONOUS MACHINES 20
Muhammad Usman Naseer, Ants Kallaste, Bilal Asad, Toomas Vaimann, Anton Rassõlkin

THE EFFICIENCY ANALYSIS OF RESONANT CIRCUITS IN HIGH-POWER SOFT SWITCHING MODE INVERTERS 27
Daria Bondarenko, Igor Voronin, Pavel Voronin, Oleg Osipov

DETERMINING THE THERMAL CONDUCTIVITY OF ADDITIVELY MANUFACTURED METAL SPECIMENS 32
Martin Sarap, Ants Kallaste, Payam Shams Ghahfarokhi, Hans Tiismus, Toomas Vaimann

DYNAMIC STABILITY ANALYSIS OF ROBOTS ON SPHERICAL BASE WITH VARIABLE LOAD MASS 36
Margarita V. Kartashova, Nikolai A. Poliakov, Elizaveta D. Moiseenkova

OPTIMIZED-FUZZY DROOP CONTROLLER FOR LOAD FREQUENCY CONTROL OF A MICROGRID WITH WEAK GRID CONNECTION AND DISTURBANCES... 42
Haider M. Jassim, Anatoliy Ziuzev

FAULT DIAGNOSIS OF THE TOOTH BELT TRANSMISSION OF CARTESIAN ROBOT 49
Siarhei Autsou, Toomas Vaimann, Anton Rassõlkin, Bilal Asad, Karolina Kudelina, Van Khang Hyunh

EVALUATION OF 5 KW CONVERTER-FED INDUCTION MOTOR LOSSES BY ANALYTICAL CALCULATION 55

Shahab Khalghani, Lassi Aarniovuori, Juha Pyrhönen

AN EXPERIMENTAL TEST BENCH FOR STUDYING SUCKER ROD PUMP 61

Samuel Isaac Tecle, Nakataev Anton Andreevich, Anatolii Ziuzev

OPTIMIZED FIELD-WEAKENING STRATEGY FOR CONTROL OF PM SYNCHRONOUS MOTORS 67

Anton Dianov

FIELD-ORIENTED CONTROL OF THE INDUCTION MOTOR AS PART OF THE SHUNTING LOCOMOTIVE POWERTRAIN CONSIDERING CORE LOSSES AND MAGNETIC SATURATION 73

Igor Zhurov, Sergey Bayda, Stanislav Florentsev

MPPT ALGORITHMS FOR WIND TURBINES: REVIEW AND COMPARISON 79

Lukin Aleksandr, Galina Demidova, Omar Alassaf, Dmitry Lukichev, Anton Rassõlkin, Toomas Vaimann

IMPLEMENTATION AND ANALYSIS OF ROLLING BEARING FAULTS CAUSED BY SHAFT CURRENTS 85

Karolina Kudelina, Siarhei Autsou, Bilal Asad, Toomas Vaimann, Anton Rassõlkin, Ants Kallaste

CYLINDRICAL BLADES MAGNUS WIND TURBINE OPTIMIZATION AND CONTROL SYSTEM 91

Omar Alassaf, Aleksandr Lukin, Galina Demidova, Gleb Kozlov, Andrey Volkhontsev, Nikolai Poliakov

ADJUSTING THE PHASE SWITCHING-ON POSITIONS OF THE TRACTION SWITCHED RELUCTANCE MOTOR IN THE LOW SPEED RANGE 96

Alexander Krasovsky, Elena Vostorgina

THE ACCURACY INCREASING OF VOLTAGE STABILIZATION ON THE INPUT OF PWM INVERTER IN THE ELECTRIC DRIVE WITH ENERGY STORAGE BASED ON SUPERCAPACITORS 102

Polyakov Vladimir, Plotnikov Iurii

ADDITIVE MANUFACTURING OF COOLING SYSTEMS USED IN POWER ELECTRONICS. A BRIEF SURVEY 108

Loránd Szabó

EQUALIZATION OF LOSSES IN A MULTILEVEL FREQUENCY CONVERTER IN FAULT CONDITION 116

Yulia Kazemirova, Alecksey Anuchin, Dmitry Aliamkin, Maxim Lashkevich, Alexandr Zharkov, Alexey Kovyazin

THERMAL CYCLING EFFECT IN A TRACTION INVERTER FOR STAR-CONNECTED AND OPEN-END WINDING PERMANENT MAGNET SYNCHRONOUS MOTORS WITH NEARLY CONSTANT LOSSES CURRENT REGULATION 122

Yousef Ali, Egor Kulik, Alecksey Anuchin, Duy Hiep Do

COMPARATIVE ANALYSIS OF DISK TYPE ELECTRICAL MACHINES DESIGNS 128

Flur Ismagilov, Vyacheslav Vavilov, Ildus Sayakhov, Alexey Zherebtsov, Razmik Dadoyan, Albina Nurieva

Author Index

2022 29th International Workshop on Electric Drives:
Advances in Power Electronics for Electric Drives
IWED2022

Moscow Power Engineering Institute, Moscow, Russia
26th– 29th of January 2022

Proceedings

IEEE
Advancing Technology
for Humanity

*Russia
Section
est. 1990*

Technical Support:

Lev Rassudov
Phone: +7 916 329 19 21
RassudovLN@mpei.ru

2022 29th International Workshop on Electric Drives: Advances in Power Electronics for Electric Drives (IWED)

Keynote Address:
GaN MHz Power Conversion:
Faster, Smaller and Tougher

Miroslav Vasić
Universidad Politecnica de Madrid
Madrid, Spain
miroslav.vasic@upm.es

Abstract—The keynote speech is focused on the challenges related with the realization of high frequency gallium nitride (GaN) based power converters. The issues related with common mode currents, dynamic channel resistance and magnetic materials selection have been presented. All the issues have been presented through illustrative examples related with industry, electric aircraft and medicine applications. The comparison between gallium nitride and silicon transistors was presented from the point of view of efficiency and power density has been presented.

Miroslav Vasić (Senior Member, IEEE) was born in Serbia, in 1981. He received the B.E. degree from the University of Belgrade, Belgrade, Serbia, in 2005, and the M.S. and Ph.D. degrees from the Universidad Politécnica de Madrid (UPM), Madrid, Spain, in 2007 and 2010, respectively. He is currently an Associate Professor at UPM. He has advised four Ph.D. thesis. He has authored or coauthored more than 75 articles in the IEEE journals and conferences. He holds five patents. His areas of interests include DC-DC converters, power converters for R.F. applications, and converter topologies optimization. He received the SEMIKRON Innovation Award for the teamwork on "R.F. Power Amplifier with Increased Efficiency and Bandwidth," in 2012. In 2015, he received a Medal from the Spanish Royal Academy of Engineering for his research trajectory as a Young Researcher. In 2016, he received the Best Young Researcher Award from UPM. Miroslav actively serves as an Associated Editor in IEEE Journal of Emerging and Selected Topics in Power Electronics and IEEE Transactions on Vehicular Technology. Since 2021 he acts as the Vice-chair of the IEEE PELS TC 10- Design Methodologies.

Keynote Address:
WBG-Devices on Modern Power Electronics and Electric Drives

Moumita Das
Indian Institute of technology Mandi
Kamand, India
moumita@iitmandi.ac.in

Abstract—The keynote report addresses wide-bandgap semiconducting devices for Power Electronics Converters and Electric Drives applications. The comparison of silicon (Si), silicon-carbide (SiC), gallium nitride (GaN) devices based on power and switching frequency was discussed. The structural comparison of Si, SiC, and GaN devices was presented. The applications in power electronics converters, AC motor drives using energy-efficient hybrid (Si+SiC/GaN) devices were considered. The advantages and challenges of using wide-bandgap devices for high-power, high-frequency applications were discussed. The hardware results use in the projects considering transistors were given as prime examples.

Moumita Das (M'19) received Ph.D. degree in the department of electrical engineering from the Indian Institute of Technology Bombay, Mumbai, India, in 2016. After PhD, she has done postdoctoral research in the School of Electrical and Electronics Engineering, University of Manchester, Manchester, U.K, and Mid Sweden University, Sweden.

She is currently working as an assistant professor at Indian Institute of Technology Mandi, India. Her research interests include charging and power management in electric vehicles, converter topologies and their modeling, design, and control with the focus on renewable energy applications using wide bandgap devices.

Keynote Address:
Exploring Inverter Topologies for Machine-tool Drives

Alecksey Anuchin
Moscow Power Engineering Institute
Moscow, Russia
anuchin.alecksey@gmail.com

Abstract—The talk presents two projects, which were carried out for machine-tool industry. The first one is the spindle drive based on a 3-level inverter topology capable of 30000 rpm with the rated power of 40 kW. The inverter unit has premium efficiency even at 16 kHz pulse-width modulation frequency. Coupled with a line filter it can be used as an active front-end as well. The performance of the proposed solution is compared with that of conventional 2-level power converters and pros and cons of both implementations are considered. The second project considers GaN switches implementation in a servo amplifier. The problems of high dv/dt, high current measurement delay and reliability issues were investigated. It was shown that novel inverter topologies and solutions can bring strong benefits to the market of machine-tool drives.

Alecksey Anuchin received the B.S., M.S and Ph.D. degree from Moscow Power Engineering Institute, Moscow, Russia, in 1999, 2001 and 2004 respectively. He is currently Head of Electric Drive Department of Moscow Power Engineering Institute. His topics of interest include control systems for power converters and electric drives, design of instructional laboratories workstations and industrial networks. Address: Krasnokazarmennaya 14, Moscow, 111250, Russian Federation.

Keynote Address:
Impedance-Source Networks: Features, Benefits and Challenges for Industrial Application

Oleksandr Husev
Tallinn University of Technology
Tallinn, Estonia
oleksandr.husev@taltech.ee

Abstract—This talk was devoted to the impedance source converters. General operation principles of the impedance-source networks along with their features and benefits are demonstrated. Brief overview of varieties of impedance-source networks and its derivation will be given. Finally, examples of pre-industrial prototypes considering multi-objective optimization will be discussed. Feasibility in the commercial application will be discussed in conclusions.

 **Oleksandr Husev** (S'10–M'12-SM'19) received the B.Sc. and M.Sc. degrees in industrial electronics from Chernihiv State Technological University, Chernihiv, Ukraine, in 2007 and 2008 respectively. He defended PhD thesis in the Institute of Electrodynamics of the National Academy of Science of Ukraine in 2012.

He is senior researcher and project leader of the Department of Electrical Power Engineering and Mechatronics, TalTech University He has over 100 publications and is the holder of several patents.

His research interests are in Power Electronics systems. Design of novel topologies, control systems based on a wide range of algorithms, including modeling, design, and simulation. Applied design of power converters and control systems and application, stability investigation.

Keynote Address:
Benefits of SiC Semiconductor Devices for Traction Applications

Mikhail Chervinsky
Wolfspeed
Moscow, Russia
mikhail.chervinsky@wolfspeed.com

Abstract—**The keynote speech is covering trends and role of Silicon Carbide (SiC) in modern power electronics. Highlighted value of technology to global environment. Provided updated summary about SiC properties and advantages at component and system levels. Presented examples of modern SiC commercial products and particular design cases. Proposed approach to compare different power semiconductor components for real designs (SiC/Si/GaN). Summarized general reliability aspects which should be considered for fair comparisons.**

Mikhail Chervinsky was born in Russia, in 1984. He received the M.S. degree from the Peter the Great St.Petersburg Polytechnic University (spbstu.ru), in 2007. He is currently an Field Application Engineer in Wolfspeed company. His areas of interests include WBG semiconductors, power converters and R.F. applications.

Virtual Labs: an Effective Engineering Education Tool for Remote Learning and not only

Lev Rassudov, Alina Korunets
Department of Electric Drives
National Research University "MPEI"
Moscow, Russia
rassyd@mail.ru

Abstract—operating real hardware is a fundamental engineering skill. For this reason, hands-on classes involving real equipment is an essential tool in engineering education. At the same time in many cases there are certain limitations related to in-person training of students: hardware availability, hardly modifiable functionality for each student and educational goal, the increased demand on remote learning and safety aspects. Modern information technologies enable various means for partially solving the mentioned issues by providing remote access, VR-technologies, digital twins, etc. The paper provides a brief overview of aspects related to introducing virtual labs into educational process, with the main focus on the requirement for the virtual hardware, such as an electric power drive system, should replicate processes with time resolution in order of several milliseconds. And the interaction of the equipment with the student should be done in a way there is a feeling of a real object, necessitating low response time in order of hundreds of milliseconds. This requires high priority of computational power allocation for mathematical model calculation as well as high communication channel bandwidth between the computation power source and the user interface. The paper presents an example of such a server deployed virtual laboratory. The functionality was inspired by that of real equipment installed at Energy saving electric power drive laboratory at the Department of Electric Drives of Moscow Power Engineering Institute. Apart from substantively replicating the functionality and feel of high-dynamic hardware real-time operation, it enables to provide each student with an individual virtual equipment set to be explored, tuned or diagnosed. Virtual labs of the kind proved to be an effective engineering education tool for remote learning, giving new opportunities in improving the in-person training process as well.

Keywords— engineering education, industry applications, distance learning, robotic systems, fault diagnosis, simulation, cloud services, digital twin.

I. INTRODUCTION

Operating real hardware is a fundamental engineering skill. For this reason, hands-on classes with real equipment during labs is an essential tool in engineering education. At the same time in many cases there are certain limitations related to in-person training of students: hardware availability, fixed functionality for each student and educational goal, the increased demand on remote learning and safety aspects. Modern information technologies enable various means for partially solving the mentioned issues by providing remote access, VR-technologies, digital twins, etc. The paper provides a brief overview of the aspects related to introducing virtual labs into educational process. The 2nd section of the paper considers the availability of hands-on courses for students, the 3rd discusses the benefits and drawbacks of remote and virtual laboratories compared to the in-person ones in terms of functionality and safety. The studying motivation aspects are considered for both virtual and real hardware in section 4. Section 5 focuses on the requirements for the virtual hardware to substantively replicate the functionality and feel of high-dynamic hardware real-time operation. The design results of a server deployed Electric drives virtual laboratory are presented. Finally, section 6 concludes the paper.

II. ACCESS TO HANDS-ON CLASSES

High complexity and cost of laboratory equipment limits the educational institutions in funding the corresponding hands-on courses. This encouraged even the leading universities if the World and some companies to cooperate within such frameworks as European networked experiments project Lila [1], iLabs by MIT [2,3], VISIR [4], "Remote-Labore in Deutschland" [5] and others for coordinating both real remote access laboratories design as well as virtual laboratories development. This way making the setups available for the students across the participating universities.

With the increased impact of information technologies on the entire humanity more attention is being paid and more funding is being provided for virtualization of not only labs, but for significant sections of curricula [6]. The demand on such remote learning technologies is constantly growing [7]. It dramatically increased with the COVID-19 pandemic [8,9] challenging academia to rapidly revolutionize the entire educational process to be continued within the shifted paradigms [10] and making virtual classes unavoidable.

The main approaches for virtualizing labs can be subdivided into two categories [11]:

- remote control laboratories;
- virtual laboratories.

Remote hardware operation in this context involves a real physical object to be operated by a human remotely via some interface. This can often include video feedback, means for control action transfer. But other technologies are becoming a trend now as well: augmented reality technologies, avatars utilization, etc. The significant drawback is that there should probably be some personnel on site, who will observe and service the installations preventing any misbehavior caused

---

The investigation was carried out within the framework of the project "Deploying Digital Twins of Robotic Systems with Cloud Technologies for Diagnostic Purposes" with the support of a grant from NRU "MPEI" for implementation of scientific research programs "Energy", "Electronics, Radio Engineering and IT", and "Industry 4.0, Technologies for Industry and Robotics in 2020-2022".

978-1-6654-6787-2/22 $31.00 © 2022 IEEE

by some interactions of the students with the controlled object. The experiments are available only when the staff is looking after the setups. The hardly scalable and modifiable set of equipment is an additional limitation in terms of access to the equipment. The remote format may in some cases reduce the feel of the students for their personal responsibility of what is happening at the site. For this reason, the available functionality should in many cases be reduced for remote control.

The other solution is a completely virtual laboratory [12,13]. It does not suffer many of the mentioned drawbacks. Server deployed labs as well as stand-alone applications can be accessed and operated by the students at any time required. There is actually no need to invest into real hardware and no corresponding maintenance costs or space is required. The solution is easier to be scaled, requiring only the corresponding IT infrastructure.

III. Real and Virtual Hardware Functionality

Implementing remote control labs is well substantiated in case of a very complicated-to-simulate hardware setup is being studied or if the hardware is being constantly updated – including firmware. This is not often the case for most of equipment used for educational purposes. The other case for implementing this technique is the ease to provide remote access to the setup via a remote desktop for example: IT and academic stuff can solve their tasks almost independently. Another benefit is that the same on-site equipment can be used for both in-person and remote education goals. It's evident that the functionality of the remote access laboratory cannot be wider than that of this laboratory on site. Moreover, some functionality can be excluded for safety or for rear necessity reasons.

Designing completely virtual labs in turn requires very deep competences both in the simulation of the equipment behavior and IT, which seems to be more challenging to organize. However, operating virtual hardware eliminates any constraints related to possible safety aspects: less protection should be foreseen and more freedom for experiments can be provided for the students. Virtual labs are not strictly bounded to real equipment setups, so the visualization can be implemented in a most convenient and transparent manner. The laboratory setup can be foreseen modifiable via a corresponding parameter set for example. Virtual labs enable to extend the functionality in comparison with real hardware and give more freedom for the tutors and students in achieving the learning objectives. This make them also an effective tool assisting the in-person training process.

IV. Studying Motivation Aspects

The motivation aspects of the students in learning to operate hardware is comprised of many factors. In the context of virtual laboratories some can be outlined:

- general access to an interactive environment (remote, virtual, on-site) for conducting experiments;
- team work;
- deeper understanding of the experiments conducted at an on-site lab;
- individual tasks.

Labs should be accessible for students independently of their studying format. For safety reasons and due to lack of hardware onsite the students traditionally operate equipment at the labs in groups. This enables to improve their team work skills, share the responsibility between each other. But each individual often requires time to get familiar with the equipment and experiments and might be interested in playing with the equipment on his own. A virtual laboratory might help him to prepare to work effectively at the lab. Providing the students with individual set of equipment stimulates them to work on their own without copying the results from each other, and may enable to satisfy their particular educational goals. It can be successfully used for self-learning as well [14]. In this terms a virtual laboratory is practically the only solution possible, as an educational institution would not handle a vast variety equipment sets.

V. Electric Drives Real-Time Virtual Laboratory

From the aforesaid it can be concluded that in many cases in-person labs assisted by virtual labs is the optimal solution for developing the fundamental engineering skills of students [15]. The requirements for the virtual lab are determined by the studying objectives and the object being virtualized [16,17, 18]. As for an electric power drive system, it should replicate processes with time resolution in order of several milliseconds. The interaction of the equipment with the student should be done in a way there is a feeling of a real object, necessitating low response time in order of hundreds of milliseconds. This requires high priority of computational power allocation for mathematical model calculation as well as high communication channel bandwidth between the computation power source and the user interface.

The paper presents an example of such a server deployed virtual laboratory. The functionality was inspired by that of real equipment installed at Energy saving electric power drive laboratory at the Department of Electric Drives of Moscow Power Engineering Institute (Figure 1). The limitation in computing power define the main focus on supporting more connections to the server, rather than implementing a 3D visualization, for example. The computing power is required to calculate the mathematical models in real time enabling the students to interact the virtual hardware almost as the real one.

The first release was a stand-alone MATLAB-based application to be installed on a PC (figure 2,a). The considered case is that someone has assembled an installation at an enterprise and the specification is missing. The task is to safely power on the installation and perform parameter determination of the equipment. After that is done, the students are requested to find in the list provided the part numbers of the hardware components installed in the setup, check if the components were selected reasonably and tune the setup. It's worth mentioning that a straight forward activation of the stand in most variants of tasks leads to failure of the simulated equipment. In real life this could have resulted in the need to repair the equipment, whereas in case of a virtual laboratory the experiment can be conducted over again. In autumn 2021 32 students of Electric machines and Electric drives training programs were using the software for performing individual tasks laboratory practicum. Each of the

(a)

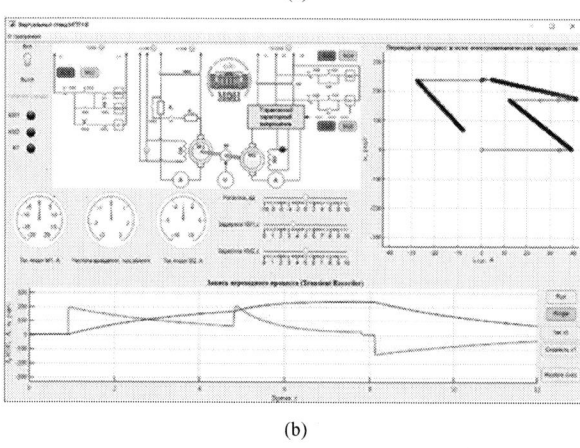

(b)

Figure 1. Energy saving electric power drive laboratory at the Department of Electric Drives of Moscow Power Engineering Institute. (a)-real setup; (b)-virtual lab, a stand-alone application.

students "burned" at least one set of virtual equipment. At the onsite labs, where protection is foreseen and many the students do not promptly understand, the protection is already functioning and affecting their results.

The stand-alone application proved to be an effective tool for education purposes. However certain technical issues arose. A locally deployed software is not convenient to be supported. Moreover, it uses local computing power of the students' PCs. 3 of the 32 students complained the computing power of their laptops was insufficient for the application to run smoothly and received a "lighter" version of the software with the reduced sampling rate of the mathematical models. Additionally, 2 students were operating remotely university PCs with the installed software package.

(a)

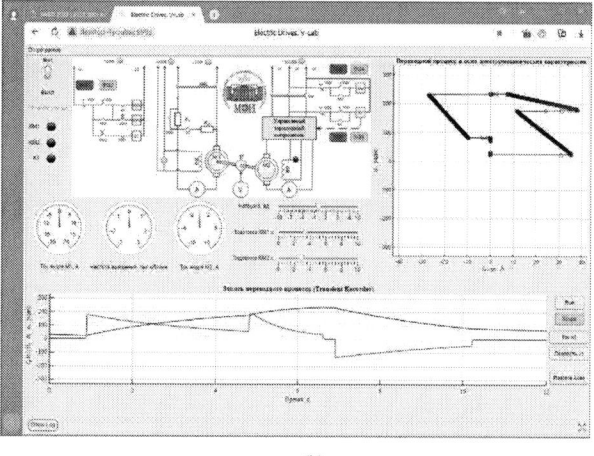

(b)

Figure 2. Virtual Energy saving electric power drive laboratory:
(a) – MATLAB Web app SERVER Interface;
(b) – Virtual lab, web application interface.

The aforementioned drawbacks forced the authors to move the virtual laboratory to a web application, making it a cross-platform solution utilizing the computing power of a server. For this MATLAB Web App Server was used (figure 2,b). The interface in web browser is presented in figure 2,c. The application functionality differs from that of a stand-alone application only in a few details. The computing power required is estimated as 1 processor core and 4GB RAM for 2 connections (students). Currently the authors are working on moving the local server deployed virtual laboratory to Moscow Power Engineering Institute IT infrastructure for it to be accessible in internet. and on the further extension of functionality including other laboratory setups virtualization and data analysis and visualization extension.

VI. CONCLUSION

The paper starts with a brief overview on labs virtualization aspects and their implementation in educational process. An example of such a server deployed virtual laboratory is presented. The functionality was inspired by that of the real equipment installed at Energy saving electric power drive laboratory at the Department of Electric Drives of Moscow Power Engineering Institute. Apart from substantively replicating the functionality and feel of high-dynamic hardware real-time operation, it enables to provide each student with an individual virtual equipment set to be explored, tuned or diagnosed. Virtual labs of the kind proved to be an effective engineering education tool for remote learning, giving new opportunities in improving the in-person training process as well.

REFERENCES

[1] T. Richter, D. Boehringer and S.Jeschke, (2011), "LiLa: A European Project on Networked Experiments". In S. Jeschke, I. Isenhardt & K. Henning, Automation, Communication and Cybernetics in Science and Engineering 2009/2010. New York. Springer International Publishing.

[2] V. J. Harward et al., "The iLab Shared Architecture: A Web Services Infrastructure to Build Communities of Internet Accessible Laboratories," in Proceedings of the IEEE, vol. 96, no. 6, pp. 931-950, June 2008, doi: 10.1109/JPROC.2008.921607.

[3] M. A. Zaman, L. T. Neustock and L. Hesselink, "iLabs as an online laboratory platform: A case study at Stanford University during the COVID-19 Pandemic," 2021 IEEE Global Engineering Education Conference (EDUCON), 2021, pp. 1615-1623, doi: 10.1109/EDUCON46332.2021.9454028.

[4] I. Gustavsson, J. Zackrisson,L. Håkansson,L. Claesson, T. Lagö, "The VISIR project — an Open Source Software Initiative for Distributed Online Laboratories", Proc. of Annual Int. Conf. on Remote Engineering and Virtual Instrumentation, 2007.

[5] T.Ortelt, and C Terkowsky. "Community Working Group Remote-Labore in Deutschland Projekte, Gemeinsamkeiten, Unterschiede.", wbv, 2020. doi: 10.3278/6004804w229.

[6] L. Montuori, M. Alcazar-Ortega, C Vargas-Salgado and P. Bastida Molina, "Methodology for the implementation of e-learning class during the COVID-19". INNODOCT 2020, Valencia, Spain, 2020. doi: 10.4995/INN2020.2020.11877.

[7] J. Bezerra "Feedback Of Engineering Students about Remote Classes During Covid-19 Pandemic", ICERI2021, 2021, Proc., pp. 754-758.

[8] S. Jackowicz and I.Sahin. "Online Education during the COVID-19 Pandemic: Issues, Benefits, Challenges, and Strategies." ISTES Organization, 2021

[9] E. Argüello. "The Impact of ''Going Virtual'' on Engineering Education During the COVID-19 Pandemic: A Student-Centered Study in Colombia". November 2021 International Journal of Engineering Education 37(6):1511–1517.

[10] L. Rassudov and A. Korunets, "COVID-19 Pandemic Challenges for Engineering Education," 2020 XI International Conference on Electrical Power Drive Systems (ICEPDS), 2020, pp. 1-3, doi: 10.1109/ICEPDS47235.2020.9249285.

[11] M.E. Auer. "Virtual lab versus remote lab". 20th World Conference on Open Learning and Distance Education, Dusseldorf, 2021

[12] K. M. Moudgalya and I. Arora, "A virtual laboratory for distance education". 2010 International Conference on Technology for Education, 2010. doi:10.1109/t4e.2010.5550036

[13] Diwakar, S. Poojary and S. B. Noronha, "Virtual labs in engineering education: Implementation using free and open source resources", 2012 IEEE International Conference on Technology Enhanced Education (ICTEE), 2012, pp. 1-4, doi: 10.1109/ICTEE.2012.6208670.

[14] N. Mikhaylov and D. Chernov. "From Virtual Lab to Virtual Development Lab". IFAC Proceedings Volumes, 45(11), 177–182, 2012. doi:10.3182/20120619-3-ru-2024.00018

[15] M. Henker and K. Kelber. "Virtualizing Electrical Engineering Teaching Labs", University of Applied Sciences (HTW) Dresden. 2021. URL: https://www.mathworks.com/company/newsletters/articles/virtualizing-electrical-engineering-teaching-labs.html.

[16] L. Rassudov, E. Akmurzin, A. Korunets and D. Osipov, "Engineering Education and Cloud-Based Digital Twins for Electric Power Drive System Diagnostics," 2021 28th International Workshop on Electric Drives: Improving Reliability of Electric Drives (IWED), 2021, pp. 1-3, doi: 10.1109/IWED52055.2021.9376395.

[17] G. Baluta, V. Horga, and C. Lazar. "Implementation of a Virtual Laboratory for Low Power Electrical Drives". 2008 13th International Power Electronics and Motion Control Conference. doi:10.1109/epepemc.2008.4635566

[18] V. Vodovozov, Z. Raud and L. Gevorkov, "Experiences with remote labs in electrical drive," 2014 55th International Scientific Conference on Power and Electrical Engineering of Riga Technical University (RTUCON), 2014, pp. 88-93, doi: 10.1109/RTUCON.2014.6998192.

2022 29th International Workshop on Electric Drives: Advances in Power Electronics for Electric Drives (IWED), Moscow, Russia. Jan 26 – 29, 2022

Numerical Analysis of an Electric High-Speed Rim-Driven Rotor

Jhon Paul C. Roque
Glyndwr University
Wrexham, UK

Yuriy Vagapov
Glyndwr University
Wrexham, UK

Robert Cameron Bolam
Glyndwr University
Wrexham, UK

Alecksey Anuchin
Moscow Power Engineering Institute
Moscow, Russia

Abstract—**Finite element analysis was conducted to study the stress behaviour in a generic electric high-revolution rim-driven rotor in relation to the number of spokes and rotational speed using the maximum distortion energy failure criterion. The paper analysed the stresses on an electric high-revolution rotor and locate the failure's potential origins. Upon analysing the FEA results, it is found that the maximum equivalent von-Mises stress is related to the square of revolution in RPM. It is concluded that it is possible to create a general formula that can predict the stresses, simplifying the development stage of an electric rim-driven rotor for high-speed applications.**

Keywords—*electric propulsion, high-speed electric motor, FEA, rim-driven rotor*

I. INTRODUCTION

Many forward-thinking organisations are directing their aviation research and development strategies towards providing the most efficient propulsion systems that lower environmental impact [1],[2]. This quest has led to an increase in attention being given to electrically powered propulsion for aircraft [3]. Conventional hub-driven propeller and fan devices have already been installed and flown on numerous test aircraft [4] and, alongside novel fan arrangements such as rim driven fan technology [5] and distributed thrust systems, offer potential solutions that could contribute towards achieving the ultimate goal of zero-emission propulsion. Electrical rim drives have been successfully implemented in the marine industry where they directly drive propellor thrusters at relatively low rotational speeds of a few hundreds of revolutions per minute.

Various numerical studies have been conducted in regard of rim-driven thrusters, propellers, and fans. For example, an open water CFD investigation on hub-type and hubless rim-driven thrusters (RDT) performance with differing hub diameter using CFX 14.0 is reported in [6]; Dubas *et al.* [7] provided the development of a CFD method and investigation of rotor-stator interaction using OpenFOAM; a FEM simulation for the proposed implementation of axial flux motor (AFM) in RDT for a marine application is discussed by Ojaghlu *et al.* [8]; a performance optimisation of RDT using RANS calculation for maximum propulsive efficiency and minimum cavitation is suggested by Gaggero [9]. These papers are essential to demonstrate the potentials and applications of rim-driven rotors. However, these numerical models are conducted for low-speed applications and are focused on the computation fluid dynamics of the fluid behaviour through and around the device. Rim driven rotors (RDR) for aerospace propulsion applications require operations at much higher rotational speeds, in the many thousands of revolutions per minute [10]-[12]. Such rotors can be incorporated within the architectures of induction, switched reluctance or synchronous AC motors. Fig. 1 shows a schematic representation of such a device, while Fig. 2 shows a commercial rim-driven thruster.

This paper aims to analyse the stresses on an electric high-speed RDR using the Static Structural program in Ansys Workbench and understand the nature of stresses and potential origins of structural failures.

II. METHODOLOGY

An accurate definition of failure is essential to approach the potential structural problem. In engineering, a material, structure or component is considered to have failed once it

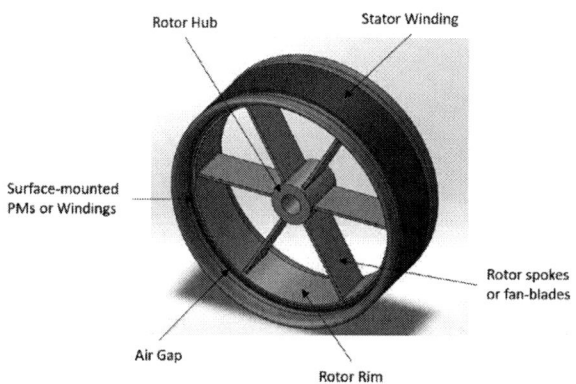

Fig. 1. Electric high-speed rim-driven device schematics.

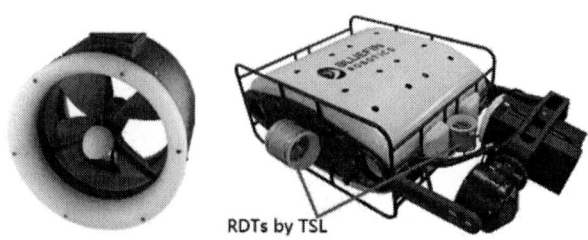

Fig. 2. Commercial Rim Driven Thruster by TSL Technology Ltd.; used in BLUEFIN HAUV [13].

This work is funded by the Welsh Government (WEFO) under the SMARTExpertise initiative (Project Reference 82321) and is supported by the European Regional Development Fund.

978-1-6654-6787-2/22 $31.00 © 2022 IEEE

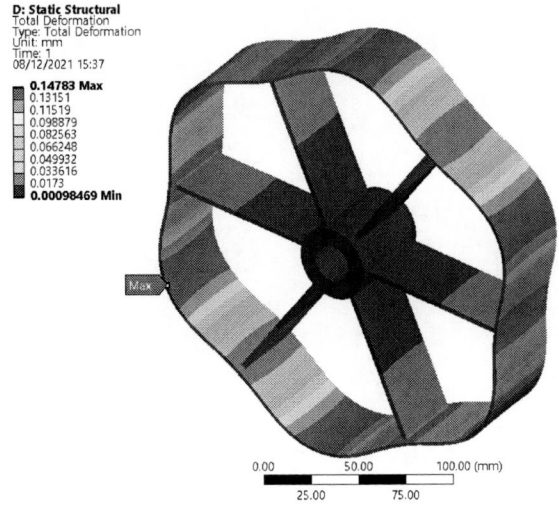

Fig. 3. Total deformation of a six-spoked rim drive at rotational speed of 14,000 RPM. (Exaggerated scale for visuals)

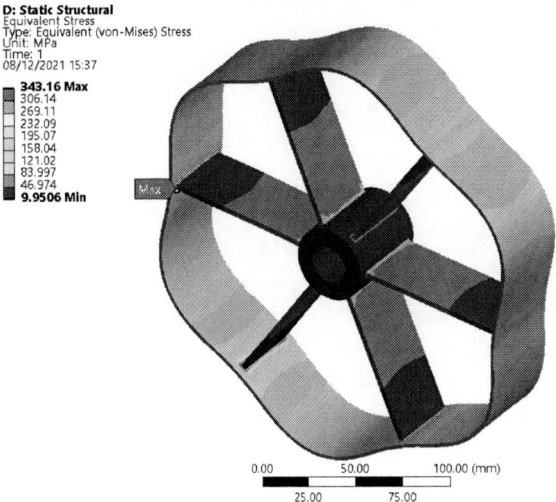

Fig. 4. Maximum equivalent von-Mises stresses of a six-spoked rim drive at rotational speed of 14,000 RPM. (Exaggerated scale for visuals)

$$\sigma_e = \frac{1}{\sqrt{2}}\left[\left(\sigma_{xx}-\sigma_{yy}\right)^2 + \left(\sigma_{xx}-\sigma_{zz}\right)^2 + \left(\sigma_{yy}-\sigma_{zz}\right)^2 - 3\left(\tau_{xy}^2 + \tau_{xz}^2 + \tau_{yz}^2\right)\right]^{0.5} \qquad (1)$$

can no longer serve its intended design functions [14]. However, in this paper, failure is associated with the physical fracture or yielding (also known as distortion and plastic strain) of materials subject to stress [15].

The type of failure depends highly on the material's physical properties, such as ductility and brittleness. Ductile materials are commonly considered to have failed as the maximum stress on the structure surpasses the ultimate yield strength; this is when the material reaches the plastic region. On the other hand, since brittle materials have no appreciable yield point, the failure on brittle materials occurs at fracture, known as "brittle fracture" [16]. This research focuses on the failure of ductile materials for electric high-revolution RDRs; thus, it is vital to use an appropriate failure theory consistent with the observation for ductile materials subject to stress.

Three-dimensional structures are subject to triaxial stress, which is composed of hydrostatic and deviatoric stresses; however, hydrostatic stresses cannot cause yielding of ductile materials; thus, a suitable failure theory should be independent of hydrostatic stress. Tresca and von-Mises failure criteria are commonly used and are proven to be consistent with the behaviour of ductile materials [15]-[17].

Due to the relatively wide elastic region of structural steel in the Ansys Material Library and the expected presence of tensile loading at the rim, hub and the length of the spokes, the von-Mises failure criterion is a better-suited failure criterion for this analysis. Furthermore, the von-Mises failure criterion is also suited for estimating within the plastic region, which is expected to occur within the simulations. Therefore, the von-Mises failure criterion is this paper's failure criterion of choice [18],[19].

The von-Mises failure criterion utilises the shear deformation as the yield mechanism of structures subject to forces; however, it does so by the principle of critical distortion energy. The critical distortion energy is the distortion energy required to cause the onset of plastic deformation. Therefore, the von-Mises failure criterion

states that yielding occurs when the effective stress equals the uniaxial yield strength. The uniaxial yield strength can be found experimentally, while the effective stress is given as follows in (1) according [18],[19], where σ_e is the effective stress, σ is the normal stress and τ is the shear stress. The first letter of the subscript represents the plane of the stress tensor while the second subscript is the direction relative to the coordinate system.

The model used in this FEA (Fig. 3) is a generic three-dimensional spoked RDR with rim thickness and rim outer diameter of 2 mm and 204 mm, respectively, and a shafted hub of thickness and outer diameter of 10 mm and 40 mm, respectively. The rim and the hub are 50 mm long, with the spokes 5mm indented from both ends of the rim and hub. The spokes have 3 mm depth, 40 mm breadth, and 80 mm high. The simulation consists of an RDR with, ns, number of spokes at variable revolution, ranging from 2 to 16 spokes and 1,000 to 14,000 RPM. The authors tracked the mass of the RDR and maximum equivalent von-Mises stress using the parameter window on Ansys Workbench.

The adaptive convergence criteria and adaptive mesh refinement controls were used to ensure the high accuracy of the simulation [20]. The adaptive convergence criteria used in this simulation tracks the maximum equivalent von-Mises stress relative error, E, given by (2), with the relative error criteria less than 2% before converge solution.

$$E > 100\% \times \left(\frac{\phi_{i+1}-\phi_i}{\phi_i}\right) \qquad (2)$$

where E is the relative error, ϕ_i is the previous tracked value, and ϕ_{i+1} is the new tracked value. In this case, the value being tracked is the maximum equivalent von-Mises stress.

The refinement depth of each refinement loop is set to two, with ten maximum refinement loops per case. This approach allowed the author to automatically refine the mesh elements around high-stress regions in the model

978-1-6654-6787-2/22 $31.00 © 2022 IEEE

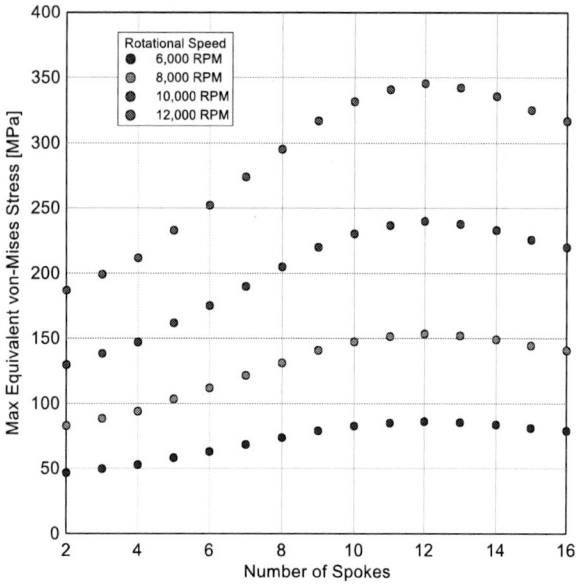

Fig. 5. Maximum equivalent von-Mises stress of rim drives with increasing number of spokes at different speeds.

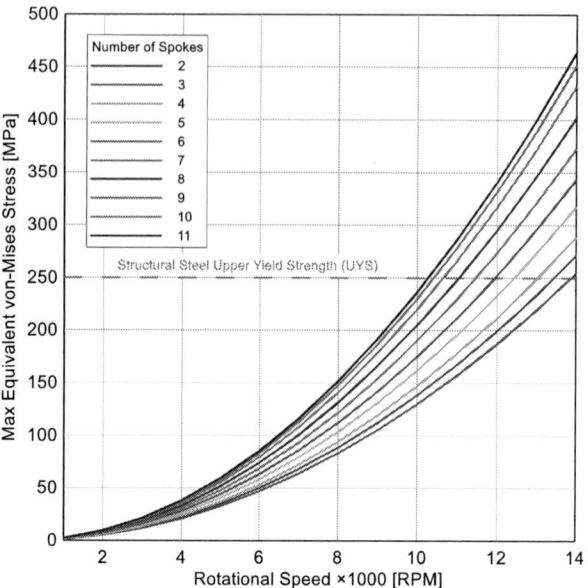

Fig. 6. Maximum equivalent von-Mises stress vs. rotational speed for various number of spokes (2-11).

while using the parameter window in Ansys Workbench for faster data gathering without compromising the simulation accuracy. The initial FEA of the model in Fig. 4 presented a logarithmic stress singularity at the sharp re-entrant corners of the rim-spoke, and hub-spoke intersect; hence, the author added blending of 1mm radius at these re-entrant corners, irradicating the stress singularities [21].

The gravity is off for the cases because the force due to gravity is negligible relative to the centrifugal force generated at high revolution. Nonetheless, the mass of each RDR for increasing numbers of spokes is shown in Table I because the mass of a rotating body is an integral factor of angular momentum that constitutes the centrifugal force [17].

III. RESULTS AND DISCUSSION

The simulation of the total deformation of a six-spoked RDR at 14,000 RPM is shown in Fig. 3. It shows that the spokeless regions of the rim are radially deforming due to rotation, which is a consistent behaviour of a rotating rim. In contrast, the length of the spokes shows low stresses in

TABLE I.   MASS OF THE RIM DRIVES
WITH INCREASING NUMBER OF SPOKES .

Number of Spokes	Mass [kg]
2	1.019
3	1.095
4	1.171
5	1.246
6	1.322
7	1.398
8	1.473
9	1.549
10	1.625
11	1.700
12	1.776

Fig. 4, suggesting that they are resisting their centrifugal load. Furthermore, Fig. 4 shows that high-stress concentrations are present at the rim-spoke intersects, which are the locations of stress singularities mentioned in section Methodology. These stress concentrations mark the failure's potential origin; therefore, the blending of these locations is essential [22].

Fig. 5 demonstrates the FEA result of the cases with increasing spoke number for different revolutions. The figure shows that a form of "disk-effect" occurs from 12 to 16 spokes. The behaviour was observed for the revolution ranging from 1,000 to 14,000 RPM; however, the behaviour is less appreciable at lower revolutions; hence, the authors omitted those results for clarity. Since the disk-effect is present in all cases with increasing spoke numbers, it is deduced that the behaviour is related to the model topology.

The disk-effect occurs when the number of spokes is sufficient to provide enough structural rigidity and causes the whole structure to behave like a solid disk, hence, the maximum equivalent von-Mises stress is reduced irrespective of the speed of revolution. However, as much as it is an interesting phenomenon, the investigation of the disk-effect is currently not within the paper's scope; hence, the authors have opted to continue with the simulations results of cases ranging from 2 to 11 number of spokes.

The maximum equivalent von-Mises stress of RDR with increasing revolution for different numbers of spokes is shown in Fig. 6. The figure shows that the maximum equivalent von-Mises is proportional to the squared of the revolution, which is a well-expected behaviour as the centrifugal force of a rotating body increases with the squared of the revolution, as shown in (3).

$$F_c = mr\omega^2 \qquad (3)$$

where F_c is the centrifugal force, m is the mass of the rotating body, r is the distance of the mass from the centre of rotation, and ω is the rotational speed in radian per second.

Furthermore, the graphs' gradient increases with a higher number of spokes, indicating that RDR with higher spoke numbers has a narrower operating range; a horizontal line extending across the graph representing the ultimate yield strength of structural steel was added for visualisation.

These findings suggest that it is possible to create a generalised formula that can predict the stresses on electric high-revolution rim-driven rotors based on the geometry, mass, and speed of revolution. However, the development of a general formula requires further work to study different permutations with the most logical approach. This is forming an approach that can potentially predict the maximum equivalent Von-Mises stresses with the RPM; however, this is still in the early stage and requires more investigation.

IV. CONCLUSION

Upon conducting this study, a method of analysing the stress behaviour of an electric high-speed RDR with increasing speed and number of spokes was developed. The initial assessment of the FEA found that the rim and the hub are deforming radially due to their centrifugal force, while the spokes resist this force. It is also found that the maximum equivalent von-Mises stress is located at the rim-spoke intersect, indicating the possible location of failure origin. Furthermore, it is found that the maximum equivalent von-Mises stress is proportional to the square of revolution, which is consistent with the nature of centrifugal force. However, further assessment of this shows that the number of spokes affects the graph's gradient, indicating that a higher number of spokes narrows the operating range of the RDR. Ultimately, because of the relationship of the maximum equivalent von-Mises stress to the revolution, it is concluded that it is possible to create a general formula that can predict the maximum von-Mises stress from the geometry, mass, and speed of revolution of an electric high-speed rim-driven device.

REFERENCES

[1] W.R. Graham, C.A. Hall, and M.V. Morales, "The potential of future aircraft technology for noise and pollutant emissions reduction," *Transport Policy*, vol. 34, pp. 36-51, July 2014, doi: 10.1016/j.tranpol.2014.02.017

[2] V. Viswanathan, and B.M. Knapp, "Potential for electric aircraft," *Nature Sustainability*, vol. 2, pp. 88-89, Feb. 2019, doi: 10.1038/s41893-019-0233-2

[3] R.C. Bolam, Y. Vagapov, and A. Anuchin, "Review of electrically powered propulsion for aircraft," in *Proc. 53rd Int. Universities Power Engineering Conference UPEC-2018*, Glasgow, UK, 4-7 Sept. 2018, pp. 1-6, doi: 10.1109/UPEC.2018.8541945

[4] R.C. Bolam, Y. Vagapov, and A. Anuchin, "A review of electrical motor topologies for aircraft propulsion," in *Proc. 55th Int. Universities Power Engineering Conference UPEC-2020*, Torino, Italy, 1-4 Sept. 2020, pp. 1-6, doi: 10.1109/UPEC49904.2020.9209783

[5] R.C. Bolam, and Y. Vagapov, "Implementation of electrical rim driven fan technology to small unmanned aircraft," in *Proc. 7th Int. Conf. on Internet Technologies and Applications ITA-17*, Wrexham, UK, 12-15 Sept. 2017, pp. 35-40, doi: 10.1109/ITECHA.2017.8101907

[6] B. Song, Y. Wang, and W. Tian, "Open water performance comparison between hub-type and hubless rim driven thrusters based on CFD method," *Ocean Engineering*, vol. 103, pp. 55-63, July 2015, doi: 10.1016/j.oceaneng.2015.04.074

[7] A.J. Dubas, N.W. Bressloff, and S.M. Sharkh, "Numerical modelling of rotor–stator interaction in rim driven thrusters," *Ocean Engineering*, vol. 106, pp. 281-288, Sept. 2015, doi: 10.1016/j.oceaneng.2015.07.012

[8] P. Ojaghlu, and A. Vahedi, "Specification and design of ring winding axial flux motor for rim-driven thruster of ship electric propulsion," *IEEE Transactions on Vehicular Technology*, vol. 68, no. 2, pp. 1318-1326, Feb. 2019, doi: 10.1109/TVT.2018.2888841

[9] S. Gaggero, "Numerical design of a RIM-driven thruster using a RANS-based optimization approach," *Applied Ocean Research*, vol. 94, pp. 1-29, Jan. 2020, doi: 10.1016/j.apor.2019.101941

[10] B. Cheng, G. Pan, and Y. Cao, "Analytical design of the integrated motor used in a hubless rim-driven propulsor," *IET Electric Power Applications*, vol. 13, no. 9, pp. 1255-1262, Sept. 2019, doi: 10.1049/iet-epa.2018.5303

[11] O. Goushcha, M. Kogler, and H. Raza, "Aerodynamic and acoustic performance of a rim driven thruster," in *Proc. AIAA Aviation 2020 Forum*, 15-19 June 2020, pp.1-11, doi: 10.2514/6.2020-2756

[12] R.C. Bolam, Y. Vagapov, R.J. Day, and A. Anuchin, "Aerodynamic analysis and design of a rim driven fan for fast flight," *Journal of Propulsion and Power*, vol. 37, no. 2, pp. 179-191, March 2021, doi: 10.2514/1.B37736

[13] X. Yan, X. Liang, W. Ouyang, Z. Liu, B. Liu, and J. Lan, "A review of progress and applications of ship shaft-less rim-driven thrusters," *Ocean Engineering*, vol. 144, pp. 142-156, Nov. 2017, doi: 10.1016/j.oceaneng.2017.08.045

[14] J. Vernon, "Failure, fatigue and creep," in *Introduction to Engineering Materials*. London: Macmillan Publishers Limited, 1992, pp. 409-418.

[15] R. Juvinall, *Stress, Strain and Strength*. New York: McGraw-Hill, 1967.

[16] R. Hibbeler, *Mechanics of Materials*, 10th Edn. Upper Saddle River: Pearson, 2018.

[17] E.J. Hearn, *Mechanics of Materials 2*, 3rd Edn. Oxford: Butterworth-Heinemann, 1997, pp. 220-299.

[18] T. Anderson, *Fracture Mechanics: Fundamentals and Applications*, 3rd Edn. Boca Raton: CRC Press, 2017.

[19] S.P. Timoshenko, and J.N. Goodier, *Theory of Elasticity*, 3rd Edn. Singapore: McGraw-Hill, 1970.

[20] Ansys, Inc, "Adaptive Convergence," Ansys, Inc, [Online]. Available: https://ansyshelp.ansys.com/account/secured?returnurl=/Views/Secured/corp/v201/en/wb_sim/ds_Convergence.html. [Accessed 29 November 2021].

[21] Z. Wu, "A method for eliminating the effect of 3-D bi-material interface corner geometries on stress singularity," *Engineering Fracture Mechanics*, vol. 73, no. 7, pp. 953-962, May 2006, doi: 10.1016/j.engfracmech.2005.10.010

[22] J.G.B. Goodno, and J.M. Gere, *Mechanics of Materials*, 9th Edn. Boston, MA: Cengage Learning, 2017.

Development of Double-motor Electric Drive for Screen Wipers

Igor Polyuschenkov
Digital Control Systems Dept.
R&P Company Rubicon – Innovation
Smolensk, Russian Federation
polyushenckov.igor@yandex.ru

Abstract—**The paper presents the materials and results of the development of a double-motor electric drive, which provides the coordinated motion of two screen wipers of a bus windshield, as the mechanical connection between them is absent. Taking into account the performances of the wiper position sensors and their location in the structure of the mechanism, as well as the absence of the possibility of reversing the direction of electric motors rotation, automatic stabilization of the electric motors speeds and positioning of the wipers in the trajectories extreme positions were used to coordinate their motion. To stabilize the speed of each of the electric motors, an automatic control with positive feedbacks on their currents is applied. The electric drive logical control during the implementation of the coordinated cyclic motion of the wipers is described, as well as the software structure of the microprocessor control system. The hardware and detailed software for this electric drive control system have been developed. During the development of the software, the tools of model-based programming and automatic generation of program code were used. The software provides automatic stabilization of the electric motors speeds, an automatic logic control for the wipers cyclic motion, input of discrete and analog signals for logic control and monitoring the electric drive serviceability. Experimental results are presented.**

Keywords—**DC machines, electric motors, closed loop systems, automatic control, position control, microcontrollers.**

I. INTRODUCTION

It is well known that there are various approaches to solving problems of technical systems motion control. Depending on the approach chosen, various technical solutions are required. This also applies to the drive of mechanisms, interconnected parts and elements of which must perform coordinated motions using electric drives. Due to the kinematic connection of mechanism elements and parts, their coordinated motion is carried out in a natural way, regardless of the structure of the electric drives that set them in motion, among which there may be even the simplest open-loop electric drives.

The use of individual electric motors makes it possible to abandon the kinematic transmission between elements and parts of mechanisms, thus simplifying the technical system design from mechanical point of view. In this case, the coordinated motion of mechanism elements and parts is carried out as a result of coordinated control of electric motors driving them as part of an electric drive system. At the same time, a method for electric drives control should be implemented that best meets the requirements of the technical specifications for accuracy, dynamics, range and functionality. To implement them, already at the stage of assembling and designing the technical system, suitable control and measuring equipment should be used to ensure full functioning of chosen control technology. However, electrical, electronic and electromechanical equipment for automotive technology is characterized by specific technical solutions, which in many respects limit the possibilities for the most advanced control methods implementation or make them impossible at all. This applies to sensors, electronic and electrical circuits, power supply circuits and other equipment. Therefore, the development of electric drives for auxiliary mechanisms of automotive technique should be carried out taking into account this circumstance.

The purpose of the project, the materials of which are given in the paper, is the development of an electric drive control system for screen wipers of automotive technique windshield, between which there is no mechanical connection. To achieve this purpose, the development and related research task is the choice and application of the suitable technical solutions. They are aimed at implementing the coordinated motion of mechanically unconnected screen wipers, as well as aimed at performing auxiliary functions with the purpose to expand the consumer performances of the developed electric drive as a full-fledged device.

It is rather difficult to give a full assessment of similar electric drive systems, including serially produced ones, from the technologies point of view. As an example, dual DC motor controller WWS2.1 from [1] can be mentioned. Their documentation does not explicitly disclose the solutions used to achieve functionality, consumer properties and characteristics. One can only come to some conclusions about their use by indirect signs – by the design and location of sensors, power converter circuits, measuring circuits and power supply circuits. Experimental studies of similar serial electric drives also allow one to come to the conclusion about the use of some technical solutions. So, the research content of this development is aimed at choosing technical solutions and evaluating their applicability depending on the design of wipers, dynamics of motion processes and so on.

Another purpose of the project development is to expand the practice of model-based design, as well as automatic code generation technology [2]. They are very promising to expand the professional capabilities of developers of various technical systems with microprocessor control [2] – [7]. For example, based on the professional education specifics and experience of engineers from various fields, model-based design and automatic code generation are tools for independent development of detailed and fully functional software [7] for diverse microprocessor control systems, including electric drives systems and motion control systems in general. So, both of named purposes are interrelated. The first purpose illustrates the achievement of the second purpose and vice versa the second purpose illustrates the achievement of the first one.

978-1-6654-6787-2/22 $31.00 © 2022 IEEE

II. Problem and Methodology of Development

A. Design and Kinematics of Screen Wipers Mechanism

The design and kinematics of the screen wipers SW1 and SW2 mechanism of the car or bus windshield with individual drives using direct current electric motors with independent excitation is shown in Fig.1. Electric motors M1 and M2 are connected to screen wipers through WG1 and WG2 worm gearboxes, as well as through crank gears. It should be noted that worm gearboxes have a property to break itself, which prevents the transition of electric motors to the generator modes and provides effective braking under action of its friction. The presence of crank gears leads to the fact that the wipers make return motion during the rotation of the electric motors shafts in one of the directions, and the reverse of their rotation directions is not required. This means that the relationship between the positions, as well as the speeds, the wipers shafts in relation to the positions and speeds of the electric motors shafts are non-linear and ambiguous. Also, Fig.1 shows the location of the PS1 and PS2 position sensors on the wipers shafts. Based on the kinematics of gears from electric motors to wipers shafts, the motion sectors of their brushes are shown in Fig.1. At the same time, the motion sector of the wiper brush from the driver side is highlighted in blue, and the motion sector of the wiper from the passenger side is highlighted in red. There is an area of overlapping these sectors. The extreme lower positions of the wipers with respect to the position sensors in Fig.1 are indicated as x_{min} and y_{min}, and the extreme upper positions are indicated as x_{max} and y_{max}. The angular distances between the extreme positions of each of the screen wipers are indicated as $\Delta\beta_1$ and $\Delta\beta_2$. The described above design of the mechanism with unconnected wipers and individual DC motors for both of them, including the performances of the position sensors and their location, are suitable for WWS2.1 application and serve as initial data for the electric drive development. In addition to the requirement to exclude the wiper intersection during motion, the following requirements are imposed:

- Continuous motion at one of two programmable speed levels, conventionally called "low" and "high".

- Intermittent operation at "low" speed with single cycles separated by pauses of programmable duration.

- The speed level choice, the choice of intermittent or continuous driving mode and the implementation of parking in home position depending on the logic control signals combination, as well as cyclic operation in conjunction with screen washer.

- Monitoring the supply current and voltage for timely response of the system to fault conditions.

- Return of wipers to home positions x_{min} and y_{min} and detection of their extreme positions when test motion.

- Optional control via the network bus interface.

The listed functions should be supplemented with the setting, configuring and access to technological information.

B. The Control System Structure of Electric Drive

Functional diagram of the screen wipers electric drive control system, which illustrates its composition and interaction with the periphery, is shown in Fig.2. Absolute encoders, having analog outputs, are used as position sensors PS1 and PS2 of screen wipers shafts. These sensors are able to measure the rotation angles of the axes in the 90 mechanical degrees sector when their output voltage signal changes in the range from 0 to 5 V. Power supply of electric motors is carried out from individual half-bridge transistor power converters, as it is shown in Fig.3. Power converters arranged according to such a scheme are used in an analogue – the electric drive WWS2.1 from [1]. The diagrams, also shown in Fig.3, illustrate the signal generation with pulse-width modulation (PWM) to control the transistors VT1 and VT2 of half-bridge power converter supplying the electric motor M1. In this case, VT1 and VT2 transistors are switched on or switched off in antiphase. In accordance with the specified diagrams when transistor VT1 is switched on, the VT2 transistor is switched off and a DC bus U_{DC} voltage is applied to the electric motor. With the switched off state of the transistor VT1, the VT2 transistor is switched on. So, the power circuit of the electric motor is shorted, and motor is in dynamic braking mode. Additional inductors L1 and L2 are presented in power supply circuits of electric motors for smoothing the pulsation of their currents. The power supply DC voltage U_{DC} level is equal to 27 V. Control of current in the power converters common DC bus is carried out using the analog-to-digital convertor (ADC) in order to protect the electric drive from exceeding the current of the alarm value occurring during a short circuit. To do this, the CS current sensor is present in the power converters scheme, shown in Fig.3. In order to prevent overloading of electric motors under the action of an external load of wipers, in the described scheme it is required to limit their currents at the level of the permissible stop level. For this, current sensors CS1 and CS2 are present in the power supply circuits of motors. This arrangement of current sensors allows one to measure the currents of motors both when they rise and when they fall along the circuits that are formed when the transistors are switched on and switched off. The currents are limited by sampling signals from sensors, comparing them with the holding current level and transistors control, depending on this comparison result.

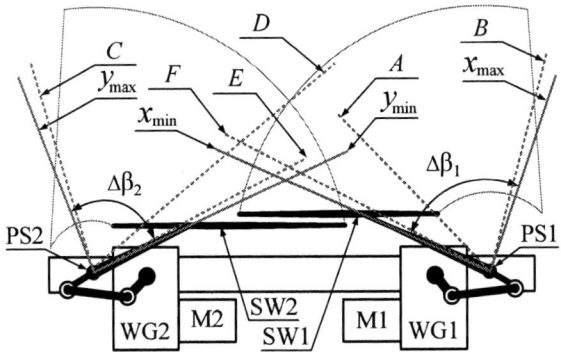

Fig. 1.   Kinematics of screen wipers mechanism.

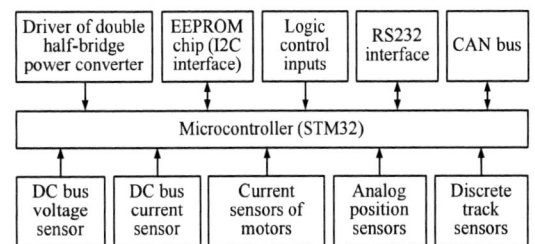

Fig. 2.   Functional diagram of screen wipers electric drive control system.

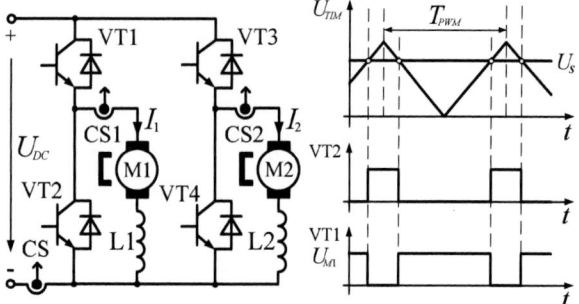

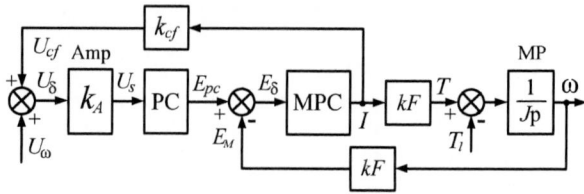

Fig. 3. Schematics of power converter and diagrams of its operation.

The U_{DC} voltage control on the DC bus is carried out in order to the timely detection of its exit permissible limits towards the decline and exceeding.

For interaction with external control devices, two digital interfaces are provided in control system. To interact with the high-level controller, the Controller Area Network (CAN) bus interface is used, the data exchange protocol on which in the message format is compatible with the format of G1939 protocol, distributed in automotive equipment control technologies. The Universal Asynchronous Receive-Transmitter (UART) digital interface is used to interact with control and verification equipment, as well as with a personal computer at the stage of the development and debugging of electric drive system. Interaction with the electrically erasable programmable read-only memory (EEPROM) chip is carried out via the Inter-integrated circuit (I2C) network bus interface. In the memory of this chip, the numerical values of the automatic control system settings and the motion modes parameters are stored in special packed format. They can be written and read.

Logic control, including the choice of the motion mode and the choice of one of preset motion speeds, as well as return of wipers to the home position, is carried out using switches connected to the microcontroller discrete inputs. Discrete track sensors are used to fix the screen wipers location in home position and to locate their lowest positions when tuning to placement of PS1 and PS2 sensors.

III. Materials of Development

A. The Automatic Control System

The need for automatic control of the coordinates of the considered screen wipers electric drive is associated with the fact that this mechanism is under the action of a reactive load arising from the brushes friction against the windshield, as well as a reactive load caused by friction in mechanical transmissions, mainly in worm gearboxes. There is also an active component of the load caused by gravity and having a variable value depending on the angular positions of the screen wipers. An obvious technical solution that allows one to synchronize the screen wipers motion by automatic control technology is a servo system [8].

The use of servo control technical solution is difficult due to the complex action of the following reasons:

- Non-linear and ambiguous relationship of wipers shafts positions, as well as their speeds, in relation to the positions and speeds of the electric motors shafts due to their connection through a crank gears.

- Use of position sensors on the screen wipers axes, not on the axes of electric motors.

- Power supply of electric motors using half-bridge transistor power converters excludes reversal of their directions of rotation.

A less perfect technical solution in comparison with the servo system can be the stabilization of the electric motors speeds with their mutual positioning at control points. However, for the reasons listed above, it is impossible to apply speed stabilization by its deviation [8], since it is difficult to unambiguously measure the speeds of electric motors connected to the axes on which the position sensors are located through a crank gears. Therefore, the stabilization of the electric motors speeds using an individual for each of them positive current feedbacks [8] was applied. This feedback reduces the change in the electric motor speed when the mechanical load changes. The use of feedback mentioned is facilitated by the fact that in power converter there are current sensors, as they are shown in Fig.3. The functional diagram illustrating automatic speed ω control with positive current I feedback of one of an independent field DC electric motor is shown in Fig.4. This functional diagram contains amplifier Amp with a gain of k_A value, a power converter PC, an MPC electric motor power supply circuit, and a mechanical part MP having an inertia J. The positive current feedback coefficient is denoted as k_{cf}. The difference between the supply voltage E_{pc} and the electromotive force E_M of electric motor is denoted as E_δ. Under the action of E_δ, the current I flows in the power supply circuit of the electric motor. The electromechanical coupling coefficient of DC motor converting the current I into its electromagnetic torque T and giving rise to the electromotive force E_M is denoted as kF. Its value depends on the magnetic flux value and the design of the DC electric motor. The rotation of the electric motor rotor is resisted by the torque of the mechanical load torque T_l.

Positioning at the control points is necessary in order to timely eliminate the misalignment of the wipers mutual position, which occurs due to the susceptibility of speed stabilization systems to mechanical load. The degree of this susceptibility is individual for each motor, since it is impossible to perfectly match the electromechanical characteristics $\omega(I)$ of electric motors with positive feedbacks on their currents using k_A and k_{cf} coefficients. In addition, the mechanical load on each of the wipers is somewhat different. Therefore, without control points positioning, the wipers coordination over time will decay under the disturbing factors action such as T_l load torque.

The power converter transfer function is as follows:

$$W_{pc}(\mathrm{p}) = \frac{E_{pc}(\mathrm{p})}{U_s(\mathrm{p})} = \frac{k_{pc}}{T_{pc}\mathrm{p}+1} \qquad (1)$$

where k_{pc} is transfer coefficient of power converter, T_{pc} is its time constant and p is Laplace operator.

The transfer function of the power supply circuit of a DC motor with independent excitation is described as follows:

$$W_{MPC}(\mathrm{p}) = \frac{I(\mathrm{p})}{E_\delta(\mathrm{p})} = \frac{\dfrac{1}{R_a}}{T_e \mathrm{p} + 1}; \; T_e = \frac{L_a}{R_a}, \qquad (2)$$

where R_a is the electric resistance of DC motor power supply circuit, L_a is its inductance and T_e is its time constant.

The static electromechanical characteristic of DC electric motor with rotor speed stabilized by action of positive current feedback is as follows:

$$\omega(I) = \omega_0 - \Delta\omega(I); \; T = kF \cdot I \qquad (3)$$

No load speed ω_0 and static differential speed $\Delta\omega$ from (3) are defined depending on other parameters as follows:

$$\omega_0 = \frac{k_A k_{pc} U_\omega}{kF}; \; \Delta\omega = \frac{R_a I}{kF}\left(1 - \frac{k_A k_{pc} k_{cf}}{R_a}\right) \qquad (4)$$

The value U_s of duty cycle of the voltage E_{pc} supplying the electric motor is calculated by the amplifier Amp depending on U_δ that is the sum of the speed reference U_ω and signal U_{cf} of positive current feedback:

$$U_s(\mathrm{p}) = k_A U_\delta(\mathrm{p}) = k_A(U_\omega(\mathrm{p}) + U_{cf}(\mathrm{p})) \qquad (5)$$

The signal U_s application when generating voltage with PWM is illustrated by the diagrams shown in Fig.3.

B. The Algorithm of Screen Wipers Coordinated Motion

Since the speed control is implemented, the operation sequence in each motion cycle is base on Fig.1 as follows:

- As a result of the previous motion cycle, both wipers are in their lowest positions x_{min} and y_{min} respectively. Driver side wiper starts motion to rise with constant speed. The passenger side wiper is stationary.

- Driver side wiper has reached position A (about 20% of the distance $\Delta\beta_1$ between its extreme positions x_{min} and x_{max}), freeing up space for passenger side wiper motion, which starts to move with constant speed.

- Rising above position A, the driver side wiper slows down depending on its current position. The higher it is in relation to position A, the lower its speed.

- In the vicinity of positions B (about 95% of the distance $\Delta\beta_1$) and C (about 95% of the distance $\Delta\beta_2$) between its extreme positions y_{min} and y_{max}), the control system fixes the passage of the wipers to their extreme positions x_{max} and y_{max} respectively. Wipers continue to move without stopping.

- Since the passenger side wiper has a constant speed, unlike the driver side wiper, it passes the uppermost position before it and continues to decrease, freeing up space for the driver side wiper.

- Having previously reduced the speed, the driver side wiper, going down from uppermost position x_{max} to A position, increases its speed depending on position. The close it is to position A, the greater its speed.

- The passenger side wiper is decreased to position D (about 20% of $\Delta\beta_2$ distance) and continues to move. Its speed decreases depending on the position. The lower the passenger side wiper, the lower its speed.

- The passenger side wiper reaches the position E (about 5% of the distance $\Delta\beta_2$) and stops when detect the signal from its discrete track sensor, waiting for driver side wiper to reach its extreme position x_{min}. Since the speed of passenger side wiper has been reduced, it stops strictly in the vicinity of y_{min} position without oscillatory component.

- Further, when descending below position A, the speed of driver side wiper again decreases depending on its current position in case it is necessary to park, if at this stage of motion such a command follows.

- The driver side wiper reaches the position E (about 5% of the distance $\Delta\beta_1$), detects the signal of its discrete sensor in x_{min} position and continues to move without stopping. Then the motion cycle is repeated.

Stopping the passenger side wiper in the lowest position y_{min} is necessary for the wipers to equalize their distances during the cycle of motion. By changing the speed of the driver side wiper, the separation of its path with the path of passenger side wiper is ensured and their intersection is avoided. The implemented logical computational process, which, depending on the measured positions of the wipers shafts, controls their motion according to the cycle described above, allows them to enter this cycle from a wide range of initial positions, which were when the electric drive was powered on. It should be noted that due to the impossibility of reversing the electric motors direction of rotation, there are "forbidden" combinations of the wipers positions, in which the entry into the motion cycle is impossible, since the crossing of the wipers is inevitable.

C. The Control System Software Development

Significant experience in software development, namely, the microcontroller computing resources distribution and assigning appropriate priorities to the control and measuring tasks, has been accumulated in the process of previous development of electric drive control systems [7]. This experience was used when developing screen wipers system as well. This applies to both hardware parts, such as STM32 microcontroller [9], and the software development tools.

The software block diagram of the screen wipers control system is shown in Fig.5. This block diagram illustrates the composition of tasks and their execution priorities. It operates like an operation system and a task manager [10]. Higher priority levels correspond to their lower numbers [9] shown in Fig.5. The software subroutines running in the background loop have minimal priority. Transitions between executions of lower-priority tasks when they are interrupted by higher-priority tasks are provided in the software, but they are not shown in Fig.5 for simplicity.

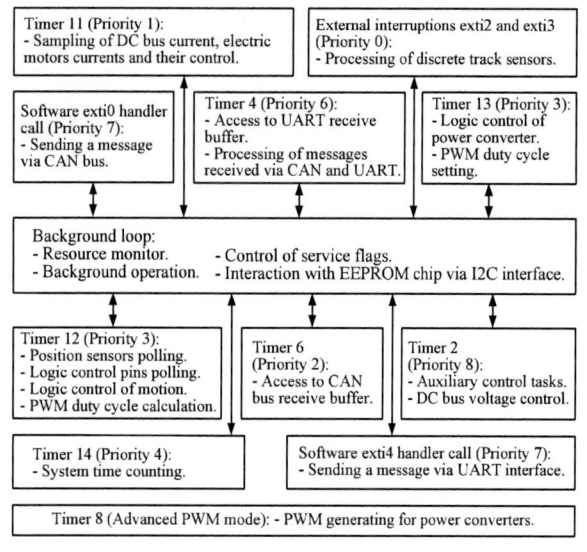

Fig. 5. The structure of wipers electric drive control system software.

The priorities of the tasks are assigned depending on the importance of their parts for the control process, taking into account the available the microcontroller computing resource. For example, the subroutine for currents sampling and their control is executed with the highest priority, since it is intended, among other things, to protect the electric drive from exceeding the current alarm value.

The software development process led to the conclusion that the amount of calculations associated with calculating the duty cycles of the voltages supplying electric motors for the electric drive automatic control of the motors speeds is relatively small. Therefore, the tasks of inputting information from analog position sensors and from discrete logic control switches, as well as the logic control process depending on the signal levels on these switches are combined with the task of automatic coordinate control in the single software subroutine executed by an interrupt from Timer 12 with a repetition rate 1.0 kHz. Computations related to receiving, processing and executing commands received in messages via digital serial interfaces UART and CAN are grouped into separate threads. Access to the UART module receive buffer and processing of the messages received via CAN queue is carried out by Timer 4 interrupt with a repetition rate of 1.0 kHz. Access to the CAN receive buffer and queuing of messages received within a waiting interval of 10 ms is also performed by a Timer 6 interrupt. The sampling and control of the DC bus voltage U_{DC} and currents of electric motors are carried out with a repetition rate of 100 Hz and 20 kHz, respectively.

Tasks executed in the background loop are assigned a repetition interval of 0.1 s, which, however, due to the lack of execution priority, deviates from this specified value, depending on the computational load of the microcontroller. The system time count, the elementary intervals of which are formed by the interrupt when the Timer 14 overflows, is required for the wipers intermittent operation mode, when there are pauses between single cycles of their motion. The sequence of these elementary intervals with a duration of 0.1 s forms pauses of a predetermined duration between the motion cycles. The mentioned above time intervals and repetition rates of the subroutines are assigned taking into account their priorities, depending on the required control

system timeliness response when detecting, capturing, processing and generating signals. The generation of the PWM signal to control transistors according to the diagrams shown in Fig.3 is carried out by Timer 8 operating in Advanced PWM mode with frequency of 25 kHz, and the inverse value T_{PWM} of the PWM period is equal to 40 µs.

When developing the software according to its structure shown in Fig.5, model-based programming and automatic code generation tools of the Waijung Blockset library [2] from the Matlab modeling system were used. Like in [7] and other developments of this paper author, these tools were used mainly as layout elements of software structure and as the hardware periphery handlers of the microcontroller – input and output ports, ADC, PWM generators and digital interfaces UART, CAN and I2C. For example, the use of timers in the Advanced PWM mode to generate PWM signals for power converters control is discussed in detail in [7]. Therefore, there is no need to re-address this issue. Mathematical and algorithmic software was developed in C language. Some model blocks used, notation on which easily allows one to compare them with the STM32 modules and with the software structure, shown in Fig.6.

Motion control of wipers through the stages of the described cycle is carried out by setting and clearing the software flags depending on the wipers position. This control has been carefully designed to avoid the collisions.

IV. EXPERIMENTAL RESULTS

The hardware parts for the electric drive control system of the screen wipers were developed and manufactured. The electronic board is placed in the case. Its design meets the ingress protection code of IP66, as well as connectors for power and measuring circuits, corresponds to automotive technology. To set the parameters of the electric drive, to set the parameters for the motion modes of the screen wipers, as well as to display the electric drive coordinates, the application for a personal computer has been developed. When using it, the electric drive is connected to a personal computer via the USB – UART communication line. Obtained using this application, experimental positions, voltages PWM diagrams and diagrams of electric motors currents are shown in Fig.7. The graphs in Fig.7 (a) correspond to the passenger side wiper, and the graphs in Fig.7 (b) correspond to the driver side wiper.

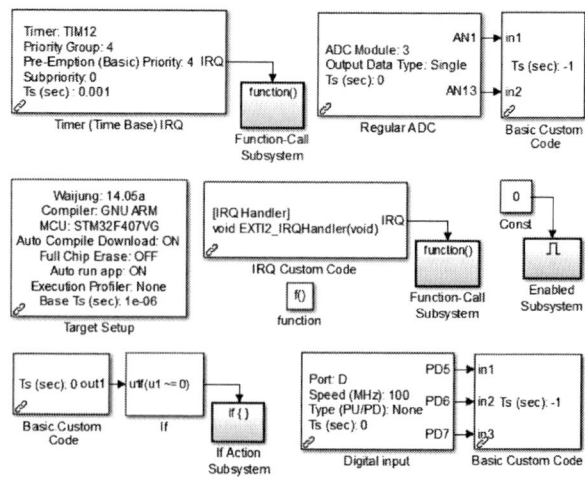

Fig. 6. Model-based elements used in control system software.

978-1-6654-6787-2/22 $31.00 © 2022 IEEE

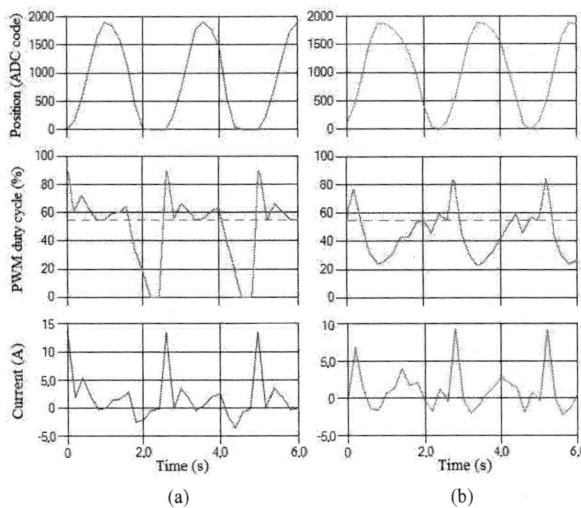

Fig. 7. Experimental diagrams of electric drive when moving in cycle.

The time axes of all sections in Fig.7 have a common reference point. The diagrams of positions are expressed in analog-to-digital conversion codes. The minimum code values on the graphs for each of the wipers correspond to their lowermost positions, and the maximum code values correspond to their uppermost positions. The graphs show the passenger wiper standstill intervals at the positioning point – in lowermost position of this wiper. When obtaining the graphs shown in Fig.7, according to (3) and (4), value of U_{ω} was set equal to 0.55, which corresponds to the electric motors supply voltages duty cycle of 55% of U_{DC} value. This level is shown in Fig.7 with a dashed line. The duty cycle values exceeding this level are the result of the positive feedbacks action on the electric motors voltages, as well as the result of voltage control depending on wipers positions. Current feedbacks automatically dose the supply voltages of electric motors in order to stabilize their speeds when exposed to a mechanical load. In this regard, there are dynamic components on the graphs of electric motors currents and voltages duty cycles. The electric drive operation at different speeds has been tested. Since the maximum possible duty cycle of the voltage supplying is equal to 95% of U_{DC} voltage, the value of electric motor speed is interconnected with the voltage reserve available for its stabilization and ensuring the dynamics during acceleration. If the set value of electric motor speed increases, this available reserve decreases. The "low" speed of the wipers is about 30 cycles per minute, and their "high" speed is about 40 cycles per minute, taking into account the accuracy of the electric motors speeds control.

The performance of various electric drive functions was tested. The motion beginning and parking of wipers from arbitrary positions, which are not "prohibited" from kinematical point of view, taking into account the electric motors irreversibility, has been tested. The electric drive interaction via CAN bus with a control device that simulates on-board computer has been debugged. Various emergency situations were simulated, including the crossing of wipers, excess currents of the emergency setting, the supply voltage going out of the permissible range. In all cases, the control system responded as foreseen. The tuning of the control system has been tested. The action of the positive current feedback k_{cf} coefficient on the accuracy of the electric motor

speed stabilizing under the action of a mechanical load, as well as its effect on the dynamics of transient processes was demonstrated. Logical control of the motion in all modes, depending on the combinations of switches was debugged.

The electric drive developed was tested for moving the windshield wipers of a passenger bus to assess the effect of the real mechanical load action. This test showed the sufficiency of the adopted technical solutions, in particular, the motors speed stabilization using positive feedbacks on their currents, as well as the use of the motions sequence of the wipers forming a cycle. The algorithm for wipers extreme positions detection using their test motion to tune the system for each installation was implemented, as well.

V. CONCLUSIONS

Results of development are formulated as follows:

- The microprocessor-based control system for the electric drive has been developed, which implements the coordinated motion of screen wipers without a mechanical connection. In addition to it, the auxiliary control functionality is implemented.

- Taking into account the design of the mechanism and the location of the position sensors, as well as their design and characteristics, stabilization of the electric motors speeds using positive current feedbacks, as well as the mutual positioning of the screen wipers at control position points, was applied to coordinate their motion. A sequence of the wipers coordinated motion in a closed cycle has been implemented.

- When developing the software, the model-based programming tools were used. The software is developed with the proper degree of detail to implement of the control process algorithms taking into account physical processes, automatic control sequences and interaction with peripheral devices.

REFERENCES

[1] Aros Electronics. [Online]. Available: www.aros.se.

[2] Waijung Blockset. [Online]. Available: www.waijung.aimagin.com.

[3] Exponenta. [Online]. Available: www.exponenta.ru.

[4] J. Krizan, L. Ertl, M. Bradac, M. Jasansky, A. Andreev, "Automatic code generation from Matlab/Simulink for critical applications", Electrical and Computer Engineering (CCECE), 2014 IEEE 27th Canadian Conference on, pp. 1–6, May 2014.

[5] K. Horvath, M. Kuslits, "Model-Based development of induction motor control algorithms with modular architecture", 2016 IEEE International Power Electronics and Motion Control Conference (PEMC), pp. 133–138, 25–28 Sept. 2016.

[6] N. He, H. Huang. Use of model-based design to teach embedded systems programming, 2017 IEEE International Conference on Electro Information Technology (EIT), Lincoln, NE, 2017, pp. 091-094. doi: 10.1109/EIT.2017.8053336.

[7] I. Polyuschenkov, "Model-oriented programming technique in the development of electric drive control system," 2019 26th International Workshop on Electric Drives: Improvement in Efficiency of Electric Drives (IWED), Moscow, Russia, 2019, pp. 1-6. doi: 10.1109/IWED.2019.8664388.

[8] A. S. Anuchin, Control systems of electric drives: Textbook for universities. Moscow: Publishing House of MPEI, 2015.

[9] STM32 Arm Cortex Microcontrollers [Online]. Available: www.st.com.

[10] Sommerville, Ian. Software engineering. – 9th ed. – Wokingham etc.: Addison – Wesley, 2011.

Initial Design Estimation of Synchronous Machines

Muhammad Usman Naseer
Dept. Of Electrical Power Engineering and Mechatronics
Tallinn University of Technology
Tallinn, Estonia
https://orcid.org/0000-0002-1682-9169

Ants Kallaste
Dept. Of Electrical Power Engineering and Mechatronics
Tallinn University of Technology
Tallinn, Estonia
ants.kallaste@taltech.ee

Bilal Asad
Dept. Of Electrical Power Engineering and Mechatronics
Tallinn University of Technology
Tallinn, Estonia
bilal.asad@taltech.ee

Toomas Vaimann
Dept. Of Electrical Power Engineering and Mechatronics
Tallinn University of Technology
Tallinn, Estonia
toomas.vaimann@taltech.ee

Anton Rassõlkin
Dept. Of Electrical Power Engineering and Mechatronics
Tallinn University of Technology
Tallinn, Estonia
anton.rassolkin@taltech.ee

Abstract—The paper presents the basic calculations, guidelines and procedures to determine the initial design parameters of a synchronous machine. This study aims to seek the adaptability of previously established design procedures for the continuously evolving design freedoms offered by additive manufacturing. An algorithm to implement this procedure has also been presented wherein necessary freedoms can be introduced to choose a specific parameter to be optimized with the associated optimization criterion. This paper also discusses the possible approaches for basic calculations of synchronous reluctance machines. The choice of a specific machine type influences various variables and procedures individually. Wherever applicable, the model can be further modified corresponding to the machine type.

Keywords—*electric machines, AC machines, rotating machines, design tools, design for experiments, design for manufacture, algorithm design and theory*

I. INTRODUCTION

With the recent emphasis on additive manufacturing (AM) of electrical machines, the already established design techniques, explicitly developed for conventional manufacturing techniques, need to be modified to consider the design improvements offered by the flexibility in design realization from utilizing AM techniques. To make it happen, the existing methods must be analyzed ~~first~~ to identify areas where the specific procedures need to be modified. This paper presents the utilization of existing schemes in designing a synchronous machine, with a few modifications allowing the selective optimization of design parameters of interest.

The synchronous machines, in general, have higher efficiencies than induction machines [1], and the construction is easier and mechanically stable due to the preference of relatively wider air-gaps. Whereas in-specific, the synchronous reluctance machines offer high torque density [2], [3] and fast response characteristics [4], [5]. This makes them the potential candidate for many industrial applications, especially small to medium scale traction applications [6], [7]. Along with their inherent utility in constant speed applications, the vast advancements in variable frequency drives have also extended their utilization in variable speed applications. Although various aspects of synchronous machine design are already well-defined, they are optimized for and only consider conventional manufacturing techniques. The fundamental aspects of the main types of electrical machine design are addressed by many authors [8]–[12]. Many research articles [13]–[22] also address improvement in

various design aspects of already designed individual electrical machines. But it is rare to find a research article describing the step-by-step approach to estimating a synchronous machine's initial design parameters. This paper describes a step-by-step process towards the said estimation utilizing the current state-of-the-art, well-presented by [9]. Furthermore, this paper presents a simplified procedure that considers the fundamental requirements to facilitate the study of interdependence and the effect of various design parameters. An algorithm for this straightforward approach is also implemented in MATLAB, allowing the selection of different design parameters to be optimized individually.

This presented algorithm, with or without some modifications, can serve various applications such as;

- a tool for beginners to grasp the fundamentals of electrical machine design

- for practicing designers, by improving the procedure further, according to their required utility

- for the researchers from allied fields of electrical machines such as power electronics-based machine control and power system quality, to estimate the basic design parameters and characteristics of any machine

With the added flexibility in selecting the fixed and optimizable parameters, the author aims to couple this procedure with the modified-winding-function-approach (MWFA) model [23]. This way, the machine can be electrically designed and analyzed from the initial stage, requiring minimum simulation runtime and computational resources to provide a better starting point for numerical analysis based final design stage (typically Finite Element Analysis).

II. DESIGN PROCESS

A. Starting Values

The commencing of the design process can be done by designating the variables of the machine design as fixed and free variables. Some variables characterizing the basics of a machine can be termed as design outcomes or selected variables such as rated power, rated terminal voltages, no. of phases, no. of pole pairs, rated rotational speed, efficiency and power factor. Since we are discussing the synchronous machines and there is no slip involved, the rated rotational speed is purely dependent on the frequency and no. of pole pair. So, they should be set accordingly.

B. Main Dimensions (D and l)

For the torque of a mechanical system rotating at n (rps),

$$T = P_{mec}/2\pi n \tag{1}$$

For synchronous machines, $n = n_{syn}$. Also, the shaft power or mechanical output of the machine is directly proportional to the rotor volume as described by (2)-(4) in [9].

$$P_{mec} = C_{mec}D^2 l_{eq} n_{syn} \tag{2}$$

$$C_{mec} = V/\sqrt{3}E_m \, \eta \cos\theta \, \pi^2 k_{w1}\hat{A}\hat{B}_\delta \tag{3}$$

$$V_r \sim \pi/4 \, D^2 l_{eq} \tag{4}$$

where V_r is indicative of rotor volume, P_{mec} is the shaft power, D is the stator inner-diameter, l_{eq} is the equivalent stack length of the machine (assuming the same height for rotor and stator), n_{syn} the synchronous speed of the machine, V is the terminal voltages of the machine, E_m is the induced emf of the stator phase winding, η is the efficiency, $\cos\theta$ is the power factor, k_{w1} is the winding factor for fundamental component, $\hat{A}$ is the peak value of linear current density and $\hat{B}_\delta$ is the peak value of air-gap flux density.

The torque being produced can also be written in the form

$$T = \sigma_{F\,tan} r_r S_r \tag{5}$$

$$T = 1/2 \, \sigma_{F\,tan} \pi D_r{}^2 l_{eq} \tag{6}$$

$$\sigma_{F\,tan} = 2T/\pi D^2 l_{eq} \tag{7}$$

also

$$\sigma_{F\,tan} = \frac{\hat{A}\hat{B}_\delta \cos\vartheta}{2} \tag{8}$$

where r_r is the rotor radius, S_r is the rotor surface area, $\sigma_{F\,tan}$ is the average tangential stress on the surface of the rotor, $\cos\vartheta$ is the spatial phase shift between the fundamental distributions of $\hat{A}$ and $\hat{B}_\delta$ (for synch. machines $\cos\vartheta = 1$) and $D \approx D_r$.

The main dimensions of the machine are calculated in the manner that from the known values of $V, \eta, \cos\theta$ and initially supposed values of $k_{w1}, \hat{A}$ and $\hat{B}_\delta$, the value of C_{mec} is calculated as described in (3). These initially supposed values will be replaced iteratively with the calculated ones from the later steps. Next, the factor $D^2 l_{eq}$ will be calculated from (2) and then the amount of tangential stress from (5) and (7).

Since the tangential stress is a function of $\hat{A}$ and $\hat{B}_\delta$, it can be selected as the optimization criterion and the values of $\hat{A}$ and $\hat{B}_\delta$ can be optimized individually within the assigned limits. The rotor material's maximum allowed mechanical stress can be taken as the optimization criterion here. Usually, a safety factor is used to not exceed the rotor's mechanical integrity. This step is further illustrated in the block diagram of Fig. 1.

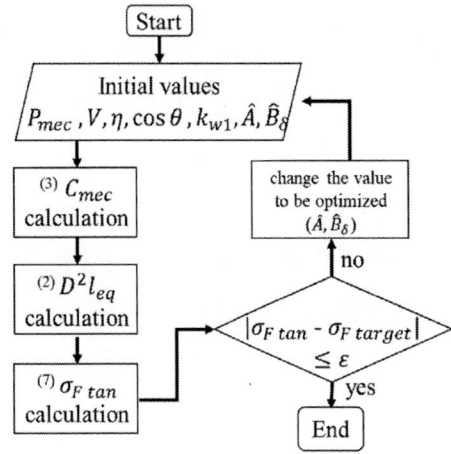

Fig. 1. Block diagram for optimizing values of $\hat{A}$ and $\hat{B}_\delta$ against the mechanical stress limit. The number on the top left corners of the blocks indicate the respective equations

The typical ratio of l_{eq} and D is defined as follows [9]

$$X = l_{eq}/D \approx \frac{\pi}{4\sqrt{p}} \tag{9}$$

where p is the number of pole pairs. Once the values of $\hat{A}$ and $\hat{B}_\delta$ have been optimized, the updated value of C_{mec} and X are utilized in (2), to estimate the main dimensions, D and l_{eq}.

C. Air gap

The air gap is defined in synchronous machines to ensure that the flux density is not reduced significantly due to the armature reaction. Applying this condition, we get [9]

$$\delta \geq 1/2 \, \alpha_{SM}\mu_o\tau_p \, A/k_C\hat{B}_\delta \tag{10}$$

where δ is the air gap width, α_{SM} is the relative pole width coefficient (for sinusoidal air-gap flux distribution $\approx 2/\pi$), μ_o is the permeability of free space, τ_p is the pole pitch and k_C is the carter factor.

D. Winding Design

The winding design commences with the selection of the number of slots per pole per phase, q_s. Ideally, a large number of slots would mean a more sinusoidal MMF distribution. But, due to the increased number of coils, the machine's price increases. So, the designer has to choose the value by making a balanced choice of slot pitch, τ_{us}.

$$Q_s = 2pmq_s \tag{11}$$

where m is the number of phases and Q_s is the total no. of stator slots.

Layered winding coupled with short-pitched coils yields a better sinusoidal distribution of current linkage, short end-windings and reduced harmonics. Also, the skewing of slots along the machine length can reduce the slotting harmonics. This fact is illustrated in Fig. 2 where the winding factors for a sample winding designed for double layer winding short-pitched with two slot pitches, $p = 2$, $m = 3$, $q_s = 4$ are shown.

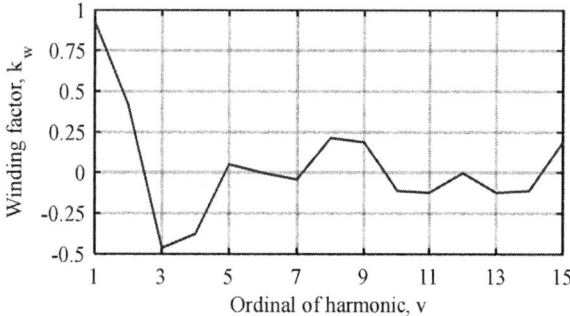

Fig. 2. Winding factors for various v^{th} order harmonics

The winding factors, namely; the distribution factor k_d, pitch factor k_p and the skewing factor k_{sq} can be calculated as below.

$$k_{dv} = \frac{\sin(v\frac{\pi}{2m})}{q\sin(v\frac{\pi}{2mq_s})} \qquad (12)$$

$$k_{pv} = \sin(v\frac{y}{y_Q}\frac{\pi}{2}) \qquad (13)$$

$$k_{sqv} = \frac{\sin(v\frac{\pi}{2}\frac{s_{sp}}{mq_s})}{(v\frac{\pi}{2}\frac{s_{sp}}{mq_s})} \qquad (14)$$

$$k_{wv} = k_{dv} \times k_{pv} \times k_{sqv} \qquad (15)$$

where v is the harmonic number, k_{wv} is the winding factor, k_{dv} is the distribution factor, k_{pv} is the pitch factor and k_{sqv} is the skewing factor. s_{sp} is the skewing of slots in terms of number of slot pitches, y_Q is the full pole pitch expressed in the number of slots, and y is the actual pitch of the coil (after short pitching). The term y/y_Q gives the relative shortening of the coil span. If different from the initial supposed value in (2), this actual winding factor can then be updated in reiteration of step-I (calculation of main dimensions).

1) Number of winding turns: The number of series coil-turns per phase, N_s can be calculated as follows,

$$N_s = \frac{\sqrt{2}E_m}{\omega k_{w1} l_{eq} \tau_p \alpha_i \hat{B}_\delta} \qquad (16)$$

$$E_m = \frac{\pi^2 k_{w1} \hat{A} \hat{B}_\delta D^2 l_{eq} n_s}{m I_s} \qquad (17)$$

$$I_s = \frac{P_{mec}}{mV \eta \cos\theta} \qquad (18)$$

where, α_i is the arithmetic average of flux density per pole and is a function of saturation. Initial value of α_i is set ≈ 0.65, assuming a low saturation level. The actual value from the magnetic voltages of different machine parts will be calculated later and updated in due sequence. In machine types without excitation field windings or permanent magnet excitation in the rotor, such as synchronous reluctance machines (SynRM), the stator winding supplies the magnetising current and should be factored in.

2) The number of conductors in a slot: Although it is advisable to connect the coils of different pole-pairs in series to avoid circulating currents from possible asymmetries, connecting the coils in parallel results in a larger number of conductors per slot and consequently less skin effect. The number of conductors in a slot z_Q is given by

$$z_Q = \frac{2amN_s}{Q_s} \qquad (19)$$

To make z_Q an even number, it may have to be rounded off, resulting in a new number of winding turns. This would necessitate calculating the new air-gap flux density corresponding to the respective number of winding turns. At this stage, the process can be reiterated with the updated value of air-gap flux density.

E. Stator-Tooth Width

To determine the stator-tooth width, the respective flux density, $\hat{B}_d'$ must be chosen corresponding to material characteristics. To avoid excessive iron losses in high-frequency machines, the lower values must be selected. The stator tooth width, b_d is calculated as follows

$$b_d = \frac{l_{eq}\tau_{us}}{k_{Fe}(l - n_v b_v)}\frac{\hat{B}_\delta}{\hat{B}_{d'}} + 0.1 \, mm \qquad (20)$$

where, k_{Fe} is the space factor for stator construction material. n_v and b_v are the number and equivalent width of ventilation ducts, respectively (if any). The ventilating ducts influence the equivalent length of the machine and the effect of field fringing on machine length due to ventilating ducts can be considered by utilizing the carter factor.

Since this procedure is optimized, originally for conventional manufacturing techniques, i.e. punched laminations, (20) has 0.1mm added to the width of stator-tooth considering the influence of punching on material crystal structure and permeability. This factor, along with the space factor k_{Fe}, should be changed accordingly when considered for other manufacturing types such as solid construction from die casting or topology optimized structure from AM.

F. Stator Slot Dimensions

With the stator-tooth width and number of conductors per slot determined, the stator slot dimensions are calculated based on the stator current I_s, allowed current density J_s and the resulting cross-sectional area of the conductors S_{cs}.

$$S_{cs} = \frac{I_s}{a J_s} \qquad (21)$$

The area of the slot S_{cus} is calculated from factoring in the slot insulation's space factor k_{cus}.

$$S_{cus} = \frac{z_Q S_{cs}}{k_{cus}} \qquad (22)$$

By suitably selecting a few basic dimensions of stator slots such as slot type, slot opening, average slot-width and insulation thickness requirement around the slot and between the layers of winding, other dimensions such as slot height and respective widths at different slot heights are calculated based

978-1-6654-6787-2/22 $31.00 © 2022 IEEE

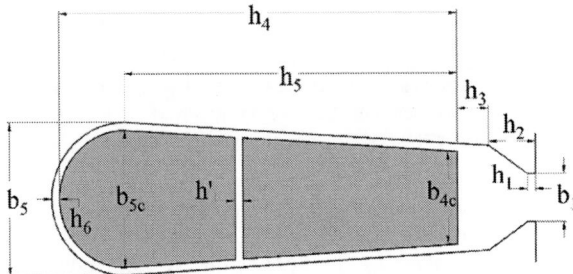

Fig. 3. Stator slot dimensions. The shaded portion represents the wound stator slot area S_{cus}.

on the total required area S_{cus}. The total slot area and wound area in a slot is illustrated graphically in Fig. 3.

G. Determining Magnetic Voltages of the Machine Parts and New Saturation Factor

To determine the saturation factor and, in turn, the updated value of α_i, the magnetic voltages of various machine parts such as stator, rotor and the air gap need to be calculated. The saturation factor k_{sat} and the value of α_i can be calculated as follows.

$$k_{sat} = \left. \widehat{U}_{m,s} + \widehat{U}_{m,r} \middle/ \widehat{U}_{m,\delta} \right. \tag{23}$$

$$\alpha_i = (1.24k_{sat} + 1) \middle/ (1.42k_{sat} + 1.57) \tag{24}$$

where, $\widehat{U}_{m,s}$, $\widehat{U}_{m,r}$ and $\widehat{U}_{m,\delta}$ are the stator, rotor and air-gap magnetic voltages, respectively. The values for these magnetic voltages can be calculated by estimating the line integral of field strength H in the respective portions of the machine.

$$U_m = \int H \cdot dl \tag{25}$$

whereas, in calculating the effective air-gap width δ_{es} considering the effect of stator slot openings, the respective carter factor k_{cs} also needs to be calculated. This effect of slot opening on the flux distribution in the air gap can be minimized by optimizing the stator tooth shape through AM, and hence this step could be eliminated.

$$dl_\delta \approx \delta_{es} = k_{cs}\delta \tag{26}$$

For various portions of the stator and rotor, since the respective apparent flux densities have already been assumed, the actual flux densities and the corresponding field strength values H to be used in (25) can be obtained by plotting (27) over the BH-curve of the material. The intersection of this plotted line with the BH-curve gives the actual flux density and the corresponding field strength H.

$$\widehat{B}_d = \widehat{B}_d' - \frac{S_u}{S_d}\mu_o\widehat{H}_{dl} \tag{27}$$

where, $\widehat{B}_d$, $\widehat{B}_d'$, S_u and S_d are the actual flux density, apparent flux density, slot area without the iron portion and the tooth area (iron portion). This process is illustrated through a sample calculation in Fig. 4 (a) and (b) for two

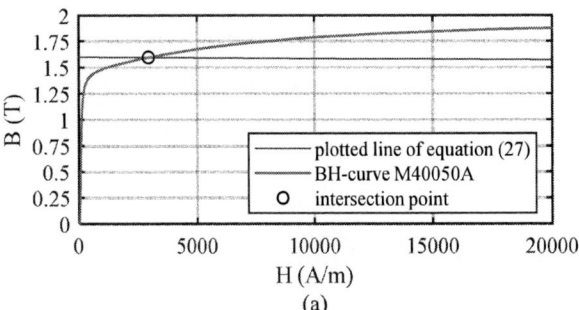

(a)

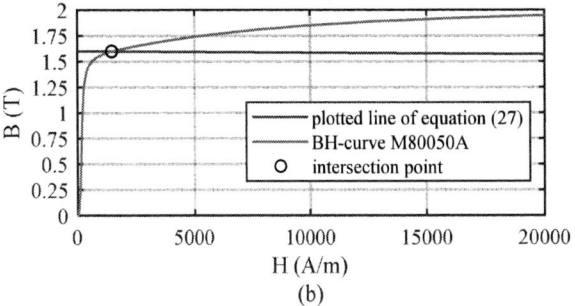

(b)

Fig. 4. Determination of actual flux density and corresponding field strength in machine parts (sample illustration for stator tooth tip), for different core materials (a) M40050A (b) M80050A

different core materials. Since this is only the initial estimation of the basic synchronous machine dimensions, the detailed design of the rotor for other machine types such as SynRMs require individual procedural formulations to calculate the actual flux density and corresponding field strength in various rotor parts. Also, in the case of the rotor of a SynRM, the number and dimensions of the flux barriers can be optimized analytically, employing MWFA at the later stage.

The SynRM's rotor houses no winding/slots, only the flux barriers; the proposed scheme for estimating the magnetic voltage of rotor portion at this stage is to calculate the slope of (27) i.e. S_u/S_d by taking the ratio of flux barrier area with the iron core area of the rotor. Here, for saturation factor estimation, the magnetic voltage of the rotor portion is assumed similar to the stator i.e. $\widehat{U}_{m,s} \approx \widehat{U}_{m,r}$.

H. Stator and Rotor Yoke heights

With the suitable selection of stator and rotor-yoke peak flux densities $\widehat{B}_{ys}$ and $\widehat{B}_{yr}$, according to the specifications of yoke material, the respective yoke heights can be calculated from the peak value of air-gap flux $\widehat{\emptyset}_m$ as follows.

$$\widehat{\emptyset}_m = \alpha_i\widehat{B}_\delta\tau_pl_{eq} \tag{28}$$

$$h_{ys} = \widehat{\emptyset}_m \middle/ 2k_{Fe}l\widehat{B}_{ys} \tag{29}$$

$$h_{yr} = \widehat{\emptyset}_m \middle/ 2k_{Fe}l\widehat{B}_{yr} \tag{30}$$

where, h_{ys} and h_{yr} are the stator and rotor yoke heights. The stator outer diameter D_{se} and outer rotor diameter D_r are calculated as follows.

978-1-6654-6787-2/22 $31.00 © 2022 IEEE

$$D_{se} = D + 2\,(h_s + h_{ys}) \qquad (31)$$

$$D_r = D - 2\delta \qquad (32)$$

III. RESULTS AND DISCUSSION

The described design scheme was implemented in an algorithm comprising a MATLAB script where the user can choose the values of free variables and the respective optimization criteria, wherever applicable. The block diagram of the algorithm is presented in Fig. 5.

Table 1 presents the calculated design parameters for the initial design estimation of 10KW, 400V, 3-phase SynRM. The winding factor k_w for various multiples of harmonics is presented in Fig. 2. Two layers of winding with short pitching of two slot pitches are considered while designing the winding. The apparent, actual flux density and the associated field strength calculation for various portions of stator-tooth is presented in Fig. 6 (a) and (b).

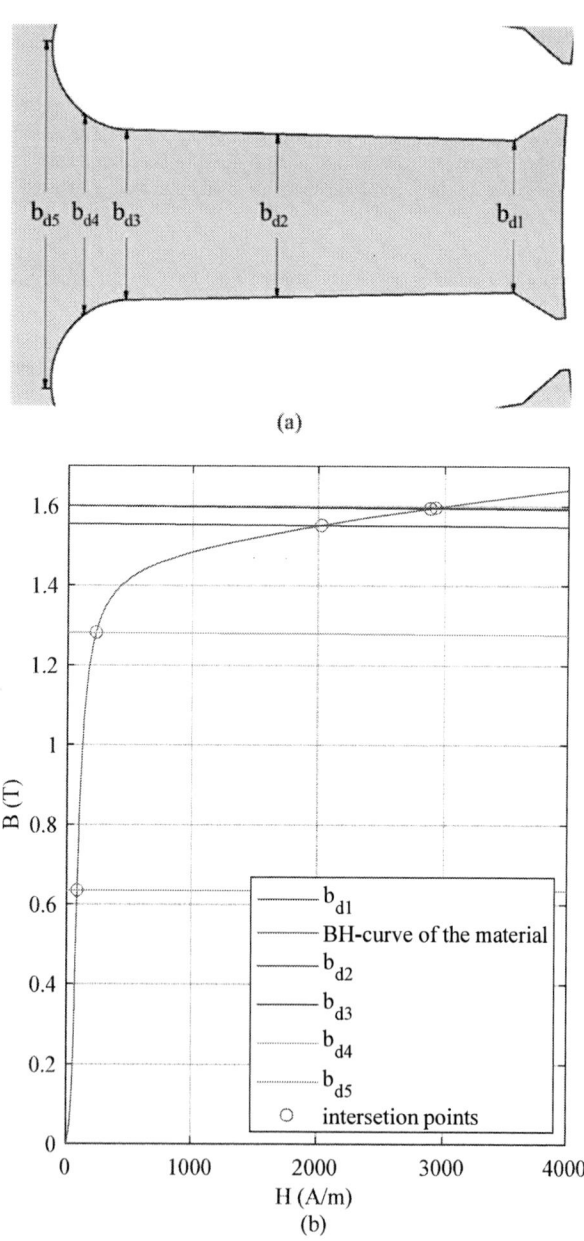

Fig. 6. (a) Stator tooth (b) Apparent and actual flux densities at respective positions, actual flux densities at the intersection point with BH-curve of the material

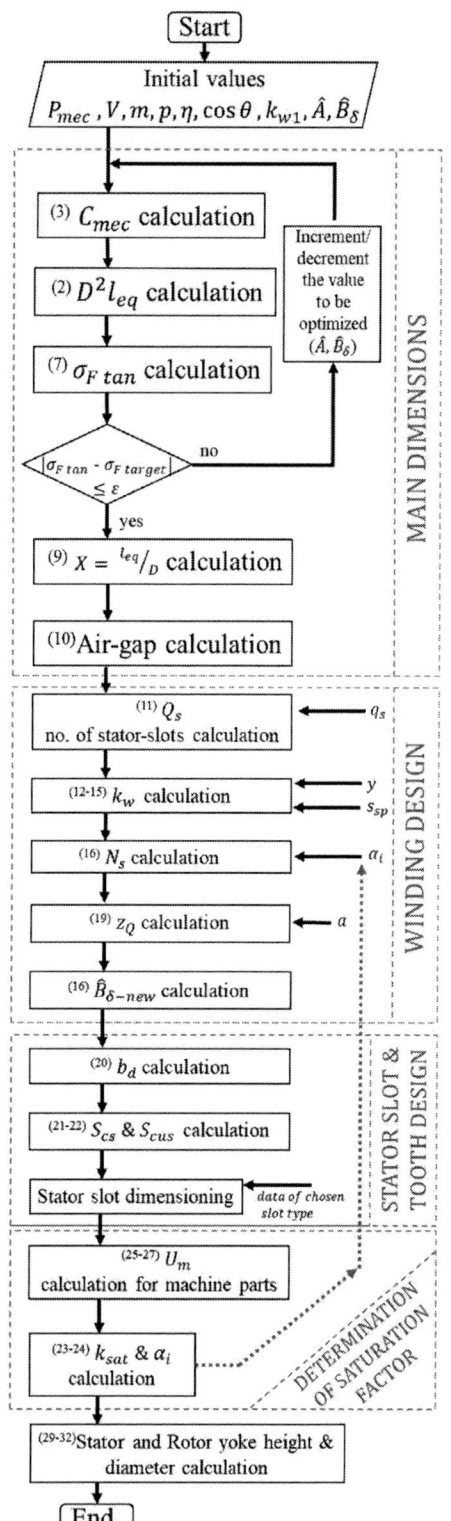

Fig. 5. Block diagram for MATLAB algorithm to estimate initial design parameters of synchronous machines

TABLE I. FIXED AND FREE VARIABLE ESTIMATION FOR INITIAL DESIGN ESTIMATION OF 10KW SYNRM

Machine's Design Parameter	Value	Unit
Machine rated power	10	kW
Terminal Voltages (L-L)	400	V
No. of phases	3	No.
No. of poles	4	No.
Frequency	50	Hz
Rotational speed	1500	rpm
Rated torque output	63.6	Nm
Peak Air-gap flux density	0.79	T
Linear current density	34.83	kA/m
Stator outer-diameter	295.82	Mm
Stator bore-diameter	170.00	Mm
Rotor Diameter	165.30	Mm
Machine stack-length	89.00	Mm
Air-gap width	2.35	Mm
No. of stator slots	36	No.
No. of winding layers	02	No.
No. of parallel paths	02	No.
Pole pitch	133.5	Mm
Slot pitch	14.83	Mm
Short pitching (no. of slot pitches)	02	No.
Stator-slot opening	3	Mm
Stator-tooth width	8.13	Mm
Stator-slot height	31.54	Mm
Stator-yoke height	31.40	Mm

IV. CONCLUSION

The design procedure's execution order suggested in [9] mainly addresses the conventional manufacturing techniques, majorly for the field-excited or permanent magnet synchronous machines. For the initial sizing of SynRMs, the corresponding areas requiring modifications or procedural formulations have been highlighted. The current state-of-the-art indicates the selection of machine constant C_{mec}, from the empirical knowledge, i.e. from the plot of C_{mec} as a function of pole power. This plot is based on the previous knowledge of well-designed machines. Since the intent here is to make the design procedure more compatible for AM techniques, where more flexible control over electromagnetic and mechanical properties of the constructed machine parts is available. In this proposed scheme, C_{mec} and the corresponding values of torque producing $\hat{A}$ and $\hat{B}_\delta$ are optimized against the tangential stress (dependent on the maximum yield strength of the printed material and the associated rotor topology).

Next, the air gap is calculated for an aimed sinusoidal distribution of air-gap flux density. The winding design comprises three slots per-pole-per-phase and a double-layered winding with short pitching of two slot pitches, for a better sinusoidal air-gap flux distribution. The corresponding winding factor calculation resulted in the fundamental component's winding factor of 0.925 whereas, the higher-order harmonic components were also reduced. Further calculation of machine dimensions and recalculation of saturation factor confirmed $\alpha_i \approx 0.65$, indicating low saturated operation of the machine at rated parameters. The procedure for magnetic voltage calculation presented in this paper is not defined for all the rotor types, such as the reluctance rotor. That is why, for the sake of this submission, it was assumed equivalent to the stator magnetic voltage. In continuation to this submission, future works include devising the analytical procedure for estimating magnetic voltage in specific rotor types such as reluctance rotor so that the saturation factor estimation can be improved.

Further, with AM, a better fill factor for winding and a better flux profile from profiled windings can be achieved. The effect of slot openings on the air gap and, resultantly, on the flux distribution over the rotor surface and the slotting harmonics can be optimized with a semi-closed, closed or specific shaped tooth profile. It would also improve the efficiency and power factor of the machine. With asymmetric design possibility from AM, leakage flux can be reduced, and effective air gap length can be increased within the same volume, yielding an improved power density. All the stated and possible improvements are only qualitative. Instead of their intuitive use in machine design, they must be quantified to formulate a complete design procedure for the advanced manufacturing techniques. This implemented algorithm will serve as the primary platform for developing improved design procedures by allowing individual section-wise quantitative modifications in the design procedure and simultaneous analysis of interdependence among design variables.

REFERENCES

[1] M. Villan, M. Tursini, M. Popescu, G. Fabri, A. Credo, and L. Di Leonardo, "Experimental comparison between induction and synchronous reluctance motor-drives," in *Proceedings - 2018 23rd International Conference on Electrical Machines, ICEM 2018*, Oct. 2018, pp. 1188–1194, doi: 10.1109/ICELMACH.2018.8506983.

[2] R. R. Moghaddam, F. Magnussen, and C. Sadarangani, "Novel rotor design optimization of synchronous reluctance machine for high torque density," *IET Conf. Publ.*, vol. 2012, no. 592 CP, 2012, doi: 10.1049/CP.2012.0256.

[3] M. N. F. Ibrahim, A. S. Abdel-Khalik, E. M. Rashad, and P. Sergeant, "An Improved Torque Density Synchronous Reluctance Machine With a Combined Star-Delta Winding Layout," *IEEE Trans. Energy Convers.*, vol. 33, no. 3, pp. 1015–1024, Sep. 2018, doi: 10.1109/TEC.2017.2782777.

[4] S. J. Kang and S. K. Sul, "Highly dynamic torque control of synchronous reluctance motor," *IEEE Trans. Power Electron.*, vol. 13, no. 4, pp. 793–798, 1998, doi: 10.1109/63.704161.

[5] B. H. Bae and S. K. Sul, "A novel dynamic overmodulation strategy for fast torque control of high-saliency-ratio ac motor," *IEEE Trans. Ind. Appl.*, vol. 41, no. 4, pp. 1013–1019, Jul. 2005, doi: 10.1109/TIA.2005.851042.

[6] N. Bianchi, S. Bolognani, E. Carraro, M. Castiello, and E. Fornasiero, "Electric Vehicle Traction Based on Synchronous Reluctance Motors," *IEEE Trans. Ind. Appl.*, vol. 52, no. 6, pp. 4762–4769, Nov. 2016, doi: 10.1109/TIA.2016.2599850.

[7] F. N. Jurca, R. Mircea, C. Martis, R. Martis, and P. P. Florin, "Synchronous reluctance motors for small electric traction vehicle," in *EPE 2014 - Proceedings of the 2014 International Conference and Exposition on Electrical and Power Engineering*, Dec. 2014, pp. 317–321, doi: 10.1109/ICEPE.2014.6969920.

[8] I. Boldea and L. Tutelea, *Electric Machines. Steady State, Transients and Design with MATLAB.* CRC Press, 2010.

[9] J. Pyrhönen, T. Jokinen, and V. Hrabovcová, *Design of Rotating Electrical Machines*. 2008.

[10] T. A. Lipo, *Introduction to AC Machine Design*. 2017.

[11] H. J.R. and T. J. E. Miller, *Design of brushless permanetn magnet machines*, vol. 732, no. 1. 2010.

[12] D. Hanselman, "Brushless permanent magnet motor design," p. 392, 2003, Accessed: Nov. 22, 2021. [Online]. Available: http://digitalcommons.library.umaine.edu/fac_monographs/231/.

[13] T. De Paula Machado Bazzo, J. F. Kölzer, R. Carlson, F. Wurtz, and L. Gerbaud, "Multiphysics Design Optimization of a Permanent Magnet Synchronous Generator," *IEEE Trans. Ind. Electron.*, vol. 64, no. 12, pp. 9815–9823, Dec. 2017, doi: 10.1109/TIE.2017.2726983.

[14] S. Vaschetto, A. Tenconi, and G. Bramerdorfer, "Sizing procedure of surface mounted PM machines for fast analytical evaluations," Aug. 2017, doi: 10.1109/IEMDC.2017.8002223.

[15] H. J. Kim, J. S. Jeong, M. H. Yoon, J. W. Moon, and J. P. Hong, "Simple Size Determination of Permanent-Magnet Synchronous Machines," *IEEE Trans. Ind. Electron.*, vol. 64, no. 10, pp. 7972–7983, Oct. 2017, doi: 10.1109/TIE.2017.2694407.

[16] C. Yang, H. Lin, J. Guo, and Z. Q. Zhu, "Design and analysis of a novel hybrid excitation synchronous machine with asymmetrically stagger permanent magnet," in *IEEE Transactions on Magnetics*, Nov. 2008, vol. 44, no. 11 PART 2, pp. 4353–4356, doi: 10.1109/TMAG.2008.2001325.

[17] M. Degano, M. Di Nardo, M. Galea, C. Gerada, and D. Gerada, "Global design optimization strategy of a synchronous reluctance machine for light electric vehicles," *IET Conf. Publ.*, vol. 2016, no. CP684, 2016, doi: 10.1049/CP.2016.0253.

[18] M. Di Nardo, G. Lo Calzo, M. Galea, and C. Gerada, "Design Optimization of a High-Speed Synchronous Reluctance Machine," in *IEEE Transactions on Industry Applications*, Jan. 2018, vol. 54, no. 1, pp. 233–243, doi: 10.1109/TIA.2017.2758759.

[19] O. Laldin, S. D. Sudhoff, and S. Pekarek, "An Analytical Design Model for Wound Rotor Synchronous Machines," *IEEE Trans. Energy Convers.*, vol. 30, no. 4, pp. 1299–1309, Dec. 2015, doi: 10.1109/TEC.2014.2366472.

[20] G. Bacco and N. Bianchi, "Design Criteria of Flux-Barriers in Synchronous Reluctance Machines," *IEEE Trans. Ind. Appl.*, vol. 55, no. 3, pp. 2490–2498, May 2019, doi: 10.1109/TIA.2018.2886778.

[21] N. Bianchi, S. Bolognani, M. Dai Pré, and G. Grezzani, "Design considerations for fractional-slot winding configurations of synchronous machines," *IEEE Trans. Ind. Appl.*, vol. 42, no. 4, pp. 997–1006, Jul. 2006, doi: 10.1109/TIA.2006.876070.

[22] N. Bernard, R. Missoum, L. Dang, N. Bekka, H. Ben Ahmed, and M. E. H. Zaim, "Design Methodology for High-Speed Permanent Magnet Synchronous Machines," *IEEE Trans. Energy Convers.*, vol. 31, no. 2, pp. 477–485, Jun. 2016, doi: 10.1109/TEC.2015.2513669.

[23] M. U. Naseer, A. Kallaste, B. Asad, T. Vaimann and A. Rassõlkin, "Analytical modelling of synchronous reluctance motor including non-linear magnetic condition," *IET Electr. Power Appl.*, Jan. 2022, doi: 10.1049/ELP2.12172.

The Efficiency Analysis of Resonant Circuits in High-Power Soft Switching Mode Inverters

Daria Bondarenko
Department of Industrial Electronics
Moscow Power Engineering Institute
Moscow, Russia
GolodnikovaDN@mpei.ru

Igor Voronin
Department of Industrial Electronics
Moscow Power Engineering Institute
Moscow, Russia
igor.p.voronin@gmail.com

Pavel Voronin
Department of Industrial Electronics
Moscow Power Engineering Institute
Moscow, Russia
voroninpa@list.ru

Oleg Osipov
Department of Electric Drives
Moscow Power Engineering Institute
Moscow, Russia
osipovoi2015@yandex.ru

Abstract—**The paper presents a comparative analysis of the efficiency of two main options of soft switching at zero current in high-power inverters. Based on the phase trajectories for both options, the maximum currents and voltages of the resonant circuit are estimated. The balance of consumed and stored energy of the resonant circuit is calculated. The advantages and disadvantages of both options are shown for the optimal choice of their application.**

Keywords— *soft switching, zero current switching, zero voltage switching, resonant circuit, power balance, energy efficiency.*

I. INTRODUCTION

Soft switching techniques originate from parallel and series resonance techniques. It is known that at a switching frequency below the resonance frequency, zero voltage conditions are achieved on switch components of the device with a parallel resonant circuit. On the contrary, a zero current condition is achieved on switch components of the device with a series resonant circuit. In comparison with pulse converters it is becomes possible to effectively reduce switching losses in resonant circuits. However, the conduction losses in this case increase due to significantly higher current value and excess energy circulating through the resonant circuit. In addition, during transmitting energy from the source to the load by the resonance method, its regulation is carried out only by the frequency-pulse modulation (FPM). This method of operation is non-linear and quite complex due to changes of load parameters. Therefore, traditional resonance technology is less effective in comparison with soft switching methods, in which resonance plays secondary role and provides commutations at zero voltage and current [1, 2, 3].

A new concept of soft switching has been proposed for powerful inverter technology. This concept states that the converter operates in a resonant mode only in short intervals of transient processes, resuming normal pulse-width modulation (PWM) operation for the rest of the time. Converters of this class are called Soft-Transition-PWM [4, 5, 6].

According to the type of soft switching, they are classified into zero voltage switching (ZVT-PWM) and zero current switching (ZCT-PWM) family. The ZVT-PWM family uses a parallel *LC* circuit, and the ZCT-PWM family uses a serial one. Resonant inductances in both versions are removed from the main circuit of the load current to the resonant circuit and operate only for its soft commutation at short intervals of switching. In order to activate the

resonance mode and synchronize it with PWM an auxiliary is used [7, 8, 9].

This paper analyzes the efficiency of two possible soft switching versions with the use of a serial resonant circuit connected to the power switches of a three-phase voltage inverter.

II. SERIAL RESONANT CIRCUT CONNECTION TO THE POWER SWITCHES

The classic ZCT version for voltage inverters requires the use of the same quantity of auxiliary and main switches, as shown in Fig. 1 for a single inverter phase. There are two possible versions of the serial resonant circuit connection to the main switches of the inverter [9, 10].

In the classic version, auxiliary switches of the same topological group (anode or cathode group) with the main switches are used. For example, with a positive load current, the resonant switching of the main switch S_1 is operated by the auxiliary switch S_{X1} and with a negative load current, the auxiliary switch S_{X2} operates the main switch S_2.

The second version requires the diagonal use of auxiliary switches in relation to the main ones. For example, when the load current is positive, the auxiliary switch S_{X2} operates the resonant commutation of the switch S_1, and when the load current is negative, the switch S_2 is operated by the auxiliary switch S_{X1}.

Both versions are invariant in relation to resonant processes in the output circuits of the main inverters switches and find practical application in modern devices.

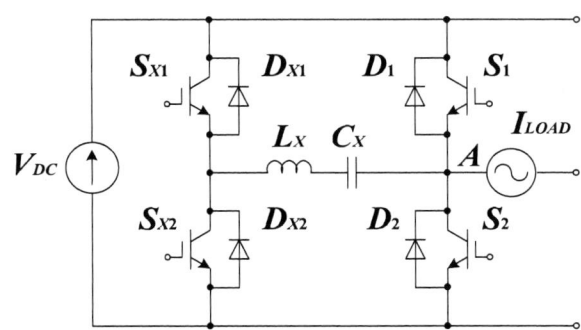

Fig. 1 Phase A of a three-phase inverter

The main switches are connected to the auxiliary block by means of a resonant circuit containing a capacitor C_X and an inductance L_X.

In the direction shown in Fig.1, the load current is commutated between switch S_2 and diode D_1. Soft switching of S_2 is provided by a serial LC circuit and an auxiliary switch S_{X2}.

The process of soft switching in the first version is considered below.

Turn-On Transition (t_0-t_1). Switch S_{X2} turns on at the moment t_0, and then a transient process begins in resonant circuit. As the value of the current I_X increases, the current in diode D_1 diverts to the auxiliary circuit. At the moment t_1, when the current I_X reaches its peak and the current in the diode D_1 drops to zero, the S_2 switch turns on at zero current condition.

Turn-On Transition (t_1-t_2). The value of the current I_X is reduced to zero, since the resonant circuit is short-circuited by switch S_2. After I_X drops to zero, the resonance process continues in the auxiliary circuit. The current I_X becomes negative, whereupon D_{X2} turns on. At this point, S_{X2} can be switched off at zero current condition. The current I_X reaches its peak value and then decreases to zero. The diode D_{X2} switches off at the moment t_2. The current I_S reaches the value of the load current, and D_{X1} switches off. The auxiliary circuit disconnects from the main circuit.

The process of soft switching off of the S_2 is as follows.

Turn-Off Transition (t_3-t_4). Before switching off S_2, S_{X2} turns on at the moment of time t_3. The transient process begins in resonant circuit again. The current I_X becomes positive and increases to a peak value. The current of switch S_2 is diverted to the auxiliary circuit. When the current I_X reaches the value of the load current, the current of switch S_2 drops to zero. As the value of I_X continues to increase, excess current flows through diode D_2. Switch S_2 turns off at zero current condition. When the value of current I_X reaches the value of the load current, the switch current I_S drops to zero at the moment t_4, diode D_2 switches off. The load current flows through the resonant circuit, charging the resonant capacitor.

Turn-Off Transition (t_4-t_5). When the resonant circuit voltage reaches V_{DC}, diode D_1 switches on. The resonance process begins in the LC circuit. As the current I_X decreases, the current in D_1 gradually increases. The value of current I_X reaches zero at the moment t_5. The voltage V_X is greater than V_{DC}, while the resonant process continues in the LC circuit.

Turn-Off Transition (t_5-t_6). The current I_X value becomes negative, diode D_{X2} is switched on, and S_{X2} turns off at zero current condition. When I_X drops to zero at time t_6, the auxiliary circuit stops resonating and disconnects from the main circuit.

The current $I_0(t)$, consumed from the input source E, is the sum of the load current pulse with a duration equal to the PWM signal and a full sinusoid wave with a resonance period. The diagrams of the main processes are shown in the Fig.2.

This type of resonant commutation is formed by parallel connection of auxiliary switches. It requires the use of 6 additional switches in a three-phase inverter. Each arm of the auxiliary circuit contains 2 diodes, so this block includes 6 diodes and 6 switches.

In the device shown in Fig.1, it is possible to use another version of the resonant commutation of the main switches using a serial LC circuit.

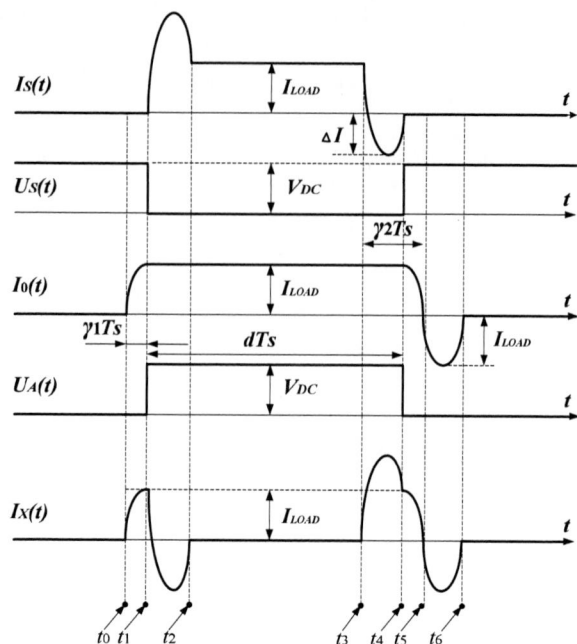

Fig. 2 Diagrams of the main switching processes of S_2

In this type of switching, the serial resonant circuit is connected to the main switch using a diagonally located auxiliary switch. Thus, the auxiliary switch S_{X2} is used to commutate the main switch S_1, and the auxiliary switch S_{X1} is used to commutate the main switch S_2.

Turn-On Transition (t_0-t_1). The S_{X2} switch is turned on at the initial moment of time t_0, and then a resonance process begins in LC circuit. The resonant current I_X is negative. It increases to a peak value and then decreases to zero, while the resonant process continues. The current I_X changes its direction and the diode D_{X1} switches on. The current I_X reaches zero again at the moment of time t_1. The S_{X2} switches off at zero current condition.

Turn-On Transition (t_1-t_2). When the value of the current I_X becomes greater than zero and continues to rise, the current in diode D_2 diverts to the auxiliary circuit. When the current I_X reaches its positive peak at t_2, and diode D_2 switches off, the main switch S_1 turns on at zero current condition.

Turn-On Transition (t_2-t_3). The current I_X decreases towards zero as the current of main switch I_S increases and reaches the load current value at t_3. Further, the current I_X quickly drops to zero, and the diode D_{X2} switches off.

Turn-On Transition (t_3-t_4). The resonance process continues in the LC circuit. The current I_X becomes negative, diode D_{X1} switches on. The current I_X reaches zero at time t_4 and the diode D_{X1} turns off. The resonant process in the LC circuit is terminated and the circuit is disconnected from the main circuit.

Turn-Off Transition (t_5-t_6). Switching off the transistor S_1 occurs as follows: before the switching off of the main switch S_1, the auxiliary switch S_{X2} is turned on again at t_5. A resonant process begins in the LC circuit. The current I_X is negative and reaches its peak, then decreases to zero. At time t_5, the main switch current I_S reaches the value of the load current and then continues to decrease.

Turn-Off Transition (t_6-t_7): Since the voltage across the resonant capacitor V_X is greater than V_{DC}, the resonant process continues in the LC circuit. The value of the current

I_X becomes positive, and the current of the main switch S_1 flows through the resonant circuit. When current I_X begins to flow through diode D_{X2}, switch S_{X2} turns off at zero current condition. At t_7, when the value of the current I_X reaches the value of the load current, the current I_S decreases to zero.

Turn-Off Transition (t_7-t_8). As the current I_X continues to increase, excess current begins to flow through diode D_1, the main switch S_1 turns off at zero current condition. The current I_X decreases to the value of the load current at the moment t_8, and diode D_1 stops conducting. Load current begins to flow through the resonant circuit, charging C_X.

Turn-Off Transition (t_8-t_{10}). When the voltage of the resonant capacitor V_X reaches zero, diode D_2 switches on. The resonance process begins again in the $L_X C_X$ circuit. The current I_X reaches the value of the load current at t_9 and then decreases. When the current I_X drops to zero, the resonant circuit is disconnected from the main circuit. Load current flows through diode D_2.

The current $I_0(t)$ consumed from the input source V_{DC} can be represented by the sum of the load current pulse (with duration equal to the PWM signal) and the full wave of a sinusoid with a resonance period. Fig.3 shows the diagrams of the processes for this type.

As in the classical version, the resonant circuit in this version works without losses, consuming energy from the power source on a positive half-wave and returning the same amount of energy to the source on a negative half-wave.

III. ANALYSIS OF THE EFFICIENCY OF THE RESONANT CIRCUIT CONNECTION OPTIONS

To estimate the effectiveness of both options considered above, their operation was analyzed in a voltage inverter with the following parameters: $V_{DC} = 400$V; the load current has a sinusoidal form with an amplitude of $I_{LOADmax} = 100$ A; the output frequency of the inverter is 400 Hz; the switching frequency is 20 kHz; the inductance of the resonant circuit is 1 µH; the capacitance of the resonant capacitor is 0.625 µF.

The trajectories of energy circulation in a sequential LC circuit have been obtained for both considered options (Fig.4).

Phase paths I_X-V_X contain 25 limit switching cycles at half-wave of sinusoidal load current. All limit cycles are stable and the conditions for soft switching at zero current are achieved at each of the 25 cycles.

The obtained results make it possible to carry out a numerical assessment of the effectiveness of the options for connecting the resonant circuit to the main switch.

For option I, the duration of the control pulses of the auxiliary transistor S_X, activating the resonant processes synchronously with the PWM control signal of the inverter, were chosen unchanged during the switching cycles:

$$\gamma_1 Tc = 2.5 \, \mu s; \; \gamma_2 Tc = 5.0 \, \mu s \qquad (1)$$

The maximum voltage across the circuit capacitor is $V_{Cmax} = 520$ V. The corresponding maximum instantaneous energy stored in the capacitor: $W_{Cmax} = 85$ mJ. It can be observed that the maximum of this energy is realized within the switching cycle at zero switch current.

Since the full wave of the resonant sinusoid does not have a constant component, it can be stated that the resonant circuit operates without energy loss. The same amount of energy is consumed on the positive half-wave, as is returned on the negative half-wave [11, 12].

It should be noted that in this mode, the amplitude of the resonant half-wave is always equal to the amplitude of the load current and therefore changes over the period on the output frequency of the inverter from zero to a maximum of 100 A. Thus, the energy consumed by the resonant circuit from the source V_{DC} changes with a change in the load current.

The maximum energy consumption W_{0max} occurs at the maximum load current and can be calculated as follows:

$$W_{0max} = 2V_{DC} I_{LOADmax} \sqrt{LC} = 63 \text{ mJ} \qquad (2)$$

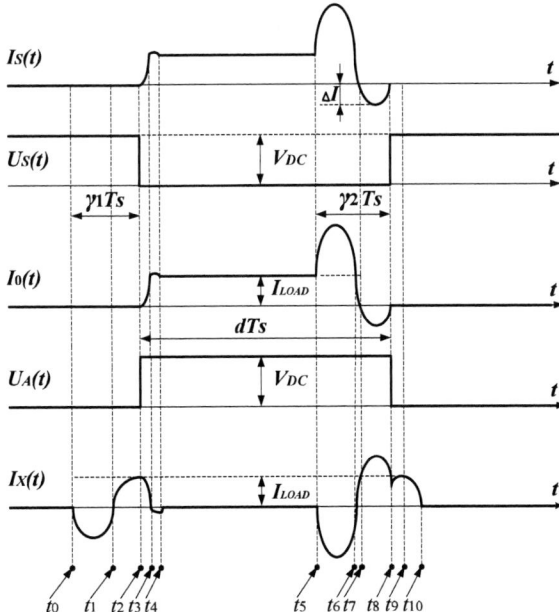

Fig. 3 Diagrams of the main switching processes of S_1

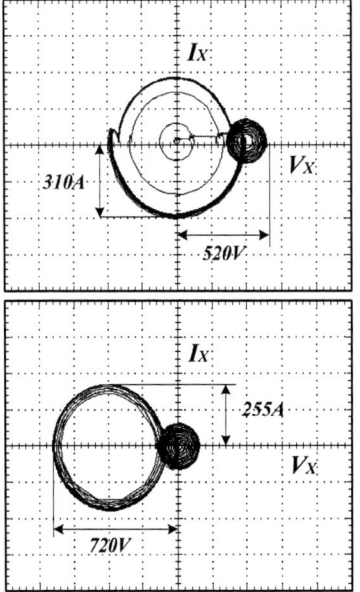

Fig. 4 Energy circulation paths of a three-phase inverter for switching options I and II (scales: 200 V/ div.; 150 A/div.)

The capacitor voltage at the maximum load current is:

$$U_c = V_{DC} - \rho I_{LOAD} = 270 \text{ V} \qquad (3)$$

where ρ is the characteristic wave impedance of the LC series circuit. In this case, the energy in the capacitor is equal to $W_C = 22$ mJ. It should be noted that the sum of the energy consumed from the source and stored in the capacitor is always equal to the maximum energy in the capacitor.

The balance of instantaneous energy in each switching cycle:

$$W_{Cmax} = W_{0max} + W_c = 85 \text{ mJ} \qquad (4)$$

The maximum current of the resonant circuit is 310 A. An auxiliary transistor is selected by the value of this current. The maximum allowable impulse current of auxiliary switch should be greater than the maximum current of the circuit.

For option II, the duration of the control pulses of the auxiliary transistor S_X, which activates the resonant processes synchronously with the PWM control signal of the inverter, do not change during the switching cycles and are equal to each other:

$$\gamma_1 Tc = 3.5 \text{ μs}; \; \gamma_2 Tc = 3.5 \text{ μs} \qquad (5)$$

The maximum voltage across the circuit capacitor within the switching cycle is $V_{Cmax} = 720$ V. The corresponding maximum instantaneous energy stored in the capacitor: $W_{Cmax} = 162$ mJ. It can be noted that this energy is almost twice as much as in the previous regime.

The current $I_0(t)$ consumed from the input source V_{DC} is exactly equal to the current of the main switch. In this mode, the resonant circuit also operates without energy loss. It should be noted that the amplitude of the resonant half-wave is always constant and equal to 255 A. Thus, the energy consumed by the resonant circuit from the source V_{DC} does not change with the load current changes, and is always equal to:

$$W_{0max} = 2V_{DC} I_{0max} \sqrt{LC} = 162 \text{ mJ} \qquad (6)$$

In contrast to the mode with 6 additional switches, in which at the moment of time t_6 a voltage across the loop capacitor occurs, but in this case, at t_6, the voltage across the loop capacitor is always zero.

The balance of instantaneous energies in each switching cycle is as follows:

$$W_{Cmax} = W_{0max} \qquad (7)$$

Thus, the energy consumed by the resonant circuit from the source V_{DC} provides the maximum charge of the capacitor within each switching cycle. The maximum current of the resonant choke is equal to 255 A, i.e. 1.2 times less than in the previous mode.

IV. CONCLUSION

One of the main advantages of the methods described above is that their resonant processes do not change the form of the PWM signals. The PWM signal at the inverter output is formed at the phase point V_A. This is ensured by

the fact that the average value of the current consumed from the source V_{DC} over the switching period is always equal to the load current.

Since the full wave of the resonant sinusoid does not have a constant component, it can be claimed that the resonant circuit works without energy loss, consuming energy from the power source at a positive half wave and returning the same amount of energy to the source at a negative half wave in both versions considered in the article.

The invariance of both versions makes it possible to decrease twice the number of auxiliary switches of the inverter. To achieve this condition the soft commutation of the main switch S_1 must be carried out by the auxiliary switch S_{X1} with a positive direction of the load current. With a negative load current, soft switching of the main switch S_2 should be carried out by the same auxiliary switch S_{X1}, which is located diagonally relative to the main switch S_2. Then only 3 auxiliary switches of the lower cathode group can be used. A similar option can be implemented using three auxiliary switches of the upper anode group.

When using the second mode, it becomes possible to reduce the impulse currents of auxiliary switches by 1.2 times. However, this doubles the maximum instantaneous energy circulating in the resonant circuit, which can lead to an increase in additional static losses in the resonant circuits of the converter. In addition, it requires the use of a resonant capacitor for an operating voltage 1.4 times higher than in the classic version.

REFERENCES

[1] J. D. van Wyk, F. C. Lee. "On a Future for Power Electronics," IEEE Journal of Emerging and Selected Topics in Power Electronics, vol. 1, no. 2, pp. 59-72, June 2013.

[2] R. H. Ashique and Z. Salam, "A Family of True Zero Voltage Zero Current Switching (ZVZCS) Nonisolated Bidirectional DC–DC Converter With Wide Soft Switching Range," in IEEE Transactions on Industrial Electronics, vol. 64, no. 7, pp. 5416-5427, July 2017.

[3] Ismail Aksoy, Hacı Bodur, Nihan Altıntaş, "ZVZCT PWM Boost DC-DC Converter," World Academy of Science, Engineering and Technology International Journal of Energy and Power Engineering Vol:9, No:8, 2015.

[4] J. Lai et al., "A Hybrid-Switch-Based Soft-Switching Inverter for Ultrahigh-Efficiency Traction Motor Drives," in IEEE Transactions on Industry Applications, vol. 50, no. 3, pp. 1966-1973, May-June 2014.

[5] Y. Chen, S. Shiu and R. Liang, "Analysis and Design of a Zero-Voltage-Switching and Zero-Current-Switching Interleaved Boost Converter," in IEEE Transactions on Power Electronics, vol. 27, no. 1, pp. 161-173, Jan. 2012.

[6] [9] E. Chu, X. Hou, H. Zhang, M. Wu and X. Liu, "Novel Zero-Voltage and Zero-Current Switching (ZVZCS) PWM Three-Level DC/DC Converter Using Output Coupled Inductor," in IEEE Transactions on Power Electronics, vol. 29, no. 3, pp. 1082-1093, March 2014.

[7] Y. P. Li, F. C. Lee and D. Boroyevich, "IGBT device application aspects for 50-kW zero-current-transition inverters," in IEEE Transactions on Industry Applications, vol. 40, no. 4, pp. 1039-1048, July-Aug. 2004.

[8] P. Sun, J. Lai, C. Liu and W. Yu, "A 55-kW Three-Phase Inverter Based on Hybrid-Switch Soft-Switching Modules for High-Temperature Hybrid Electric Vehicle Drive Application," in IEEE Transactions on Industry Applications, vol. 48, no. 3, pp. 962-969, May-June 2012.

[9] L. Solero, D. Boroyevich, Y. P. Li and F. C. Lee, "Design of resonant circuit for zero-current-transition techniques in 100-kW PEBB applications," in IEEE Transactions on Industry Applications, vol. 39, no. 6, pp. 1783-1794, Nov.-Dec. 2003.

[10] Voronin P.A., Voronin I.P. "High Power Converters with Resonant Switching on the Alternating-Current Side," Russian Electrical Engineering, 2015, vol. 86, n.4. – pp. 180 – 186.

[11] Voronin I., Voronin P., Osipov O. "The Efficiency Analysis of Resonant Circuits in High-Power Converters of Electrical Energy with Soft Switching Mode," 28th International Workshop on Electric Drives: Improving Reliability of Electric Drives, IWED - Proceedings, 2021, 9376381.

[12] Voronin P.A., Voronin I.P. "Development of the Topology and Analysis of the Efficiency of Voltage Inverters on Composite Resonant Switches," Russian Electrical Engineering, 2017, vol. 88, n.8. – pp. 544-550.

Determining the Thermal Conductivity of Additively Manufactured Metal Specimens

Martin Sarap
Department of Electrical Power
Engineering and Mechatronics
Tallinn University of Technology
Tallinn, Estonia
martin.sarap1@taltech.ee

Ants Kallaste
Department of Electrical Power
Engineering and Mechatronics
Tallinn University of Technology
Tallinn, Estonia
ants.kallaste@taltech.ee

Payam Shams Ghahfarokhi
Department of Electrical Power
Engineering and Mechatronics
Tallinn University of Technology, Riga
Technical University
Latvia, Estonia
payam.shams@taltech.ee

Hans Tiismus
Department of Electrical Power
Engineering and Mechatronics
Tallinn University of Technology
Tallinn, Estonia
hans.tiismus@taltech.ee

Toomas Vaimann
Department of Electrical Power
Engineering and Mechatronics
Tallinn University of Technology
Tallinn, Estonia
toomas.vaimann@taltech.ee

Abstract—**This paper presents a method for experimentally determining the thermal conductivity of additively manufactured specimens. Selective laser melting was used to fabricate two metal rods – one from an aluminum alloy (AlSi10Mg) and the other from silicon steel (Fe-3.7%w.t.Si) powder. For measuring the thermal conductivity, an experimental setup is designed and the measurements are carried out. According to the experimental results, the thermal conductivities of the aluminum and steel rods are 107.8 ± 3.95 and 25.97 ± 0.86 W/m/K respectively.**

Keywords—Thermal conductivity, three-dimensional printing, electric machines

I. INTRODUCTION

Additive Manufacturing (AM), more commonly known as 3D printing, is a novel manufacturing technique for producing three-dimensional objects via the subsequent fabrication of thin layers. Due to its advantages in quick prototyping, abilities in creating complex shapes without significant limitations and a wide selection of materials, AM is a major driving force of design innovation in many fields. One of those is the field of electrical machines, where various novel solutions have been realized with AM [1]. Many of these are based on the ability of 3D printing silicon steel, which has been shown to produce additively manufactured parts with good magnetic properties [2]. In addition, AM can be beneficial for thermal engineering, as novel cooling methods for electrical machines can significantly benefit from the design freedom offered by it [3]. While the magnetic properties of additively manufactured silicon steel have been investigated [4], its thermal characteristics have previously not been measured. An effective design of an additively manufactured electric machine might also require the printing of a material with high thermal conductivity, of which the aluminum alloy AlSi10Mg has received the most attention in the AM space [5].

There are several different AM methods, each with their own advantages and disadvantages, which can be used to print different types of materials [6]. These methods are often divided into three groups based on the method of forming the part – material extrusion, vat photopolymerization and Powder Bed Fusion (PBF). For manufacturing the specimens described in this paper, a PBF method called Selective Laser Melting (SLM) was used as it is the most common AM method for producing electrical machine parts. With this method, layers of metal powder are melted by a high-powered laser to create the three-dimensional object. The process of melting the powder creates a potentially porous and non-uniform part with unknown physical properties. These can depend on the particle size of the powder, printing orientation of the part, laser power and other printing parameters, which is why it is important to measure the physical parameters of the finished objects. For thermal engineering, the most important parameter is the thermal conductivity of the material, which is why the main objective of this paper is to experimentally determine the thermal conductivity of the additively manufactured specimens.

II. MEASURING THERMAL CONDUCTIVITY

The thermal conductivity of a material describes its ability to transfer heat. In the construction of electrical machines, a higher thermal conductivity is beneficial, as it allows for more efficient heat transfer from the internal heat generating areas, resulting in a lower operating temperature. As temperature is a critical parameter when designing electrical machines, the thermal conductivity of the material used needs to be known. In a typical electric machine, much of the heat generated must travel through the soft magnetic material, which is generally a poor thermal conductor, meaning that it makes up a large part of the total thermal resistance and that the thermal performance of the electric machine is heavily dependent on the thermal conductivity of the soft magnetic material. For this reason, knowing the thermal conductivity of the soft magnetic material is crucial for achieving optimal thermal designs

The thermal conductivity of a material can be measured in several different ways: guarded-comparative-longitudinal heat flow technique [7], Lee's disk and Searle's bar methods [8], by modulated temperature differential scanning calorimetry [9], transient plane source method [10], transient flash method, laser flash method and many others. The various methods are suitable for different conductivity values, sample shapes and states of matter. Some require proprietary test apparatus or known reference samples. For determining the thermal conductivity of the AM samples, a longitudinal heat flow technique is used, as it is the most suitable method for measuring solid materials with relatively high conductivities.

This technique consists of heating the specimen (ideally a completely uniform rod) from one end and cooling it from the other end to achieve a temperature gradient over the length of the sample, which depends on the heat flux through the specimen and the thermal conductivity of the material. Ideally all of the heat would flow axially through the specimen stack (heater-specimen-cooler), as only the total heat input can be accurately measured. Although it can never be eliminated, the non-axial heat flow can be reduced by maximizing the non-axial thermal resistance. Practically this can be done by covering the specimen with a layer of insulation. The non-axial heat flow can be further reduced by minimizing the axial thermal resistance, which can be done by reducing the contact resistances in the specimen stack and by increasing the effectiveness of the cooling.

With this method, knowing the heat flow through the isolation is the main challenge, as it's not directly measured. Therefore, it needs to be approximated with the use of a mathematical model, which is calibrated using temperature measurements of the heat sink in addition to thermal imaging of the insulation surface. Since the thermal resistance of the insulation is in any case several orders of magnitude higher than the specimen stack axial resistance, the heat flow through the insulation will be minimal and even a large relative uncertainty in its value will have a small impact on the final accuracy of the measurement. Still, this source of error is considered, which is why minimizing the thermal resistance between the heating element and ambient is important.

III. EXPERIMENTAL SETUP AND PROCEDURE

The method described in the previous chapter is realized for the purpose of experimentally determining the thermal conductivity of the additively manufactured specimens. A schematic for the experimental setup is pictured in Fig. 1.

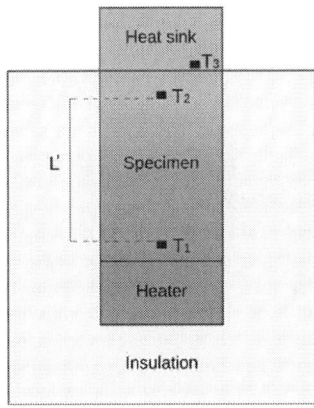

Fig. 1. Schematic of the measuring setup

T_1, T_2 and T_3 depict the locations of the temperature sensors. T_1 and T_2 measure the temperatures of the hot and cold side of the specimen which are used to calculate the thermal conductivity of the material λ (W/m/K) through Fourier's law of conduction:

$$\lambda = \frac{Q_s \cdot L'}{\Delta T \cdot A_s} \qquad (1)$$

where Q_s (W) is the heat flow through the specimen, L' (m) the distance between the temperature sensors, A_s (m^2) the area of the specimen and ΔT (°C) the temperature difference between T_2 and T_1.

To heat the specimen, a current is passed through a resistor connected to one end of the specimen, allowing the total heat generated to be accurately measured with a digital multimeter. T_3 is measuring the temperature of the heat sink (the sensor is placed next to the specimen, on the copper cold plate) to help estimate the thermal resistance of the system. The insulation surface temperature is measured with a thermal camera and is denoted as T_{ins} in the results. The force pressing down on the heat sink assures adequate contact to reduce the thermal resistance between the heat sink and the top part of the specimen.

Q_s is found by subtracting the heat losses from the total heat generated. As the heat flow through the specimen is not directly measured it must be approximated through the use of a mathematical model. For this reason, the setup was modelled using 3D finite element analysis. As the temperature of the heat sink and the surface of the insulator body are measured, these can be simulated as constant temperature boundary conditions. The only unknown input parameters are the contact resistances on the heater and cooler surfaces, which determine the absolute temperature of the test sample. Since these temperatures were measured, the model can be fitted to the temperatures (by changing the contact resistances) to achieve a relatively simple and accurate simulation of the system. The fitting of the contact resistance also accounts for any leaks and imperfections in the isolation, as the relative insulation losses depend only on the ratio of the thermal conductivities of the paths through the specimen and through the insulation. Using this method, the heat flow through the specimen Q_s was found to be 96 and 94 % of the total heat input for the aluminum and steel samples respectively.

A. Test Samples

Two specimens (Fig. 2.) were manufactured with SLM. Both were printed horizontally, meaning that the thermal conductivity was measures in the transverse direction in relation to the printing process. Both tests were done without annealing the specimens to measure the as-built properties of the materials. The first specimen was printed from an aluminum alloy (AlSi10Mg) widely used in AM for its good physical and mechanical properties in addition to easy printability. The thermal conductivity of this material (after annealing) is given in the manufacturer's data sheet as 130-150 W/m/K at 20 °C [11]. Another supplier of the same alloy gives a value around 100 W/m/K before heat treatment and around 160 W/m/K after [12]. The thermal conductivity of this material has been measured by other researchers as well - Sélo et al. used a similar method and measured it to be 115 ± 9 W/m/K in the transverse direction [13].

The second specimen was printed from a Fe-3.7%w.t.Si powder with good magnetic properties, which together with the material's exact composition were measured by Tiismus et al. [14]. The thermal conductivity of steel depends heavily on its composition and no data was provided with the supplied material. For comparison, a commercially available 3.0 % Si electrical steel is used, with a given thermal conductivity value of 28 W/m/K [15]. The lengths of the aluminum and steel samples are 120.2 and 89.4 mm respectively and both were given a precise 19.5 mm diameter on a lathe. While a longer specimen is preferable for more accurate results, the length was limited by the manufacturing process used. The printing parameters for both specimens are presented in table I.

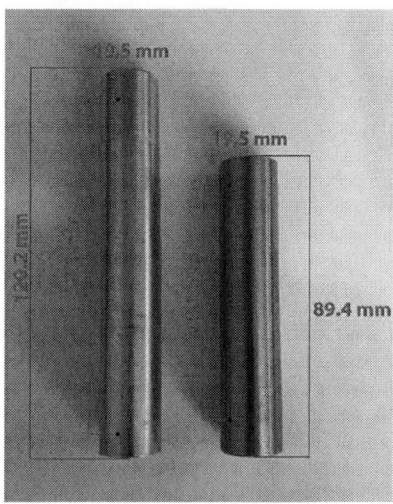

Fig. 2. Additively manufactured specimens.

Table I. Printing parameters

Parameter	Value
Layer thickness	50 μm
Hatch distance	170 μm
Laser Power	350 W (primary) / 100W (secondary)
Scanning velocity	1.15 m/s (primary and secondary)
Scan strategy	Striper
Environment	Nitrogen
Oxygen content	~0.1 %

The test samples have radially placed holes for the sensors, which are dimensioned such that the sensors can be firmly inserted without any noticeable slack. As the parts were 3D-printed, the holes could be made rectangular, allowing for a greater certainty in the position of the working part of the thermal sensor. The holes were placed 10 mm from the ends of the rods (this was reduced with post-processing) to reduce the effects of any non-axial heat flow, as the exact temperature of the sensor is highly sensitive to the surrounding geometry. Aside from the diameter, the most crucial dimension is the distance between the temperature sensors. The opposing edges of the two rectangular holes were measured with calipers to be 98.1 and 68.1 mm apart for the aluminum and steel specimens respectively. The ends of the test rods were polished, as a smoother surface (in addition to thermal paste) allows for a lower thermal contact resistance [16] and as a result reducing the total axial thermal resistance of the specimen stack.

B. Equipment Used

The temperature sensors used were PT100 resistance temperature detectors (RTD) with an accuracy class of F 0.1 in the range 0 °C to 150 °C. Aside from the accuracy, the main advantage of the sensor is its small size (0.8x1.25 mm), which helps reduce its impact on the total heat flow. The temperature measurements were recorded using a precision data logger. A 20 W resistor powered through a digital multimeter was used as the heat source while a high-performance CPU cooler was used as the heat sink. For insulation, a cube of polystyrene (λ = 0.037 W/m/K [17]) with a side length of 300 mm was used. The test setup is pictured in Fig. 3.

Fig. 3. Photo of the measurement setup

IV. THE ANALYSIS OF THE EXPERIMENTAL DATA

The measurements were performed in a thermally stable room with minimal airflow to keep both the active cooling of the test sample and passive cooling of the insulation surface constant. The system was given several hours to reach equilibrium, as any effects of heat capacity were undesirable. Temperature readings at several different heating powers were recorded to measure the any possible temperature dependence of the thermal conductivity. The results for the measurements are presented in tables II and III.

Table II. Measurement results for the aluminum (AlSi10Mg) specimen.

Q (W)	T_1 (°C)	T_2 (°C)	T_3 (°C)	T_{ins} (°C)
5.22	43.68	28.28	22.41	22.33
10.52	64.85	33.88	22.61	22.35
15.02	82.57	38.69	22.87	22.40

Table III. Measurement results for the steel (Fe-3.7%w.t.Si) specimen.

Q (W)	T_1 (°C)	T_2 (°C)	T_3 (°C)	T_{ins} (°C)
0.97	31.92	23.61	22.30	22.22
2.02	42.16	25.13	22.38	22.33
4.85	69.55	29.18	22.30	22.10
6.09	80.82	30.27	22.27	22.74
7.00	89.37	31.61	22.43	22.22

A. Uncertainty Analysis

The uncertainty of the thermal conductivity value is found by calculating the upper and lower limits. This is done simply by using the uncertainty values of the measurements to calculate the maximum and minimum conductivity. Most of the uncertainty comes from the accuracy of the temperature sensors, which is defined by their accuracy class in the standard IEC 60751 and is expressed as $T_\varepsilon = \pm 0.1 + 0.0017|t|$. The large constant in the expression causes the relative uncertainty of the measurement to decrease with the increase of temperature, meaning that the thermal conductivity can be more accurately measured at higher heating powers.

The uncertainty for the heat flow through the specimen $Q_{s\varepsilon}$ (W) is found by calculating the mathematical model using the lower and upper uncertainty limits of the measured temperatures and dimensions. The uncertainty value for both specimens was calculated to be 0.5 % of the total input power.

The dimensions of the specimens were measured with calipers and the dimensional uncertainties of the specimens L'_ε (m) and $A_{s\varepsilon}$ (m²) are taken from these measurements. The uncertainty of the input power Q_ε (W) is derived from the DC voltage and current uncertainties of the digital multimeter [18]. The minimum and maximum values for the thermal conductivity are then calculated through formula 2:

$$\lambda = \frac{(Q_s \pm Q_{s\varepsilon} \pm Q_\varepsilon) \cdot (L' \pm L'_\varepsilon)}{((T_1 \mp T_{1\varepsilon}) - (T_2 \pm T_{2\varepsilon})) \cdot (A_s \mp A_{s\varepsilon})} \quad (2)$$

V. RESULTS AND DISCUSSION

The measurements were conducted at different input powers so that the thermal conductivity values for different temperatures could be measured. The specimen temperature for each measurement $\overline{T}$ (°C) is defined by the average value of the measured temperatures T_1 and T_2. The calculated thermal conductivity values at the measured temperatures together with the calculated uncertainties for both specimens are presented in tables IV and V.

Table IV Measured thermal conductivities for the aluminum specimen

$\overline{T}$ (°C)	λ_{Al} (W/m/K)
35.98	106.83 ± 6.46
49.37	107.05 ± 4.50
60.63	107.80 ± 3.95

Table V Measured thermal conductivities for the steel specimen

$\overline{T}$ (°C)	λ_{Fe} (W/m/K)
27.77	25.01 ± 2.28
33.65	25.37 ± 1.44
44.96	25.46 ± 1.01
49.37	25.73 ± 0.96
55.55	25.81 ± 0.89
60.49	25.97 ± 0.86

The thermal conductivity values of both specimens showed small increases with temperature, which is consistent with measurements done on similar materials [19], [20], although this increase is mostly within the measurement uncertainty. The measurement uncertainties are reduced with the increase of temperature as the temperature sensor uncertainty includes a relatively large constant part. If the most accurate measurement is used, the thermal conductivities of the aluminum and steel samples can be given as 107.8 ± 3.95 and 25.97 ± 0.86 W/m/K respectively. The result for the aluminum sample is in line with the manufacturer provided data and results obtained by other researchers while the thermal conductivity of the steel sample is comparable to a similar commercially available material

VI. CONCLUSION

Aluminum and steel specimens were additively manufactured with SLM and an experimental setup suitable for measuring the thermal conductivity of additively manufactured metal rods was described. The transverse directional steady state thermal conductivity values of the pre-annealed specimens were experimentally determined and the results were presented together with the calculated uncertainties. It was found, that the thermal conductivities of the specimens were similar to comparable materials, although further testing is required to determine the effect of orientation and other printing parameters on the thermal conductivity of an AM part.

REFERENCES

[1] M. U. Naseer, A. Kallaste, B. Asad, T. Vaimann, and A. Rassõlkin, "A review on additive manufacturing possibilities for electrical machines," Energies, vol. 14, no. 7. MDPI AG, Apr. 01, 2021. doi: 10.3390/en14071940.

[2] H. Tiismus, A. Kallaste, A. Belahcen, A. Rassõlkin, T. Vaimann, and P. S. Ghahfarokhi, "Additive Manufacturing and Performance of E-Type Transformer Core," Energies 2021, Vol. 14, Page 3278, vol. 14, no. 11, p. 3278, Jun. 2021, doi: 10.3390/EN14113278.

[3] P. S. Ghahfarokhi et al., "Opportunities and Challenges of Utilizing Additive Manufacturing Approaches in Thermal Management of Electrical Machines," IEEE Access, vol. 9, pp. 36368–36381, 2021, doi: 10.1109/ACCESS.2021.3062618.

[4] H. Tiismus, A. Kallaste, A. Belahcen, T. Vaimann, A. Rassõlkin, and D. Lukichev, "Hysteresis measurements and numerical losses segregation of additively manufactured silicon steel for 3D printing electrical machines," Applied Sciences (Switzerland), vol. 10, no. 18, pp. 1–15, 2020, doi: 10.3390/APP10186515.

[5] N. T. Aboulkhair, M. Simonelli, L. Parry, I. Ashcroft, C. Tuck, and R. Hague, "3D printing of Aluminium alloys: Additive Manufacturing of Aluminium alloys using selective laser melting," Progress in Materials Science, vol. 106, p. 100578, Dec. 2019, doi: 10.1016/J.PMATSCI.2019.100578.

[6] F. Calignano et al., "Overview on additive manufacturing technologies," Proceedings of the IEEE, vol. 105, no. 4, pp. 593–612, 2017, doi: 10.1109/JPROC.2016.2625098.

[7] "Standard Test Method for Thermal Conductivity of Solids Using the Guarded-Comparative-Longitudinal Heat Flow Technique, ASTM E1225 - 20, 2020."

[8] "MATERION PERFORMANCE ALLOYS Thermal Conductivity Lee's Disk Method Searle's Bar Method".

[9] "Standard Test Method for Thermal Conductivity and Thermal Diffusivity by Modulated Temperature Differential Scanning Calorimetry, ASTM E1952-17, 2017".

[10] "Hot Disk | Technology." https://www.hotdiskinstruments.com/technology/ (accessed Nov. 23, 2021).

[11] SLM Solutions, "Material Data Sheet Al-Alloy AlSi10Mg/EN AC-4300/EN AC-AlSi10Mg."

[12] "EOS Aluminium AlSi10Mg Material Data Sheet Metal Solutions".

[13] R. R. J. Sélo, S. Catchpole-Smith, I. Maskery, I. Ashcroft, and C. Tuck, "On the thermal conductivity of AlSi10Mg and lattice structures made by laser powder bed fusion," Additive Manufacturing, vol. 34, p. 101214, Aug. 2020, doi: 10.1016/J.ADDMA.2020.101214.

[14] H. Tiismus et al., "Ac magnetic loss reduction of slm processed fe-si for additive manufacturing of electrical machines," Energies, vol. 14, no. 5, Mar. 2021, doi: 10.3390/en14051241.

[15] Si 3 and A. A 0 4%, "Chemical composition, typical Electrical Steel, Thin Non-Oriented Grades Coating SURALAC ® 7000".

[16] Y. A. Cengel, "Heat Transfer: A Practical Approach."

[17] I. Gnip, S. Vejelis, and S. Vaitkus, "Thermal conductivity of expanded polystyrene (EPS) at 10 °C and its conversion to temperatures within interval from 0 to 50 °C," Energy and Buildings, vol. 52, pp. 107–111, Sep. 2012, doi: 10.1016/J.ENBUILD.2012.05.029.

[18] "A Greater Measure of Confidence 6½-Digit USB Digital Multimeter", Accessed: Dec. 03, 2021. [Online]. Available: www.keithley.com

[19] S. Liu, S. Chang, H. Zhu, J. Yin, T. Wang, and X. Zeng, "Effect of substrate material on the molten pool characteristics in selective laser melting of thin wall parts," International Journal of Advanced Manufacturing Technology, vol. 105, no. 7–8, pp. 3221–3231, Dec. 2019, doi: 10.1007/S00170-019-04540-1.

[20] B. Groschup, A. Rosca, N. Leuning, and K. Hameyer, "Study of the Thermal Conductivity of Soft Magnetic Materials in Electric Traction Machines," Energies 2021, Vol. 14, Page 5310, vol. 14, no. 17, p. 5310, Aug. 2021, doi: 10.3390/EN14175310..

978-1-6654-6787-2/22 $31.00 © 2022 IEEE

2022 29th International Workshop on Electric Drives: Advances in Power Electronics for Electric Drives (IWED), Moscow, Russia. Jan 26 – 29, 2022

Dynamic Stability Analysis of Robots on Spherical Base with Variable Load Mass

Margarita V. Kartashova
Faculty of Control System and
Robotics
ITMO University
Saint-Petersburg, Russia
fawntiog@gmail.com

Nikolai A. Poliakov
Faculty of Control System and
Robotics
ITMO University
Saint-Petersburg, Russia
polyakov_n_a@itmo.ru

Elizaveta D. Moiseenkova
Faculty of Control System and
Robotics
ITMO University
Saint-Petersburg, Russia
y_moiseenkova@inbox.ru

Abstract— **Article explores the possibility of maintaining the equilibrium stability of the ball balancing robot (also known as a BallBot) with forced and free movement. This type of robots is initially unstable; therefore, it is necessary to create a control system that provides resistance to external influences. That requires to include into the analysis parameters of load's moment of inertia and rotational torque of the sphere base. A mathematical model of the sphere-based robot system was developed based on the augmented model of the inverse pendulum. A linear state controller with an internal model of free movement of the object has been obtained. Verification of the developed control system is performed using simulation in MATLAB / Simulink. The initial results of the study, including balancing in place and moving along the trajectory, are presented.**

Keywords— *robot control, robot motion, mobile robots, linear feedback control systems, control system synthesis, motion control, motion planning, path planning.*

I. INTRODUCTION

A ball-balancing robots are an excellent example of dynamically stable robots that might be developed with the framework of a modern approach to robotics [1]. Thus, this type of robots is initially unstable, and their main stable position is provided with an operating principle of the inverted pendulum, they are much more resistant to external influences, because they can easily bend over and restore their balanced state. Since unicycle robots are statically unstable their main disadvantage is constant need to produce some oscillation to ensure balance [2][3]. This is accompanied by small but regular movements from the originally set trajectory.

Such robots can be used as human assistants in many areas [4]: from desktop robot- companions to more complex, human high structures suitable for work in medical organizations or production. Nowadays there are few prototypes that may perform specific tasks. By now, from a constructive point of view, the most complete robots are Rezero and BallBot. Rezero [5] is supposed to be made a street robot photographer or equipped with a manipulator, and a fully functional robot assistant BallBot [3][6].

Research have shown that this class of robots, robots on a spherical base, is very stable. Scientists believe such robots to be an excellent tool for moving heavy loads, especially with these robots coordinated work for moving platforms with the load in concert (Fig. 1, a). To date, there is no research on how such robots work with a load, but at the same time, it is obvious that among the undeniable advantages of such a system is its high maneuverability and compactness, because such robots occupy a vertical rather than horizontal position. This feature will allow them to easily maneuver in people crowd and avoid obstacles.

At the same time, when turning on wheels in a limited space, there may be a number of difficulties associated with the need to provide a turn path. Therefore, if Segway robots are linked to such a task [7], it will be necessary to install an additional drive that provide robot's body rotation (Fig. 1, b).

Today, there is no public research on the operation of robots on a spherical base with a load, despite the obvious advantages of such a design in this matter. For analysis purposes test weight is introduced. Test weight is an additional load in relation to the structural elements of the robot, in general, with variable overall dimensions and weight. The purpose of this work is to study the control system of a robot on a spherical base with a test weight load in conditions of its variable parameters and evaluation of the drive requirements based on performed experiments.

II. STRUCTURE OF THE DESIRED SYSTEM. PROBLEM STATEMENT

For realization of this work, it is necessary to create a test model to verify the equilibrium maintenance of the robot based on such parameters as distance between centers of mass, moment of inertia of the balancing weight, mass, and radius of the spherical base. For proof-of-concept purposes the ball-based robot model must demonstrate advantages of dynamic stability and resistance to external disturbing influences.

- It is required the model to maintain the stability when moving along a given trajectory. The equations of

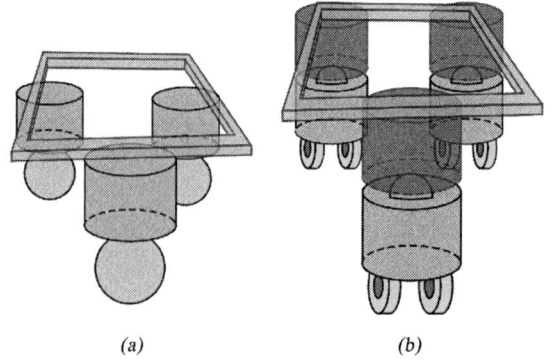

(a)　　　　*(b)*

Fig. 1. The difference between the structural models of group work of robots on a spherical base and Segway robots

978-1-6654-6787-2/22 $31.00 © 2022 IEEE

motions for a spherical base robot will be defined using a simplified test weight model, with it being determined by variable moment of inertia.

- To synthesize a linear controller to control both free and forced motion of an object, it is required to simplify the mathematical model to a linear system of equations.

- It is necessary to form criteria for generating task signals for trajectory movement. The requirements for the drive will be determined for further design.

The block diagram representing the required control system is shown in Fig. 2.

III. Synthesis Of The Simplified Mathematical Model

The simplified model is a combination of rigid hollow sphere with a certain radius, mass and cylinder balancing on this sphere. Fig. 3 shows a schematic representation of such a structure in XZ plane coordinates. The control actions are rotational moments $Torque_x$ and $Torque_y$ (Fig. 2), that are fed to orthogonally rotating rollers rotating a spherical base (not shown in Fig. 3).

Consider a simplified model in the XZ coordinate plane.

Mathematically, the equations of motion for the simulation model are based on an improved model of an inverted cart-pendulum system [8][9]. In contrast to the inverted pendulum model, where a rigid sphere acts as the base, in this model the balancing load is not a material point, but a test weight model with a certain moment of inertia.

Lagrange equations were used to get nonlinear equations in "state-space" form [8][10]:

$$
\begin{cases}
\dfrac{d}{dt}\dfrac{dL}{d\dot{q}_1} - \dfrac{dL}{dq_1} = Q_1 \\[2mm]
\dfrac{d}{dt}\dfrac{dL}{d\dot{q}_2} - \dfrac{dL}{dq_2} = Q_2
\end{cases}
\tag{1}
$$

where $q_1 = \theta_x$, $q_2 = x_x$ are generalized coordinates, Q_1, Q_2 are generalized forces; $L = W_k - W_p$ is the Lagrange Equation, where W_k is the kinetic energy, W_p is the potential energy.

Equation for total kinetic energy:

$$
W_{k\,base} = \frac{1}{2}Mv_{1x}^2 = \frac{1}{2}M\dot{x}_x^2
\tag{2}
$$

The kinetic energy of the pendulum W_{kpend} is calculated as the sum of rotational and translational motions:

$$
W_{k\,pend} = \frac{1}{2}mv_{2x}^2 + \frac{J\dot{\theta}_x^2}{2} =
$$

$$
= \frac{1}{2}m(\dot{x}_x^2 - 2|OA|\cdot\dot{x}_x\dot{\theta}_x cos\theta_x +
\tag{3}
$$

$$
+|OA|^2\dot{\theta}_x^2) + \frac{J\dot{\theta}_x^2}{2}
$$

where M is the mass of the moving base, m is the mass of the balancing weight, v_1, v_2 are linear velocities, J is the moment of inertia of the pendulum, L is the distance between the centers of mass.

The moment of inertia of the pendulum is present in the formula in its final form after applying the Huygens-Steiner theorem [11].

For a balancing cylinder, assuming that the axis of rotation passes through its end, the resulting moment of inertia will be calculated as follows:

$$
J = \frac{1}{3}mh^2 + \frac{1}{4}mr^2
\tag{4}
$$

where h is the high of the cylinder, r is the radius of the cylinder.

Total kinetic energy of the system:

$$
W_k = \frac{1}{2}(M+m)\dot{x}_x^2 - m|OA|\dot{x}_x\dot{\theta}_x cos\theta_x +
$$

$$
+\frac{1}{2}m|OA|^2\dot{\theta}_x^2 + \frac{J}{2}\dot{\theta}_x^2
\tag{5}
$$

Potential energy of the system:

$$
W_p = mg|OA|cos\theta_x
\tag{6}
$$

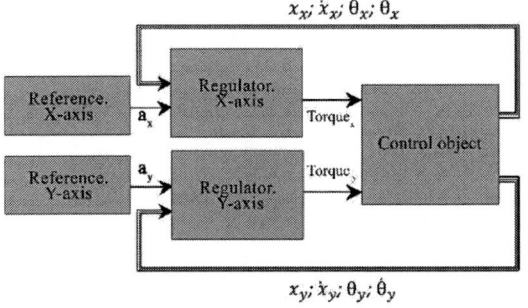

Fig. 2. The block diagram representing the required control system

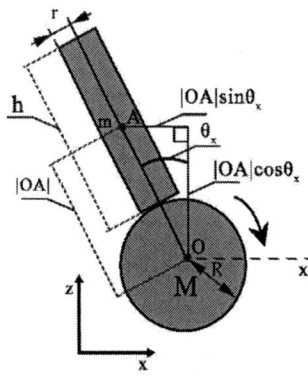

Fig. 3. Simplified model of the robot on a spherical base, used to derive equations of motion

978-1-6654-6787-2/22 $31.00 © 2022 IEEE

Making allowance for the Lagrange Equation:

$$L = \frac{1}{2}(M+m)\dot{x}_x^2 - m|OA|\dot{x}_x\dot{\theta}_x cos\theta_x +$$

$$+\frac{1}{2}m|OA|^2\dot{\theta}_x^2 + \frac{J\dot{\theta}_x^2}{2} - mg|OA|cos\theta_x \quad (7)$$

The equations of motion of the system will take the form:

$$\begin{cases} \dot{x}_1 = \dot{x}_x \\ \dot{x}_2 = \ddot{x}_x = \dfrac{S_1 + S_2 + 2Torque \cdot J/R}{H(H - cos(2\theta_x) + 2M \cdot L) + 2J(m+M)} \\ \dot{x}_3 = \dot{\theta}_x \\ \dot{x}_4 = \ddot{\theta}_x = \dfrac{S_3 + Torque \cdot Hcos\theta_x/R - L \cdot M \cdot \dot{\theta}_x \cdot \dot{x}_x \cdot sin\theta_x}{H^2(1 - cos\theta_x^2) + J(m+M) + M \cdot H \cdot L} \end{cases} \quad (8)$$

where:

$$H = |OA|m;$$

$$G = \dot{x}_x \cdot sin(2\theta_x);$$

$$S_1 = HL\left(-2\dot{\theta}_x^2 H \cdot sin\theta_x - G \cdot \dot{\theta}_x + 2 \cdot Torque/R\right);$$

$$S_2 = H(G \cdot H \cdot \dot{\theta}_x + g \cdot sin(2\theta_x) -$$
$$- 2 \cdot J \cdot \dot{\theta}_x \cdot sin\theta_x);$$

$$S_3 = Hsin\theta_x\begin{pmatrix} g \cdot m + M \cdot g - \dot{\theta}_x \cdot \dot{x}_x + \\ +m \cdot \dot{\theta}_x \cdot \dot{x}_x - m \cdot \dot{\theta}_x^2 \cdot cos\theta_x + M \cdot \dot{\theta}_x \cdot \dot{x}_x \end{pmatrix}$$

The friction between the sphere and the surface is given as viscous damping with parameters: normal contact stiffness $\mu_{cs} = 10^8 \text{ N} \cdot \text{m}$; damping factor $\mu_d = 10^2 \text{ N/(m/s)}$; kinetic friction coefficient $\mu_k = 0.5$; static friction coefficient $\mu_s = 0.5$ [12] [13].

IV. LINEARIZATION

As can be seen above, the resulting dynamical system is nonlinear since it contains trigonometric ratios. Since the design and modeling of such systems is too complicated to implement in practice, or rather requires enormous computing power to account for all possible influencing factors and system behaviors, in the case of a nonlinear dynamic system to simplify it was used its linearization [14].

Linearization is a way of reducing a system of nonlinear equations of the form:

$$\dot{x} = f(x) \quad (9)$$

to the form:

$$\dot{x} = A \cdot x \quad (10)$$

where A is matrix of numerical coefficients, x is a state vector for further analysis and modeling.

Since the linearized system shows a simplified model of behavior, for its linearization it is necessary to choose some fixed point or neighborhood set of points for which the properties of the system will be determined.

The fixed point is chosen so that the equality is fulfilled:

$$f(x^*) = 0 \quad (11)$$

In this case, linearization is made around fixed point $\theta = 180°$ with the allowance for the equilibrium of the system.

By linearizing the system, a series of partial derivatives in the form is obtained:

$$\frac{df}{dx}\bigg|_{x*} = \left[\frac{\partial f_i}{\partial x_j}\right] \quad (12)$$

The state value matrix shows the values of the function $\dot{x} = f(x)$ in neighborhood of a point x^*; these values are as approximated as possible to the values of the system $\dot{x} = A \cdot x$.

V. TRAJECTORY CONTROLLER CALCULATION & TRAJECTORY MOTION PROBLEM STATEMENT

It is advisable to use the modal control method when designing a system [15]. If it is taken into account that all state variables can be measured with sensors (e.g., an accelerometer, a gyroscope, an encoder), the variables of the controlled system are fulfilled, where each of the state variables is determined in a unified manner and the controller coefficients are set uniquely [16].

In this case the linear regulator used to control the robot represents a state regulator. The method of a state regulator construction is based on creating feedback matrix, which forms a control action in such a way that the error of movement along the x coordinate is reduced to 0. The block diagram of such a system for the x coordinate is shown in Fig. 4 (a).

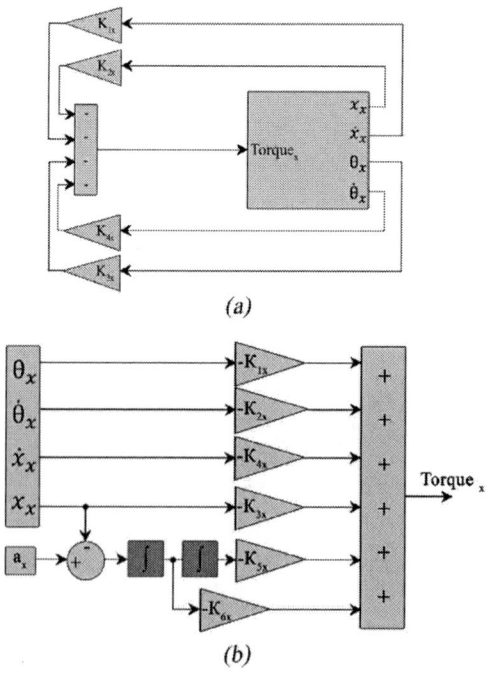

(a)

(b)

Fig. 4. Block diagram of state controller: (a) – state controller; (b) – controller with internal model

Provided that the plant model is as follows:

$$\begin{cases} \dot{x} = Ax + Bu \\ \quad y = Cx \end{cases} \qquad (13)$$

where u is the control vector; x is the state vector, A is called the input (or control) real- or complex-valued matrix, B is an output real- or complex-valued matrix, C is an output real- or complex-valued matrix. [16].

Disturbance model takes the form:

$$\dot{e} = Ae - Be \qquad (14)$$

State controller looks like:

$$u = Ke \qquad (15)$$

However, this type of a controller allows to operate only the free motion of the object, without taking into account the errors in setting its parameters, which can lead to the final instability of the system.

To control the forced motion of an object, it is necessary to upgrade the state regulator with an internal model. After that it will take into account external influences.

To control the forced motion of an object, it is necessary to upgrade the state regulator to one with an internal model that it will take account of external influences. There will be an increase in the order of the transfer function (2nd type) in terms of disturbance, which, in turn, means that zero steady-state error will be provided at a linearly varying speed. A block diagram of such a system is shown in Fig. 4 (b).

State-space model takes the form:

$$\begin{cases} \dot{x} = Ax + Bu \\ \dot{z} = \Gamma z + L\varepsilon \\ \quad y = Cx \end{cases} \qquad (16)$$

where Γ is the matrix defining the required dynamic properties of the control system; $\varepsilon = g - y$ is the error vector; g is the input signal vector; y is the output signal vector; H is the output matrix of the reference model.

Thus, final variant of the control regulator with internal model will look like:

$$\begin{aligned} u &= -K_1 x - K_2 z \\ K &= [K_1 \, K_2] \end{aligned} \qquad (17)$$

and represent a matrix of matrices consisting of feedback control coefficients.

VI. TRAJECTORY MOTION

When it is necessary to move an object, the end coordinate on the plane is set. And the control system perceives the distance between the start and end positions as an error. In a real system, the torque developed by the electric drive is limited, which is reflected in the limited speed of the system.

For such working conditions, trajectory planers are created: they allow the object not to perceive the final and

initial coordinates separately from each other but connect them with a certain trajectory. The trajectory motion is realized by introduction of motion limits in terms of the maximum permissible speed and acceleration, which makes it possible to ensure the robustness of the system with an electromagnetic converter based on the choice of an electric machine.

According to the source [17], the following concept of generating a task signal during trajectory motion was applied. It consists of 2 conventional parts: a trajectory generator, which sets the final and initial coordinates in 2-dimensional space, and a block responsible for the physical component of the object, the so-called trajectory planner.

The trajectory generator outputs the signal a_g, which can be expressed by the formula:

$$a_g(t) = a_{g0}(t) + \omega_g t \qquad (18)$$

where a_g – this is the desired spherical base coordinate, a_{g0} – initial coordinate of the spherical base, $\omega_g = \frac{a_g(t)}{dt} = v_1$ – spherical base linear velocity.

In this case, it is considered a continuous system; eventually it should be used a discrete system on the base of encoder with the period-based method measures of speed and average filter as presented in article [18, 19]. In the future, it is planned to conduct a more comprehensive study related to this issue.

The state matrix describing the equation would look like:

$$\begin{bmatrix} \dot{a}_g \\ \dot{\omega}_g \end{bmatrix} = \begin{bmatrix} 0 & 1 \\ 0 & 0 \end{bmatrix} \begin{bmatrix} a_g \\ \omega_g \end{bmatrix} \qquad (19)$$

Next, the generated signal a_g goes to the trajectory planner. This part is necessary because direct use of the generator signal can lead to unacceptable system accelerations and speeds outside the preset limit values.

Signal $a_g(t)$, passing through the trajectory planer, is converted into a signal $a_z(t)$. The state matrix describing this signal has the form:

$$\begin{bmatrix} \dot{a}_z \\ \dot{\omega}_z \end{bmatrix} = \begin{bmatrix} 0 & 1 \\ 0 & 0 \end{bmatrix} \begin{bmatrix} a_z \\ \omega_z \end{bmatrix} + \begin{bmatrix} 0 \\ 1 \end{bmatrix} \upsilon \qquad (20)$$

where $\omega_z = v$, υ – acceleration of signal $a_z(t)$.

During all these transformations an error $e(t)$ occurs in trajectory planning, which can be expressed as the difference between the outgoing signals:

$$e = a_z(t) - a_g(t) \qquad (21)$$

The task of the signal υ is to get rid of this error in the shortest possible time, while satisfying the following conditions:

$$|\upsilon| < \upsilon_{max}; \ |\omega_z| < \omega_{max} \qquad (22)$$

where υ_{max} and ω_{max} – limiters of the maximum permissible acceleration and speed.

The state matrix describing system errors will take the form:

$$\begin{bmatrix} \dot{e} \\ \dot{\omega}_e \end{bmatrix} = \begin{bmatrix} 0 & 1 \\ 0 & 0 \end{bmatrix} \begin{bmatrix} e \\ \omega_e \end{bmatrix} + \begin{bmatrix} 0 \\ 1 \end{bmatrix} \upsilon \qquad (23)$$

where $e = a_z(t) - a_g(t)$; $\omega_e = \omega_z - \omega_g$.

Further, using the Pontryagin maximum principle for a stationary system and taking the initial error as zero, a function describing the signal formation error can be obtained:

$$e = \frac{\omega_e |\omega_e|}{2\upsilon_{max}} \qquad (24)$$

As a result, by introducing the constraints υ_{max} and ω_{max}, the system operates at a given speed.

These limits are part of the base position control loop and affect the rate of the reference change. Therefore, speed reference limits affect the speed system output, but are not limit values for it.

VII. RESULTS OF MODELING

Several tests were conducted to assess the quality of the designed model.

With the parameters indicated in Table 1, the trajectory graphs and actual coordinates were obtained (Fig. 5). It illustrates the desired and actual trajectory of the system. This modelling was carried out with a speed limit in trajectory planner $v_{max} = 3$ m/s.

It is shown that there is significant overshoot in the actual trajectory. It can be the result of the impossibility of a comprehensive regulation of the movement limits (acceleration, speed) depending on the section of the path; for example, if it is possible to smoothly decrease the maximum permissible acceleration closer to the braking section, this overshoot could be reduced.

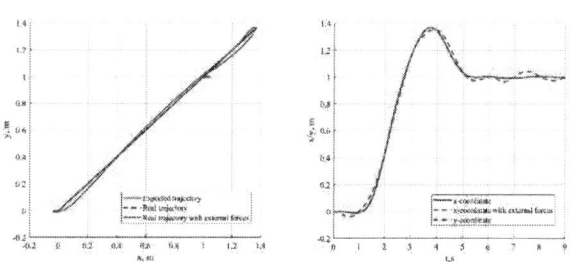

Fig. 5. Comparison of real and expected trajectories

TABLE I. SYSTEM PARAMETERS

Parameter	Value
Base radius R, m	0.24
Cylinder radius r, m	0.05
Base mass M, kg	5
Cylinder mass m, kg	10
Distance between centers of mass $|OA|$, m	0.6
Cylinder height h, m	0.72

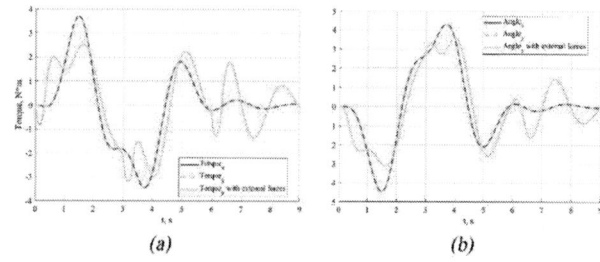

Fig. 6. (a) Rotational torques along the axes (graph $Torque_x$ with external forces is the same as graph $Torque_x$); (b) Deflection angles of the balancing load along axes (graph $Angle_x$ with external forces is the same as graph $Angle_x$)

Also, such deviations from the desired path are a consequence of linearization. However, the coordinates of the start and end points in the expected and real trajectories are the same.

Fig. 6(a) and (b) show the rotational torques along the axes, as well as the deflection angles of the balancing load model from the vertical position. The maximum deflection angle is $\theta_{max} = 4.3°$, the maximum torque is $Torque_{max} = 3.6$ N m.

Graphs were obtained under the disturbing influences represented by impulse shocks displacing the pendulum ($F_{external} = 2$ N). They are shown on fig. 5 and 6 as parameters with external forces. This force acts along the x-axis, hence the y-axis rotational moments will be greater to compensate this effect.

Also, studies were carried out on the dependence of the maximum rotational moment on the mass of the test weight model. Fig. 7 shows the graphs of the dependence of the maximum torque along the x and y axes. Because the original system is symmetric, when moving along a straight line with the same final coordinates along the axes, the maximum moments along the axes coincide. As can be seen from the graphs, with an increase in mass, the rotational moment increases proportionally, this dependence is linear−. Therefore, the requirements for the drive will increase with increasing mass and, accordingly, the moment of inertia of the test weight model.

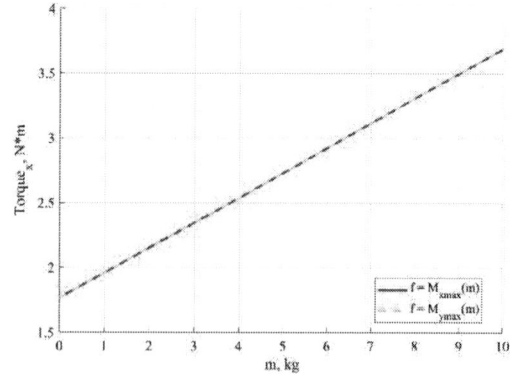

Fig. 7. Dependence of the maximum rotational moment $Torque$ on the mass of the test weight model along the axes

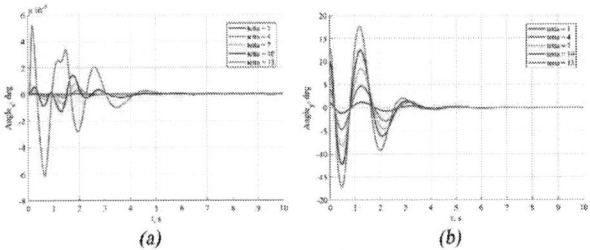

Fig. 8. (a) $Angle_x$ of the balancing load with different initial deflections; (b) $Angle_y$ of the balancing load with different initial deflections

A change in the load mass with a step of 1 in the range from 1 to 10 kg entails a change in the deflection angle in the range from 0.02 degrees along both axes. It can be noted that in the general case, with an increase in the inertia of the balancing object, it deviates more and more from the vertical, which is logical, since with an increase in mass, its force of attraction to the earth increases. However, with an increase in the mass of the cargo model, the maximum allowable angle to maintain a balanced state decreases for the same reasons.

Graphs in Fig. 8, for example, of the dependence of the angles along the x and y axes in time are presented for various initial deflection of a test weight model with a mass $m = 10$ kg. With an increase in the initial deflection, the oscillation of the transient process increases, and when $\theta_0 = 16°$ is reached, the object loses its stability (not shown on the graph). The limiting angle of deflection of the test weight model under the given weight and size conditions is $\theta_0 = 13°$.

VIII. CONCLUSION

At this stage of work, the resulting model is enough to experimentally determine the requirements for the drive based on certain load parameters. It was concluded that with an increase in the mass of the test weight model, the maximum torque that rotates the base increases, and therefore a more and more powerful drive will be needed. Also, studies were carried out on the influence of periodic impulse disturbing forces: the object remains stable, which indicates a high robustness of the system. The trajectory movement test was carried out; it demonstrates the presence of overshoot in a real system, which is most likely a consequence of the simplification of real nonlinear equations of motion and the impossibility of complex regulation of the limiters in the position loop.

In the future, it is planned to synthesize a universal platform for calculating drive parameters, which will simplify and unify the development process for such a class of robots.

The problem of balancing with a load can be considered as the problem of transferring a large heavy object of complex shape and with the non-standard center of mass location. This task can be solved by using a group of robots.

To sum up, it can be said that studies of the influence of load parameters (mass, moment of inertia, center of gravity) on the parameters of the robot control system could become the basis for creating standard solutions for choosing design parameters based on the load requirements. Therefore, solving the problem of balancing with a variable load, as well as the methodology for choosing the optimal drive, systematizes the development of a hardware platform for robots on a spherical base.

REFERENCES

[1] Wolfram Burgard, Cyrill Stachniss, Maren Bennewitz, Kai Arras, «Robot Control Paradigms», 2007.

[2] R. Hollis, «Ballbots», Scientific American, p. 72–78, October 2006.

[3] T. B. Lauwers, G. A. Kantor, and R. L. Hollis, «A Dynamically Stable Single-Wheeled Mobile Robot with Inverse Mouse-Ball Drive», в Conf. on Robotics and Automation, 2006.

[4] A.I.Alexan, A.R.Osan, S.Oniga, «Personal assistant robot,» в Design and Technology in Electronic Packaging, 2012.

[5] Maria Vittoria Minniti, Farbod Farshidian, Ruben Grandia, Marco Hutter, «Whole-Body MPC for a Dynamically Stable Mobile Manipulator», 2019.

[6] Tom Lauwers, George Kantor, and Ralph Hollis, «One is Enough!», в 12th Int'l Symp. on Robotics Research, 2005.

[7] S. N. R. Loomo, «LOOMO,» segway.com, 2021. [Internet Resourse]. Available: https://www.segwayrobotics.com/#/loomo.

[8] C. D. Green, "Equations of Motion for the Cart and Pole Control Task," 2020.

[9] R. M. M. Karl Johan Astrom, «Feedback Systems: An Introduction for Scientists and Engineers», Princeton University Press, 2008.

[10] B. Butenin, «Introduction to Analytical Mechanics», pp. 56-59, 1971.

[11] T. S.M., «Short Course in Theoretical Mechanics», High school, pp. 268-269, 1995.

[12] Y. G. I. E. Dorian A Hanaor, «Static friction at fractal interfaces», Tribology International, 2016.

[13] B. N. V. A. I. Persson, «Theory of rubber friction: Nonstationary sliding», Physical Review, April 2002.

[14] Steven L. Brunton, J. Nathan Kutz, «Data-Driven Science and Engineering: Machine Learning, Dynamical Systems, and Control», Cambridge University Press, 2017.

[15] V.V. Grigoriev, N.V. Zhuravleva, G.V. Lukyanov, K.A. Sergeev, «Synthesis of automatic control systems by the modal control method», 2007.

[16] I. Miroshnik, «Automatic control theory», Saint Petersburg, 2005.

[17] Sergey Lovlin, Artur Abdullin, Madina Tsvetkova, Aleksandr Mamatov, «Real-Time Optimal Trajectory Planning for Precision Tracking Systems with Dynamic Constraints», 26th International Workshop on Electric Drives: Improvement in Efficiency of Electric Drives (IWED), Jan 30 – Feb 02 2019

[18] Alecksey Anuchin, Valentina Astakhova, Dmitry Shpak, Alexandr Zharkov, Fernando Briz «Optimized Method for Speed Estimation Using Incremental Encoder», 19th International Symposium Power Electronics

[19] Alecksey Anuchin, Anton Dianov, Dmitry Shpak, Valentina Astakhova, Ksenia Fedorova «Speed Estimation Algorithm with Specified Bandwidth for Incremental Position Encoder» in Proc. 17th Int. Conf. on Mechatronics - Mechatronika (ME), 2016

978-1-6654-6787-2/22 $31.00 © 2022 IEEE

Optimized-Fuzzy Droop Controller for Load Frequency Control of a Microgrid with Weak Grid Connection and Disturbances

Haider M. Jassim
Ural Power Institute
Ural Federal University
Yekaterinburg, Russian Federation
khdzhassim@urfu.ru

Anatoliy Ziuzev
Electric drives dept.
Ural Federal University
Yekaterinburg, Russian Federation
a.m.zyuzev@urfu.ru

Abstract—Microgrids have been used to manage the electricity grid with a large number of integrated distributed generators. Without a proper control and management mechanism, the Microgrid efficiency and reliability may plummet which will influence the operation of critical loads connected to the grid. A Fuzzy-Droop controller is proposed to maintain the active and reactive power balance in the Microgrid and, hence, maintain frequency at the nominal value. A particle swarm optimization algorithm is implemented over the proposed design to locate the best gain values of the utilized controllers according to a certain cost function. The proposed control scheme is applied to the diesel generator and battery system units which will guarantee a rapid and sufficient response to disturbances in the network. Furthermore, a weak connection between the Microgrid and the main electricity grid is assumed, which will add more challenges to the proposed control technique. The effectiveness of the design is evaluated and compared to other control schemes.

Keywords—Microgrid, Load frequency control, Fuzzy controller, Droop controllers, Distributed generators.

I. INTRODUCTION

The increased integration of distributed generators (DGs) and the penetration of new technologies like electric vehicles (EVs) have challenged the conventionally controlled distribution network. The control and management systems are required to optimize the performance of the distribution network by maintaining a specific level of voltage and frequency while attenuating the effects of grid disturbances. The changes in frequency level need to be regulated and reduced rapidly to meet certain standers set by the power distributors. By patriating the electricity grid into smaller Microgrids, the distribution network can be managed in a more efficient manner. Distributed generators like diesel engines and photovoltaic (PV) solar panels could be easily integrated to support the network while a cost-effective operation of battery-on-grid systems and EV's can be ensured. These Microgrids could be ranged from tens to hundreds of Megawatts depending on their application site, and they can be operated with weak or no connection to the main grid [1,2]. Diesel generators and battery systems has the potential to emulate the inertial response provided by the main grid in the presence of grid disturbances. The power could be injected back to the network which allows regulating the voltage and frequency levels to their nominal values [2,3,4]. The grid controller is responsible for managing grid resources, during islanding and weak connection modes, and providing sufficient network stability by maintaining the active and reactive powers which imply a stable voltage and frequency. By sustaining these variables at an acceptable level, a safe and reliable operation of critical loads like industrial and hospital machines can be guaranteed [4].

Droop controllers have been employed to regulate the frequency and voltages of different types of Microgrids by injecting and suppressing power from the grid when necessary. Their hierarchical architecture allows for more flexibility and rapidness in response to grid events. Fuzzy droop controllers, on the other hand, have been utilized for the purposes of maintaining stability and performance of the network when unpredictable disturbances and uncertainties are in presence. The structural nature of the fuzzy controllers enables them to handle higher disturbances and a wider range of uncertainties in the system model. A sectional droop controller with dynamically changing gains has been implemented to regulate the frequency of a Microgrid [5]. This controller has been integrated with a fuzzy-based pitch controller of an on-site wind turbine to provide better performance at energy generation and stabilization levels. Another research had the focus on testing a wider range of distributed generators connected to a DC Microgrid [6]. A supercapacitor has been used to enhance the grid performance. However, the objective was to maintain the state of charge (SoC) of the integrated battery system at a certain level to reduce the economic impact of the Microgrid technology. A fuzzy logic system has been used to tune the parameters of the sliding mode controller in order to obtain a fast correction for reactive power levels in the system [7]. The resulted droop controller has been used to adapt the maximum power tracking strategy to improve the generated power and allocate resources in the Microgrid. This is advantageous for reducing the influence of switching to islanding mode when losing the connection to the main grid. A similar idea has been explored by [8], where a fuzzy-sliding mode controller has been employed to minimize the effect of the nonlinearities in the Microgrid architecture. The research in [9], suggested the combination of the H_∞ controller with the fuzzy logic controller to increase the robustness of the system to disturbances and modeling errors with load perturbation. In the research, modeling has been performed for each distributed generator which compensated for modeling the whole grid. A generalized droop control scheme has been proposed to manipulate the values of active and reactive power to regulate the voltage and frequency simultaneously in AC Microgrid [10]. The research investigates the utilization of a fuzzy rule system to tune the generalized droop control scheme and optimize the operation of the Microgrid based on the variations in the power and frequency levels. A three-level fuzzy droop control scheme has been proposed by [11] to ensure power-sharing in case of various grid catastrophic events. The first level has been used to control the voltage and current supplied to the load while the second and third levels compensated the active and reactive power to achieve a stable voltage and frequency grid-wide. The importance of maintaining the values of voltage and frequency at a certain level measured on Microgrid loads has been highlighted by [12], where a fuzzy logic droop controller has been employed to minimize the fluctuation in voltage level for data centers load. The research emphasizes the significance of small changes in these parameters on the functionality of such critical loads.

In this research, a two-level droop controller for load frequency regulation in a Microgrid is developed. The primary level is implemented as a classical PID controller while the fuzzy logic techniques are employed in the secondary level. The parameters of the two-level controller scheme are to be tuned, and the performance of the grid is to be optimized using the particle swarm optimization (PSO) method. The focus of this research is to stabilize and rapidly regulate the frequency of a critical load connected to the grid. The time response of the frequency deviation should follow certain international standers and maximum limits should not be breached. This will be guaranteed by the formulated cost function which will guide the optimization procedure. It is also assumed that the Microgrid is connected to technologies like battery systems, diesel generators, PV solar panels, electric vehicles (EVs). Two droop controllers are developed in a decentralized manner to manipulate the power produced by the battery system and generator unit which inject inertia into the system and maintain the frequency level. While the DC-DC converter of the PV panels and the EVs charging unit are controlled by fuzzy logic controllers. The aggregated model of the Microgrid depends on the connected technologies, their penetration levels, and the type of load being serviced.

In the next section, the overall description of the Microgrid architecture will be examined. Section III will discuss the utilized controller schemes implementation to the Microgrid system, while the results will be presented in section IV. Section V will provide the conclusion.

II. STRUCTURE OF MICROGRID

In this section of the research, the dynamical model of each part of the Microgrid will be presented. It should be mentioned that the Microgrid has been structured according to the work in [13], where they fused their design with a wide verity of technologies and each segment having a certain contribution to the final model of the Microgrid. A general illustration of the Microgrid is shown in figure (1). In the figure, the Microgrid is connected to the main electricity grid by a switch which is assumed to be proportional. The state of grid connection (GC) can vary from fully connected to fully isolated which indicates that the Microgrid is operating in islanding mode. The Microgrid is assumed to be DC in nature and power convertors are being utilized between its components.

A. Diesel Generator Model

The diesel generator is considered an essential supplier of electrical power to Microgrid loads in case of a weak or no connection to the main grid. The centrifugal governor of the turbine can be modeled as a first-order transfer function with a time constant [13]:

$$\Delta P(s) = \frac{1}{1+T_{die}\,S} \qquad (1)$$

where T_{die} is the time constant of the diesel governor and ΔP is the change in the diesel power. As this constant decreased in value, the governor will respond more quickly to the disturbances in the Microgrid. The actual turbine of the generator can also be modeled as a first-order transfer function [13]:

$$PG(s) = \frac{1}{D+H\,S} * (\Delta P(s) + P_{d,L}(s)) \qquad (2)$$

where H represents the inertia of the turbine and it depends on the material and shape used in its construction. On the other hand, D represents the damping factor of the turbine. While the $P_{d,L}$ encapsulates all the load and disturbances that affect the unit. The sole operational function of the diesel generator is established by controlling the fuel injection valve, the speed of the turbine can be regulated and, thus, the output power (PG) can be manipulated to restore balance into the power system.

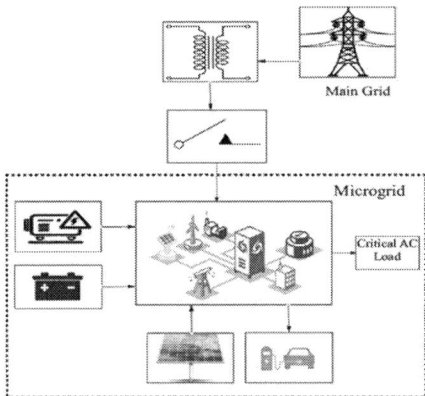

Fig. 1. General Microgrid architecture.

B. Battery System Model

Similar to the diesel generator unit, battery systems are utilized for supporting the grid when a disturbance affects the quality of the supplied power. Although they are more limited in terms of power capacity, battery systems can have a rapid response to events using their reserved energy. To achieve a functional operation in a Microgrid environment, battery systems are connected to a bidirectional power electronic convertor which enables them to absorb and eject power. The power controller part of the battery system can be modeled as [13]:

$$PB(s) = \frac{1}{1+T_{bat}\,S} \qquad (3)$$

Considering that T_{bat} is the time constant for battery system response. To emulate the physical operation of the battery system, it is essential to consider two other parameters. The maximum installed capacity $P_{Max}^{insalled}$ which represents the maximum power value that can be stored in the battery reserve. While the $P^{discharge}$ represents the maximum power discharge value can be achieved by the power electronic devices without demining the internal switches. Factoring these parameters into the model of the battery system, the output power can be as follows:

978-1-6654-6787-2/22 $31.00 © 2022 IEEE

$$PB^{out} = \begin{cases} -PB, & PB(s) \leq P_{Max}^{insalled} - P^{discharge} \\ PB, & PB(s) \geq -P_{Max}^{insalled} - P^{discharge} \\ PB(s), & otherwise \end{cases} \quad (4)$$

The values of the maximum capacity and maximum discharge parameters depend on the topology of the employed battery system. For practical application, these extremes should be avoided to prevent the overheating of the power electronic elements since the control signal will be supplied in form of a PWM signal. A deep discharge of the battery should also be prevented because such an event may lead to a significant degradation in the battery state of health (SoH).

C. PV Solar panels Model

The integration of renewable energy sources in the Microgrid provides the opportunity to sustain a large share of the grid in the situation when the main grid cannot provide enough energy. PV solar panels can be modeled using the five-parameter model for output current manipulation [13]. However, for this work, it is beneficial to consider only the DC-DC power convertor model. Owing to the fact that, the main objective of this research is to harness the power of the PV solar cells by controlling the converter. The PV solar cells then are considered as a current source with the produced signal being fed directly to the DC-DC converter. The convertor will manage the power passed to the grid according to the provided control signal. The following DC-DC buck converter model is adapted from [14] after making appropriate changes:

$$V_{out} = \frac{s\frac{1}{C}}{s^2 - \frac{1}{RC}s + \frac{1}{LC}} I_{in} \quad (5)$$

where R, C and L represent the resistor, capacitor, and inductor coefficients respectively.

D. Electric Vehicles Model

Electric vehicles represent the recent challenge for electricity grid. They can be viewed as a variable DC load with unpredictable connection and disconnection periods. There is significant research in the field to establish a control scheme for managing the charging periods and energy of the EVs. However, in this work, we are more concerned with the effect of the EV's DC charger on the Microgrid and best strategy to control its operation. The dynamical model of the DC charger maybe represented by two consecutive blocks. First block is the DC-DC charger which can be modelled as in equation (5). The second stage is the power booster which can be modelled as [15]:

$$V_{out}^{booster} = \frac{s\frac{1}{L}}{s^2 + \frac{1}{L}s + \frac{1}{LC}} I_{out}^{buck} \quad (6)$$

Where the output current of the DC-DC converter will act as an input to the booster unit. The two units will operate synchronously, and their operation will be controlled by synchronized switches.

E. The Agregated Model of the Microgrid

The complex architecture of the Microgrid depends on the combination of the four previously demonstrated technologies. Each component of the Microgrid participates by its penetration value and the actual produced or consumed power during the operation. These values are fused in one coherent whole that connects them to the consumer represented by the AC critical load in fig. 1. Each component is controlled by a specific control mechanism to regulate its operation as required. Output and input values of all components are normalized then converted to penetration percentage. Furthermore, the output of this combined system is the frequency deviation from the nominal rated frequency, and the task of all controllers is to manipulate the output power of their associated systems to reduce the deviation to zero. Hence, the output voltage of each component is applied to the Microgrid connected load and the frequency is measured. The connection of these components to the Microgrid is achieved by a DC/AC convertor. While the connection of the Microgrid to the main grid can be specified as a percentage but in the actual physical system, it might be specified indirectly as a result of certain faults, disturbances, or by operator choice.

III. CONTROLLER SCHEME DESIGN

For a Microgrid with distributed generators, it is required to construct a decentralized controller to handle the variations in each subsystem and maintain the stability of the overall system. Therefore, each technology utilized by the Microgrid including the diesel generator, battery system, DC-DC converter of the PV panels, and the EVs charger unit is needed to be controlled by a separate controller. However, the main objective of all designed controllers is to regulate the frequency measured at the critical load. A particle swarm optimization PSO algorithm will be employed for this purpose. Consequently, the tuning mechanism of the distributed controllers will be conducted in a centralized fashion with certain goals set in the cost function. As previously illustrated, the most vital parts in the Microgrid are the diesel generator and the battery system. The diesel generator provides a sufficient power when needed while the battery system is utilized for fast reaction to power change in the grid. Thus, the fuzzy droop controllers will be specifically applied on these units while other segments (PV, EVs) will be controlled by regular Fuzzy controllers.

A. Fuzzy-Droop controller

The fuzzy droop controllers are responsible for stabilizing the Microgrid frequency after the occurrence of disturbances. These controllers consist of two segments operating simultaneously on the system. The primary PID controller acts to restore the active balance in the grid when a frequency deviation is sensed across the AC load. While the secondary Fuzzy-Like (PI+P) controller is to be triggered if a large deviation is sensed at the output due to disturbances of generation unit outages. The secondary load frequency controller will inject inertia into the Microgrid and restore the nominal value of frequency. The fuzzy droop controller structure can be shown in fig. 2 where an illustration of the diesel generator control mechanism is demonstrated. after being integrated into the Microgrid, the output of the diesel generator could be considered as the frequency deviation in the grid from the nominal rated frequency (Δf). An identical structure is employed in the control of the battery system due to its capability in supplying and absorbing power at high speed to restore balance in the network. However, it should be mentioned that because of the current limitation in battery technology, the real-world implementation of such a

Microgrid will rely heavily on the diesel generator since it can provide high power for longer periods.

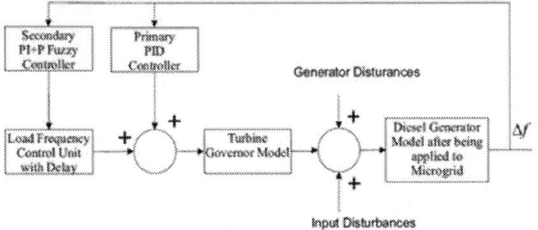

Fig. 2. Fuzzy Droop controller applied to generator model.

B. Type 1 Fuzzy Logic controllers

Fuzzy logic controllers are utilized because of their ability to handle uncertainties and disturbances that occur during system operation. Type 1 fuzzy logic mechanism is built on the concept of converting the crisp input values to the fuzzy domain by fuzzification process. This process utilizes input membership functions to project the input values into that domain. Then output values of the fuzzy mechanism will depend on the firing of fuzzy rules built in the inference engine. This value will, again, be projected through an output membership function to produce the final output. Input and output variables are crisp numerical values which corresponds to physical signals utilized by the system and can be transferred to control signals using actuators. Figure (3) demonstrates the input membership functions, output membership functions, and the fuzzy surface of the fuzzy logic controller used in this research. The first input corresponds to the frequency deviation while the second input corresponds to the rate of change in that deviation. Table (1) shows the rule system followed to design the PI+P-like fuzzy controller.

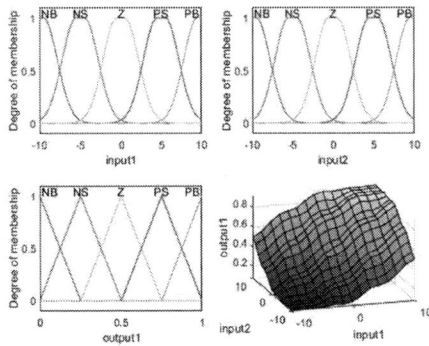

Fig. 3. Type 1 fuzzy logic controller.

For the electric vehicle charger and the PV system, represented by the DC-DC converter, this will be the only controller in operation. While, as discussed previously, for the diesel generator and the battery system it will be considered as the secondary controller.

TABLE 1. FUZZY RULES

$\Delta f / \Delta \dot{f}$	NB	NS	Z	PS	PB
NB	NB	NB	NB	NS	Z
NS	NB	NB	NS	Z	PS
Z	NB	NS	Z	PS	PB
PS	NS	Z	PS	PB	PB
PB	Z	PS	PB	PB	PB

C. Type 2 Fuzzy Logic

A more advanced fuzzy logic system has been introduced recently to cope with the increased uncertainty in high nonlinear systems. Although the type 2 fuzzy logic controller is quite similar to its type 1 counterpart, the main difference is that type 2 employs a higher dimension membership function [16]. The extra dimension represents the magnitude of the uncertainty which adds a tremendous amount of calculation in the fuzzification process. This will also require a type reduction process in the defuzzification process before calculating the output of the fuzzy system [17]. In this research, a comparison between the operation of type 1 and type 2 fuzzy logic controllers will be conducted based on the application to the constructed Microgrid model. The designed type 2 PI+P controller is equivalent to the type 1 structure illustrated in part B of this section. The only difference is that the utilized membership functions will have thickness representing the uncertainty. The Karnik–Mendel algorithm is employed as a type-reduction mechanism. A detailed explanation about the fuzzy logic type 2 systems can be found in [18]. We will be confined with the current description due to the page limitation.

D. Optimization of controllers

After setting up all controller modules for each segment of the Microgrid, a particle swarm optimization (PSO) method is implemented over the entire design to find the best value of each controller gains. These values will guarantee an optimal operation of the Microgrid, and the achievement of the optimization goals set by the operator. These goals are formulated as a cost function which is expressed as follows:

$$cost = a \int |\Delta f| + b \int \forall (\Delta f > 0) + c \int \forall (\Delta f > 0.15) \quad (7)$$

where a, b and c are penalties for each part of the cost function representing their contribution. The first part of the equation will act to minimize the overall error of the time response. The second part will ensure that the optimized controller will not produce a large overshoot in the frequency, while the last part of the equation is added to emphasize the importance of not violating the standers set by the grid operator. In this case, the (0.15 H_z) represents the most restricted steady-state deviation value allowed by the European Union countries. The comprehensive structure of the controller scheme being applied to the Microgrid segments is demonstrated in fig. 4. Where the controllers applied to the diesel generator and the battery system are the fuzzy droop controller, while the controllers manipulating the PV and EVs charger models are the fuzzy logic controllers.

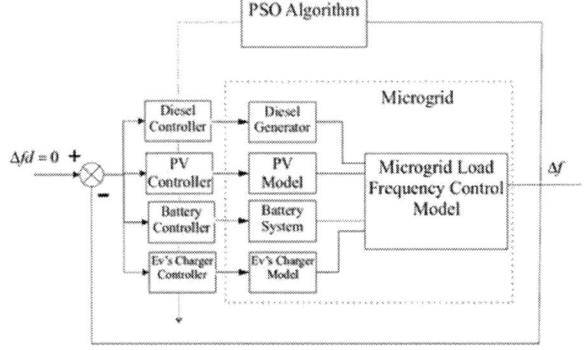

Fig. 4. Overall optimal controller scheme.

IV. SIMULATION AND RESULTS

The designed controllers are integrated into the system model and the optimization procedure is then invoked. The PSO algorithm will attempt to minimize the cost function in equation (7) by manipulating the controller parameters of the diesel generator, battery system, PV's DC-DC converter, and the EVs charger modules. Although the optimization is conducted in a centralized manner, the controllers are connected in a decentralized structure in which the operation of one controller does not affect other controllers. The penetration level and operational setting of the Microgrid are illustrated in the table below:

TABLE 2. MICROGRID OPERATIONAL SETTING

Part of the system	Parameter Name	Value
Diesel Generator	penetration level	40%
	T_{die}	2s
	H	2 Kg m^2
	D	1 Ns/m
Battery System	Penetration level	40%
	T_{bat}	0.02s
	Maximum discharge power	20%
	Maximum capacity	100%
PV Module	Penetration level	25%
EVs Module	Penetration level	25%
The Main Grid	Connection strength	20%
Disturbance	Demand disconnection	25%
	Disconnection at time	50s
	Generator outage	25%
	Generation outage at time	1000s

Other parameters, especially the PV panel DC-DC converter and the EV charger, can be selected according to practical implementations in the literature [14], [15]. However, their exact values will not affect the outcome of the research since an optimization algorithm is being employed. In this section, we will compare the results of the implementation of PID, Fuzzy PI+P, and Fuzzy-Droop controllers. The structure of the proposed controller shown in fig. 2 employs a fuzzy PI+P controller as a secondary controller while a PID controller has been implemented as the primary controller. The best cost function minimization outcome has been achieved by the Fuzzy-Droop controller structure with the result of the optimization process being demonstrated in fig. 5. next part of this section will analyze the effectiveness of the proposed design by comparing the results to other controllers, then a comparison between the utilization of the fuzzy type 1 and fuzzy type 2 technologies will be conducted.

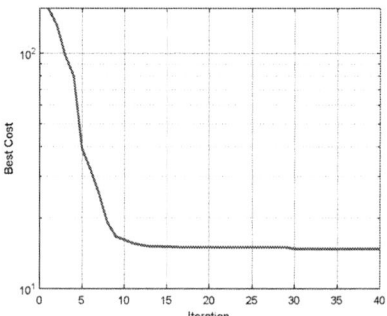

Fig. 5. PSO optimization for Fuzzy-Droop scheme.

A. Fuzzy Type 1-Droop Controller

As being illustrated previously, the optimization procedure is conducted over the entirety of the design for all utilized controllers. Figure (6) demonstrates the time response of the system's utilized technologies concerning the application of PID-only controllers. As can be seen in the figure, the control mechanism is activated when the system encounters the demand disconnection disturbance at the simulation time (50s). Similarly, when the generation outage is invoked at the simulation time (1000s) the controller attempt to stabilize the load frequency of the Microgrid by extracting more energy from the diesel generator than its actual installation capacity.

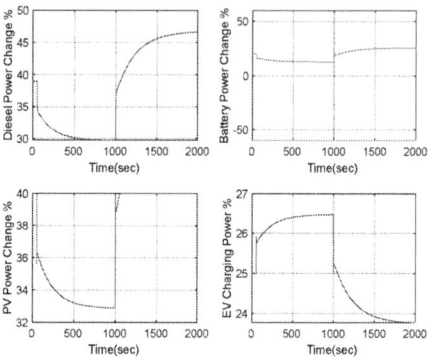

Fig. 6. Responses of Microgrid to PID-only control scheme.

Due to the stiffness of the response and the exhaustion of resources, the PID-only controller scheme fails to meet the system requirements. The Microgrid's technologies response to fuzzy PI+P controller scheme is exhibited in fig. 7. The controller completely fails to maintain the function of the Microgrid and reach a saturation state in all technologies. This is due to the fact that large errors have overwhelmed the fuzzy system and extended behind the reach of its defined input ranges shown in fig. 3. The saturated control action has led to the deformation of the test scenario such that the generator outage test is not being demonstrated.

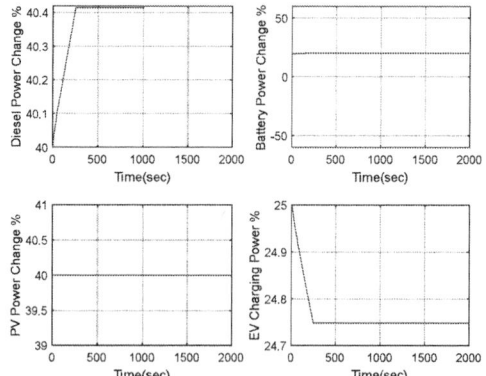

Fig. 7. Responses of Microgrid to fuzzy PI+P control scheme.

On the other hand, the implementation of the Fuzzy-Droop controller seems to produce a smooth transition in the response of all modules while maintaining system resources. As demonstrated in fig. 8, during disturbances, the generator unit has been utilized within the acceptable assumed installation capacity. The scheme guarantees the maximum usage of PV penetration capacity during the entire simulation time. While the EVs charging is not severely affected as in the case of the PID-only scheme.

978-1-6654-6787-2/22 $31.00 © 2022 IEEE

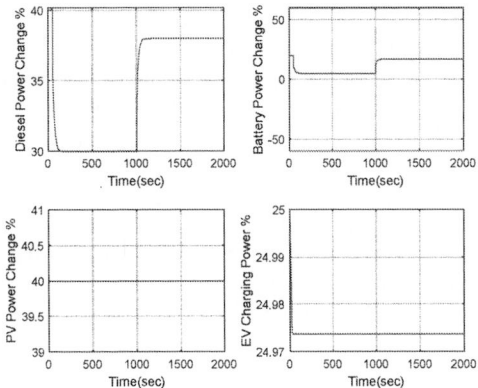

Fig. 8. Responses of Microgrid to Fuzzy-Droop control scheme.

Finally, the comparison of the load frequency responses of the three control schemes is shown in fig. 9. As can be inferred from the figure, the Fuzzy-Droop controller successfully maintains the frequency nominal value and eliminate the error resulting from disturbances. The controller produces lower overshoot and undershoot values in response to the disturbances and restabilizes the Microgrid in a shorter transition time. It should be mentioned that the outcome of the controller lies within the standers of the European Union energy operators.

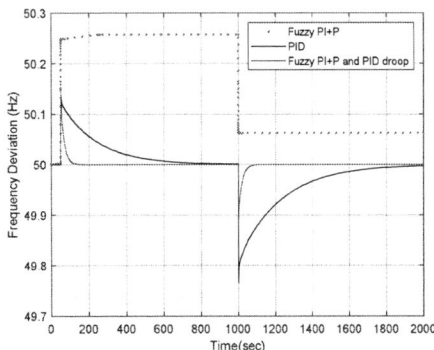

Fig. 9. Comparison of load frequency response of the three controller schemes.

B. Fuzzy Type 2-Droop controller

The fuzzy type 1 technology in the secondary controller has been replaced by the fuzzy type 2 technology. An experiment has been done to establish which is the more rigorous and robust controller to be implemented on the proposed Microgrid design. Again, the fuzzy type 2 mechanism will be employed by the Fuzzy-Droop controllers acting on the diesel generator and the battery system. Moreover, it will be implemented in the fuzzy controllers that are working on the PV's DC-DC converter and the EVs charger modules. The nature of the fuzzy type 2 systems allows them to deal with model changes and high levels of disturbances which will be handled as uncertainty in the design of the system. The time response of the fuzzy type 2 controllers is quite similar to their type 1 counterparts, with a slight improvement in the response. Therefore, it will not be demonstrated in this work. However, the main advantage of the fuzzy type 2 system is when a parametric change is encountered in the Microgrid. The time delay in the load frequency control unit shown in the figure is changed by 25% of its value. While the time constant of the diesel governor

unit is changed randomly with 50% of its value demonstrated in table 2. Figure (10) exhibits the changes in the response of fuzzy type 1 system and the occurred results from fuzzy type 2. A clear advantage for the utilization of fuzzy type 2 technology can be observed. The fluctuation in the frequency deviation response of the fuzzy type 1 system is eliminated by the utilization of fuzzy type 2 technology. Moreover, fuzzy type 2 controller produces less overshoot and settling time than the fuzzy type 1 implementation. Though, it should be mentioned that the computation time for fuzzy type 2 inference mechanism is much larger than it is in fuzzy type 1 which limits the application of such systems to devices with high computation capabilities.

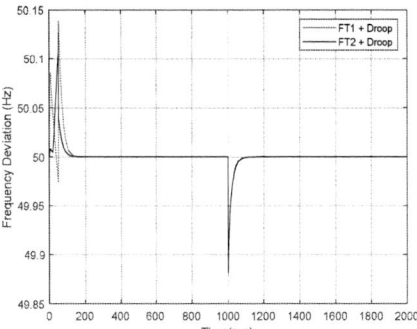

Fig. 10. Frequency deviation response of fuzzy type 1 and type 2 systems.

V. Conclusion

In this research, a Fuzzy-Droop control scheme has been proposed for load frequency control of a Microgrid. The studied Microgrid design accounts for the most common technologies integrated into the current practically implemented Microgrids. Furthermore, the Microgrid is assumed to have a weak connection to the main grid which resulted in a weak inertial response in case of disturbances. The Fuzzy-Droop control mechanism has been applied to the diesel generator unit and the battery system since they represent the actual drivers of the system regarding the maintenance of the active and reactive power balance in the network. A comparison between the operation of different controller structures and the proposed control scheme has been conducted, and the superiority of the proposed design has been demonstrated. A fuzzy type 2 technology implementation is compared to its type 1 counterpart to illustrate the effect of the uncertainties on the results of the load frequency control problem. It has been found that the fuzzy type 2 system has a smoother frequency recovery than the fuzzy type 1 mechanism.

REFERENCES

[1] A. Ziuzev and H. M. Jassim, "Power Hardware-in-Loop Implementation for Power Grids and Devices: Report and Review," in *2021 XVIII International Scientific Technical Conference Alternating Current Electric Drives (ACED)*, 2021, pp. 1–6.

[2] K. Benjamin, L. Robert, and I. Toshifumi, "A look at microgrid technologies and testing, projects from around the world," *IEEE Power energy Mag.*, vol. 6, no. 3, pp. 41–53, 2008.

[3] X. Lyu, Z. Xu, J. Zhao, and K. P. Wong, "Advanced frequency support strategy of photovoltaic system

considering changing working conditions," *IET Gener. Transm. Distrib.*, vol. 12, no. 2, pp. 363–370, 2018.

[4] Y. Hamdaoui and A. Maach, "Smart islanding in smart grids," in *2016 IEEE Smart Energy Grid Engineering (SEGE)*, 2016, pp. 175–180.

[5] U. Datta, J. Shi, and A. Kalam, "Primary frequency control of a microgrid with integrated dynamic sectional droop and fuzzy based pitch angle control," *Int. J. Electr. Power Energy Syst.*, vol. 111, pp. 248–259, 2019.

[6] A. M. Yasin, "Energy Management of a Stand-Alone DC Microgrid Based on PV/Wind/Battery/Diesel Gen. Combined with Supercapacitor," *Int. J. Renew. Energy Res.*, vol. 9, no. 4, pp. 1811–1826, 2019.

[7] Y. Zhu, Z. Wang, Y. Liu, H. Wang, N. Tai, and X. Jiang, "Optimal control of microgrid operation based on fuzzy sliding mode droop control," *Energies*, vol. 12, no. 19, p. 3600, 2019.

[8] Y. Mi, H. Zhang, Y. Fu, C. Wang, P. C. Loh, and P. Wang, "Intelligent power sharing of DC isolated microgrid based on fuzzy sliding mode droop control," *IEEE Trans. Smart Grid*, vol. 10, no. 3, pp. 2396–2406, 2018.

[9] L. Sedghi and A. Fakharian, "Voltage and frequency control of an islanded microgrid through robust control method and fuzzy droop technique," in *2017 5th Iranian joint congress on fuzzy and intelligent systems (CFIS)*, 2017, pp. 110–115.

[10] S. Ahmadi, S. Shokoohi, and H. Bevrani, "A fuzzy logic-based droop control for simultaneous voltage and frequency regulation in an AC microgrid," *Int. J. Electr. Power Energy Syst.*, vol. 64, pp. 148–155, 2015.

[11] H. R. Baghaee, M. Mirsalim, and G. B. Gharehpetian,

"Performance improvement of multi-DER microgrid for small-and large-signal disturbances and nonlinear loads: Novel complementary control loop and fuzzy controller in a hierarchical droop-based control scheme," *IEEE Syst. J.*, vol. 12, no. 1, pp. 444–451, 2016.

[12] N. L. Diaz, T. Dragičević, J. C. Vasquez, and J. M. Guerrero, "Intelligent distributed generation and storage units for DC microgrids—A new concept on cooperative control without communications beyond droop control," *IEEE Trans. Smart Grid*, vol. 5, no. 5, pp. 2476–2485, 2014.

[13] S. Sakkas, "Control of a DC Microgrid," Delft University of Technology, 2018.

[14] A. J. Forsyth and S. V Mollov, "Modelling and control of DC-DC converters," *Power Eng. J.*, vol. 12, no. 5, pp. 229–236, 1998.

[15] K. Colak, M. Bojarski, E. Asa, and D. Czarkowski, "A constant resistance analysis and control of cascaded buck and boost converter for wireless EV chargers," in *2015 IEEE Applied Power Electronics Conference and Exposition (APEC)*, 2015, pp. 3157–3161.

[16] N. N. Karnik and J. M. Mendel, "Centroid of a type-2 fuzzy set," *Inf. Sci. (Ny).*, vol. 132, no. 1–4, pp. 195–220, 2001.

[17] M. Y. HASSAN, H. I. ALI, and H. M. JASSIM, "Hybrid H-infinity fuzzy logic controller design," *J. Eng. Sci. Technol.*, vol. 15, no. 1, pp. 1–21, 2020.

[18] J. M. Mendel, R. I. John, and F. Liu, "Interval type-2 fuzzy logic systems made simple," *IEEE Trans. fuzzy Syst.*, vol. 14, no. 6, pp. 808–821, 2006.

Fault Diagnosis of the Tooth Belt Transmission of Cartesian Robot

Siarhei Autsou
Department of Electrical Power
Engineering and Mechatronics
Tallinn University of Technology
Tallinn, Estonia
siaaut@ttu.ee

Toomas Vaimann
Department of Electrical Power
Engineering and Mechatronics
Tallinn University of Technology
Tallinn, Estonia
toomas.vaimann@taltech.ee

Anton Rassõlkin
Department of Electrical Power
Engineering and Mechatronics
Tallinn University of Technology
Tallinn, Estonia
anton.rassolkin@taltech.ee

Bilal Asad
Department of Electrical Power
Engineering and Mechatronics
Tallinn University of Technology
Tallinn, Estonia
bilal.asad@taltech.ee

Karolina Kudelina
Department of Electrical Power
Engineering and Mechatronics
Tallinn University of Technology
Tallinn, Estonia
karolina.kudelina@taltech.ee

Van Khang Hyunh
Department of Engineering Sciences
University of Agder
Grimstad, Norway
hyunh.khang@uia.no

Abstract— **Condition monitoring of any electromechanical equipment is an essential tool to extend lifetime of the device and provide reliable work of the whole system. Due to friction mechanical subsystem is usually the weakest part of electromechanical equipment that. Early-stage diagnosis of that type of faults may save the equipment from serious failures. In this paper the model of tooth belt transmission of the Hirata cartesian robot is presented. Transmission operational principles and mathematical model reviewed from the point of fault diagnosis of the system. Importance of fault diagnosis of the transmission and diagnostics methodology is highlighted. Applied fault diagnostic method is based on Fast Fourier Transform (FFT) method. The simulation results of faulty tooth belt transmission are analyzed.**

Keywords—Fault diagnosis, motion analysis, robot motion, Fast Fourier Transforms, power transmission.

I. Introduction

Nowadays it is hard to find the industrial mechanism that does not include mechanical transmission. There are several types of transmission which is used in industrial robotics [1], [2]. Each type of mechanical transmission has its advantages and disadvantages, which depends on application, however, the main task of any type is to transfer the mechanical power and move different parts of the system [3]–[5]. According to Radzevich et al. [5] transmissions are subdivided to teeth, belt, worm, and chain gears.

Current work is based on the study of transmission used in a cartesian robot. Industrial cartesian robots consist of orthogonal axes, that move in different ways relatively to each other. This type of robots is designed to work with special attachments and details. They are used for conveyor systems, construction technological processes and pick and place operations inside closed area. Smooth and quiet movements of the parts robot provided by right work transmission of a cartesian robot [6], [7]. That type of industrial robot usually uses worm and belt gears [8]. Typically, worm gears are used to move the tool of cartesian robot, which increases accuracy of the tool [9], [10], and belt gears are used to move the biggest parts of the robot. Using the belt gear reduces the damage of construction in case of mechanical faults of transmission [11], [12].

In this paper a tooth belt transmission of Hirata cartesian robot is considered. Studied transmission consists of two pulleys and belt. To reduce the speed of the motor first pulley (connected to motor) is smaller than the second one (connected to load). In this case, transmission is increasing the torque on the second pulley, which is respectively able to move a bigger mass[13]. The main faults occurring during tooth belt transmission operation, according to [14], [15], are over/under tension of the belt, overheating of belt and pulleys, belt break and tooth pulley/belt damage. All of these faults are not able to damage the robot, but can bring to stochastic and nonlinear character to work process. The damages of belt or pulleys lead to reducing accuracy, speed and other important features of the robot. The diagnosis of faults can reduce the risks of above consequences and minimize the possibility of occurring faults. [14], [15]. For fault diagnosis of electromechanical system typically fast Fourier transform (FFT) and continuous Wavelet transform (CWT) is used. However, nowadays some advanced techniques like fuzzy logic (FL) or machine learning (ML) algorithms are applied [16]–[18].

The structure of the paper is as follows. Section II presents Hirata cartesian robot used for case study. Section III describes in details the tooth belt transmission (TBT) of Hirata robot. Equations used to develop the studied robot kinematic model (including gear model) are presented. Moreover, TBT transfer function is discussed. Also, in this section, there is presented description of bearing and corresponding faults that can occur during operation. Section IV describes TBT faults simulation and discusses implementation of fault diagnosis methods to such type of industrial robot.

II. Description of the Research Object

Cartesian robots are widely used in industry and they may have different structure, complex mechanical construction, and several levels on different axes [6]. The general sketch of a cartesian robot is presented in Fig. 1. As it can be seen from the sketch, the system has several mechanical gears that may be counted as the weak parts of the robot transmissions, because accuracy of the robot will be reduced if some of the gear fails. TBT has several advantages that help to reach the higher accuracy in such systems, e.g. absence of slippage and over tension, small load on the shafts and constant gear ratio [14], [15].

The research leading to these results received funding from the EEA/Norway Grants 2014–2021, Industrial internet methods for electrical energy conversion systems monitoring and diagnostics.

978-1-6654-6787-2/22 $31.00 © 2022 IEEE

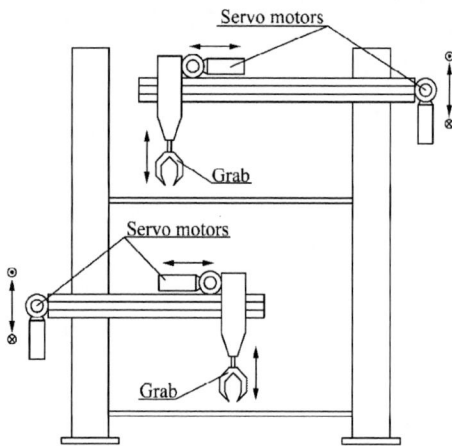

Fig. 1. Sketch of the caresian robot.

Hirata cartesian robot is a standard type of this class robots. It has three orthogonal axes, which can work with special attachments (light and medium weight) [19]. The illustration of Hirata cartesian robot is presented in Fig. 2(b). The view of TBT installed on the Hirata cartesian robot is presented in Fig. 2(a). The smaller pulley is driving pulley, another one is driven pulley.

(a)

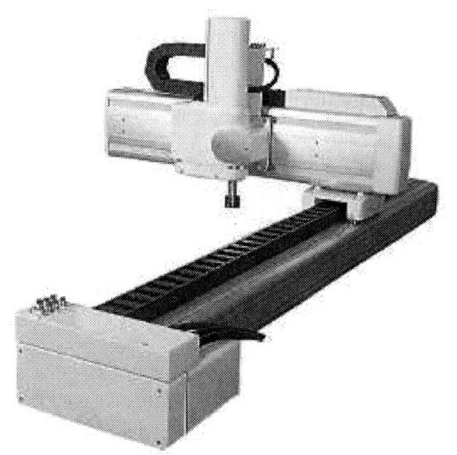

(b)

Fig. 2. The view of Hirata certesian robot (a) and view of the TBT (b).

The main parameters used for transmission mathematical model development are presented in Table 1.

TABLE I. MAIN PARAMETERS TBT OF HIRATA CARTESIAN ROBOT

Name of parameter	Value of parameter	
	Value	Dimension
Gear ratio	2	
Driving pulley diameters	75	mm
Driven pulley diameters	150	mm
Distance between the axes of the pulleys	150	mm
Length of belt	400	mm
Width of belt	20	mm

III. TOOTH BELT TRANSMISSION MODEL AND MECHANICAL FAUTS

To understand the nature of faults in TBT, corresponding model is developed. This model is based on TBT power-torque calculation [20], [21] and gear modelling examples [22]–[25]. To estimate the model reaction on mechanical faults, real-value experimental datasets from the faulty bearings are used.

A. Tooth Belt Transmission

The sketch of TBT used for model development is presented in Fig. 3, where ω_1, ω_2 – rotation speed of the driving and driven pulleys respectively, M_1, M_2 – torque of the driving and driven pulleys respectively, d_{o1}, d_{o2} – outer diameters of the driving and driven pulleys respectively, d_{n1}, d_{n2} – inner diameters of the driving and driven pulleys respectively, F_n – tension force, and F_t – tangential force.

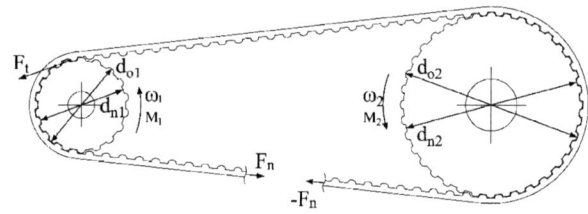

Fig. 3. Sketch of the TBT used for model development.

The transfer function of TBT is as follows:

$$W(s) = \frac{\omega_1(s)}{M_2(s)}. \tag{1}$$

As seen from (1), the input parameter is rotation speed of the driving pulley and the output parameter is torque of the driven pulley. Considering that

$$\omega_1(t) = \frac{d\varphi_1}{dt}, \tag{2}$$

then

$$W(s) = \frac{\varphi_1(s)}{T_2(s)}. \tag{3}$$

Torques influenced of driven pulley are presented in Fig. 4.

978-1-6654-6787-2/22 $31.00 © 2022 IEEE

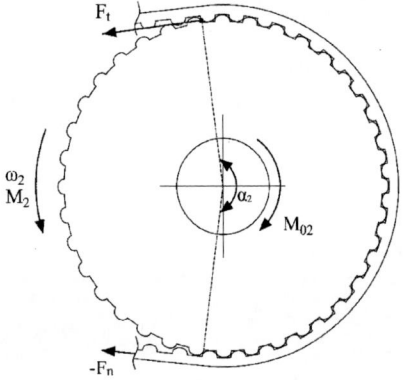

Fig. 4. The cheme of torques acting on driven pulley, where α_2 – the angle of belt girth.

The equation of torques acting on driven pulley by established movement is:

$$T_2 + T(F_t) - T_{02} = 0. \qquad (4)$$

T_2 is obtained from equation (4):

$$T_2 = T_{02} - T(F_t), \qquad (5)$$

where T_{02} – friction forces torque, $T(F_t)$ – tangential force torque. The tangential force can calculate the following way:

$$F_t = \frac{10^3 * P}{v_1}. \qquad (6)$$

Based on the equation (6) the expression for tangential force torque is obtained:

$$T(F_t) = \frac{9.55 * 10^3 * P * u}{n_2} \qquad (7)$$

or

$$T(F_t) = \frac{3.2 * 10^2 * P * u * \pi}{\omega_1}, \qquad (8)$$

where P – power of motor, u – gear ratio, n_2 – number of revolutions, w_1 – angular velocity of the belt on the driving pulley. The friction torque is summary tension forces of driving and driven lines of transmission. In this case, the friction torque is integral over all elements of the belt contacted with pulley.

$$T_{02} = r_{o2} \int_0^{\alpha_2} F_n d\alpha \qquad (9)$$

Considering that during the belt running on the pulley, the summary force (F_u) and centrifugal force (F_c) are occurring.

$$F_u = 1,1 \ldots 1,3 F_t \qquad (10)$$

$$F_c = \rho A v_2^2 \qquad (11)$$

where ρ – density of the belt, A – section area of the belt, v_2 – linear velocity of the belt on the driven pulley.

Based on equations (9)-(11) the final equation for friction torque is:

$$T_{02} = r_{o2}\alpha_2(1,1F_t + \rho A v_2^2 + F_0) \qquad (12)$$

where F_0 – the starting tension force. The F_0 can be calculated as follows:

$$F_0 = b(F_y + q_m v_2^2) \qquad (13)$$

where F_y – beginning tension of belt (table value); b – width of belt; q_m – mass of 1 meter of belt with width 1mm (table value).

The angle of belt girth can be calculated like this:

$$\alpha_2 = \pi + 2\arcsin\left(\frac{r_{o2} - r_{o1}}{L}\right) \qquad (13)$$

where L – distance between the axes of pulleys. The linear velocity of the belt can be calculated:

$$v_2 = \omega_2 r_{n2} \qquad (15)$$

Considering that:

$$u = \frac{\omega_1}{\omega_2} \qquad (16)$$

In the result we have:

$$v_2 = \frac{\omega_1 r_{n2}}{u}. \qquad (17)$$

The final equation for the torque of the driven pulley:

$$T_2 = A\omega_1^2 + B\omega_1 + C \qquad (18)$$

where

$$A = \frac{r_{n2}^2}{u^2}(\rho A r_{o2}\alpha_2 + q_m b)$$

$$B = \frac{r_{n1}}{Pu(1.1 * 10^3 - 3.2 * 10^2 \pi r_{n1})}$$

$$C = bF_y$$

In the result the transfer function of TBT is:

$$W(s) = \frac{s}{As^2 + Bs + C} \qquad (19)$$

B. Bearings

Faults occurring in TBT also influent on bearings of the robot, and vice versa. For example, displacement of the pulley's centers leads to rising load to bearings. Additional load to bearings extends influence on friction in the bearings, that lead to drying of lubrication and appearing rust [16], [26]. Also, smaller damages bearings lead to displacement normal work of TBT. The sketch, where faults can occur in bearing, presented in Fig. 5. The highlight area in the sketch is places where bearing has additional friction and load during the displacement of pulley's centers. These faults lead to occurrence of non-desirable disturbances and stochastic behavior of robot operation. Broken bearings reduce the efficiency as well.

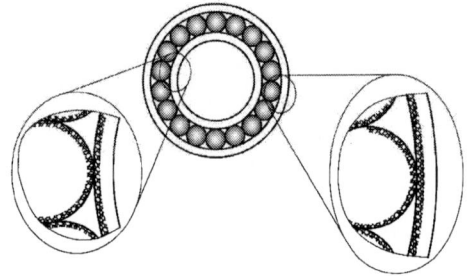

Fig. 5. Sketch of bearing faults.

In the literature, there are also discussed different faulty conditions of bearings, such as high temperature, attendance of currents on the motor shaft, cage damages, and different mechanical damages [3], [16], [26]. The diagnosis of bearing disturbances, where body damage of bearing by lubricant absences and rust occurred by micro damage's, is presented in Kudelina et al [16]. These bearing damages have an influence on the pulleys of TBT and can lead to unexpected fault.

IV. DIAGNOSIS MECHANICAL PARTS

A. Bearing Diagnosis

For bearings' analysis, in this work, fast Fourier transform and continues Wavelet transform were used. In Fig. 6, there are presented FFT spectra of faulty bearings.

(a)

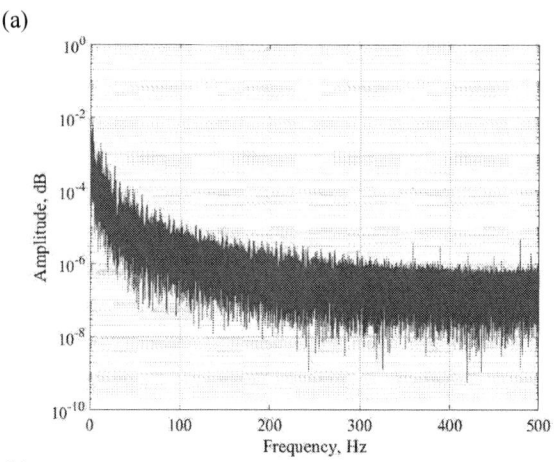

(b)

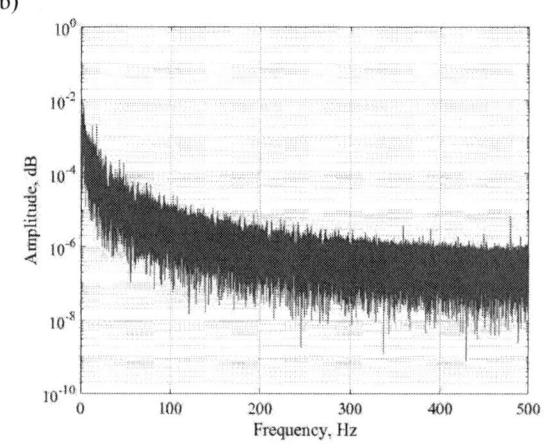

Fig. 6. Amplitude spectrum of the bearing output signal by FFT: (a) no lubricant case, (b) rusted bearing.

The bearing amplitude spectrum has a lot of disturbances, which influent on the normal work bearing. This disturbance lead to wear of bearing and possibility of failure the TBT. However, FFT method can't show magnitude of the disturbances, in this case need to use another diagnosis method.

Analysis by CWT method is presented in Fig. 7.

(a)

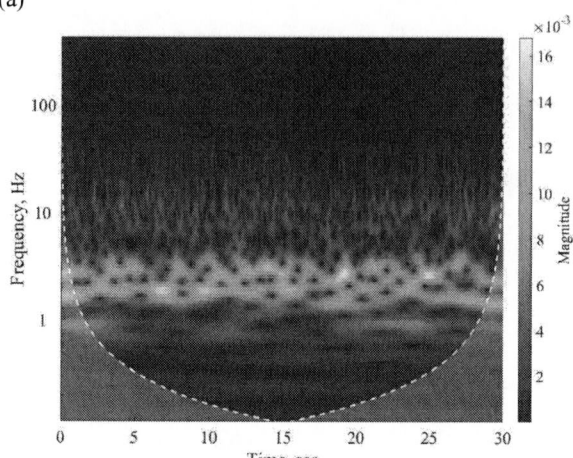

(b)

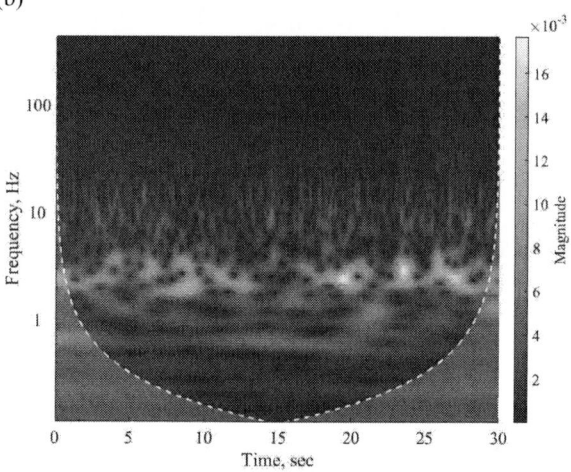

Fig. 7. Magnitude spectrum of the bearing outbut signal by CWT: (a) no lubricant case, (b) rusted bearing.

There can be seen the same structure of disturbances. Using CWT method, it can be said, in which development stage is the fault as well as the failure type can be prognosed.

B. TBT Diagnosis

During the continuous duty, different damages and faults of TBT may occur. For example, in belt or tooth of pulley, misalignment of shafts, faults of the control system [22], [24]. These types of faults have a nonlinear character and add the disturbance to output signal. Other damages, like wear of tooth belt or pulley are adding the additional harmonics to output signal. In this case we can find the reason of the faults in the power system [20], [22], [24], [27]. Also, the bearing fault lead to disturbances in the TBT. But based on the spectrum of output signal it is possible to decide, which moment the TBT can break down. For diagnosis the Fourier Fast Transform (FFT) method is used. Model sketch of TBT is presented in Fig. 8.

978-1-6654-6787-2/22 $31.00 © 2022 IEEE

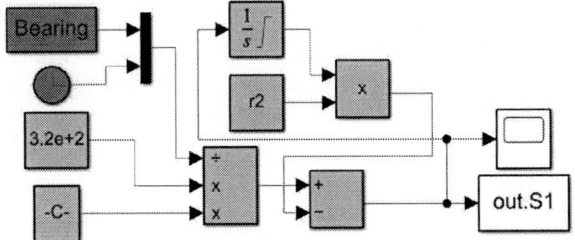

Fig. 8. The model sketch of TBT, where green area is model of the TBT and red area is faulty bearing signal.

During continuous duty, different damages and faults of TBT may occur. For example, in case of belt or tooth of the pulley, shaft misalignment or control system error can be presented [22], [24]. These faults' types have a nonlinear character adding disturbances to output signal. Other damages, such as wear of tooth belt or pulley add additional harmonics to output signal. In this case, the reason of these faults in the power system can be found [20], [22], [24], [27]. Also, bearing fault lead to disturbances in TBT itself. However, based on the spectrum of output signal, it is possible to detected, in which moment TBT is to be broken down. For diagnosis, the Fourier Fast Transform (FFT) method was used.

The results of modelling TBT with FFT diagnosis method is presented in Fig. 9 (a, b). Graph shows the output signal with noise (blue line) and without (red line).

(a)

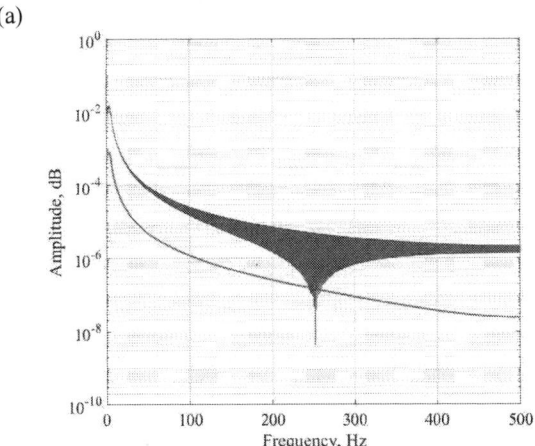

(b)

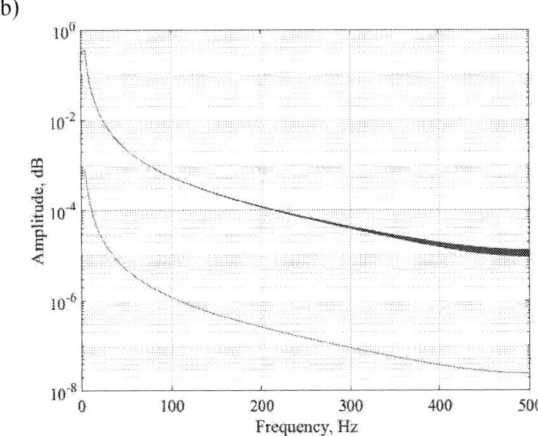

Fig. 9. Amplitude spectrum of the output signal by FFT method: (a) damage transmission with bearing in no lubricant case, (b) damage transmission with rusted bearing.

These disturbances are occurred by the bearing's damages. In the amplitude of spectra, it can be seen significant deviations from the normal work of the TBT, as shown in Fig. 9 (a). In this case, these disturbances can lead to unexpected break down of TBT due to additional load to pulleys. As presented in Fig. 9 (b), in the amplitude spectrum, it is seen smaller disturbances on TBT. However, regarding the character of these disturbances, it can be suggested that by time these faults will develop and lead to more serious consequences.

For validation of obtained results, CWT method was used, which is shown in Fig. 10 (a, b).

(a)

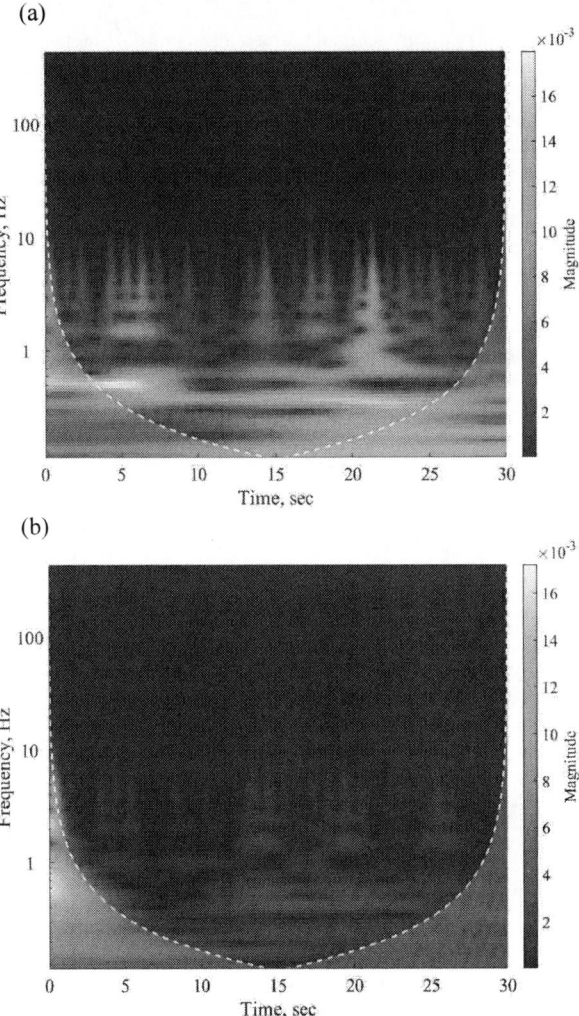

(b)

Fig. 10. Magnitude spectrum of the output signal by CWT method: (a) damage transmission with bearing in no lubricant case, (b) damage transmission with rusted bearing.

As seen from graphs, similar results can be observed. In the Fig. 10 (a), there are many disturbances in output signal. It proofs the fact that TBT can break down in sooner time. In the Fig. 10 (b), there are small amplitude disturbances. Based on the character of disturbances, it can be concluded that bearings' damages influent on TBT too. Results of modelling gives information about how long the belt transmission can work under these conditions.

V. CONCLUSION

This paper presents the diagnosis importance of the smallest damages in a robot's transmission system. As can be seen from the results, damage causes have significant disturbances in the operation of the transmission, which can lead to the failure of the entire mechanism. Two diagnosis methods (FFT and CWT) were presented. Both of these methods show good results for diagnosis of faults in the mechanical part of the robot. Based on the modelling results, it can be said that CWT method is more visual than FFT method.

At the same time, data spectral analysis helps by FFT helps to get the two-dimensional representation of the output signal, where frequency and amplitude are independent values. Regarding CWT, it is possible to estimate time of disturbances' appearance. Therefore, spectral analysis is a good filtering instrument that helps to get information about noises in output signals.

Based on real-value results of bearings' diagnosis, additional work tine of TBT can be suggested and expensive restore of the robot in case of TBT break can be prevented. As further work, advanced diagnostic methods will be considered, which allow assessment of not only mechanical damage, but also failures occurring in the control system of robot.

REFERENCES

[1]   V. Vullo, *Gears Volume 1: Geometric and Kinematic Design*, vol. 1. Springer Nature, 2020.

[2]   Isak Karabegović, *Industrial Robots: Design, Applications and Technology*. Nova Science Publishers, Incorporated, 2020.

[3]   Joseph R. Davis, *Gear Materials, Properties, and Manufacture*. ASM International, 2005.

[4]   D. W. Dudley, *Handbook of Practical Gear Design*. CRC Press, 1994.

[5]   S. P. Radzevich, *Dudley's Handbook of practical gear design and manufacture: Third Edition*. CRC Press, 2016.

[6]   Harry Colestock, *Industrial Robotics: Selection, Design, and Maintenance*. McGraw-Hill, 2005.

[7]   R. K. Miller, *Industrial Robot Handbook*. Springer Science & Business Media, 1989.

[8]   I. Dudás, *The Theory and Practice of Worm Gear Drives*. Penton Press, London, 2004.

[9]   P. Henson and S. Marais, "The utilization of duplex worm gears in robot manipulator arms: A design, build and test approach," *2012 5th Robot. Mechatronics Conf. South Africa, ROBMECH 2012*, pp. 3–6, 2012, doi: 10.1109/ROBOMECH.2012.6558461.

[10]   R. Tadakuma *et al.*, "The gear mechanism with passive rollers: The input mechanism to drive the omnidirectional gear and worm gearing," *Proc. - IEEE Int. Conf. Robot. Autom.*, pp. 1520–1527, 2013, doi: 10.1109/ICRA.2013.6630772.

[11]   S. Zhang, "Model predictive control of operation efficiency of belt conveyor," *Proc. 29th Chinese Control Conf. CCC'10*, pp. 1854–1858, 2010.

[12]   X. Cao, X. Zhang, Z. Zhou, J. Fei, G. Zhang, and W. Jiang, "Research on the monitoring system of belt conveyor based on suspension inspection robot," *2018 IEEE Int. Conf. Real-Time Comput. Robot. RCAR 2018*, pp. 657–661, 2019, doi: 10.1109/RCAR.2018.8621649.

[13]   Charles J. Boner, "Gear and transmission lubricants," *Ind. Lubr. Tribol.*, vol. 50, no. 1, pp. 121–131, 1998, doi: 10.1108/ilt.1998.01850aad.001.

[14]   K. Ma, X. Wang, and D. Shen, "Design and Experiment of Robotic Belt Grinding System with Constant Grinding Force," *Proc. 2018 25th Int. Conf. Mechatronics Mach. Vis. Pract. M2VIP 2018*, 2019, doi: 10.1109/M2VIP.2018.8600899.

[15]   A. G. Katsioula, Y. L. Karnavas, and Y. S. Boutalis, "An enhanced simulation model for DC motor belt drive conveyor system control," *2018 7th Int. Conf. Mod. Circuits Syst. Technol. MOCAST 2018*, pp. 1–4, 2018, doi: 10.1109/MOCAST.2018.8376636.

[16]   K. Kudelina *et al.*, "Bearing fault analysis of bldc motor for electric scooter application," *Designs*, vol. 4, no. 4, pp. 1–18, 2020, doi: 10.3390/designs4040042.

[17]   A. Jnifene and W. Andrews, "Experimental study on active vibration control of a single-link flexible manipulator using tools of fuzzy logic and neural networks," *IEEE Trans. Instrum. Meas.*, vol. 54, no. 3, pp. 1200–1208, 2005, doi: 10.1109/TIM.2005.847136.

[18]   K. Kudelina, T. Vaimann, B. Asad, A. Rassõlkin, A. Kallaste, and G. Demidova, "Trends and challenges in intelligent condition monitoring of electrical machines using machine learning," *Appl. Sci.*, vol. 11, no. 6, 2021, doi: 10.3390/app11062761.

[19]   R. L. Timings, *Newnes Mechanical Engineer's Pocket Book*. Newnes, 2004.

[20]   O. KROL and V. SOKOLOV, "Research of toothed belt transmission with arched teeth," *Diagnostyka*, vol. 21, no. 4, pp. 15–22, 2020, doi: 10.29354/diag/127193.

[21]   D. Wojtkowiak, K. Talaśka, D. Wilczyński, J. Górecki, and K. Wałęsa, "Determining the power consumption of the automatic device for belt perforation based on the dynamic model," *Energies*, vol. 14, no. 2, 2021, doi: 10.3390/en14020317.

[22]   A. Tonoli, N. Amati, and E. Zenerino, "Dynamic modeling of belt drive systems: Effects of the shear deformations," *J. Vib. Acoust. Trans. ASME*, vol. 128, no. 5, pp. 555–567, 2006, doi: 10.1115/1.2202153.

[23]   K. Kubas, "A model for analysing the dynamics of belt transmissions with a 5pk belt," *Arch. Motoryz.*, vol. 67, no. 1, 2015, doi: 10.14669/AM.VOL67.ART25.

[24]   B. Stojanovic, S. Tanasijevic, and N. Miloradovic, "Tribomechanical systems in timing belt drives," *J. Balk. Tribol. Assoc.*, vol. 15, no. 4, pp. 465–473, 2009.

[25]   S. M. Merghache and M. E. A. Ghernaout, "Experimental and numerical study of heat transfer through a synchronous belt transmission type AT10," *Appl. Therm. Eng.*, vol. 127, pp. 705–717, 2017, doi: 10.1016/j.applthermaleng.2017.08.079.

[26]   K. Kudelina, B. Asad, T. Vaimann, A. Rassolkin, A. Kallaste, and D. V. Lukichev, "Main Faults and Diagnostic Possibilities of BLDC Motors," *2020 27th Int. Work. Electr. Drives MPEI Dep. Electr. Drives 90th Anniv. IWED 2020 - Proc.*, 2020, doi: 10.1109/IWED48848.2020.9069553.

[27]   B. Stojanovic, N. Marjanovic, and M. Blagojevic, "Failure analysis of the timing belt drives," no. February 2015, 2011.

Evaluation of 5 kW Converter-Fed Induction Motor Losses by Analytical Calculation

Shahab Khalghani
Laboratory of Electric Drives Technology
Lappeenranta University of Technology
Lappeenranta, Finland
shahab.khalghani@student.lut.fi

Lassi Aarniovuori
Power Electronics, Machines and Power System Group
Aston University
Birmingham, UK
l.aarniovuori@aston.ac.uk

Juha Pyrhönen
Laboratory of Electric Drives Technology
Lappeenranta University of Technology
Lappeenranta, Finland
juha.pyrhonen@lut.fi

Abstract— **An analytical calculation model is constructed to determine the losses of a 5-kW squirrel cage induction machine at multiple operation points at different speeds. The losses are analyzed and compared with measured results to emulate the converter supply according to IEC standard loss segregation. The operation area includes load torque values below 125% of the rated torque and speeds below the rated operation point. The model utilizes motor design data, including geometry details and material parameters, to predict power losses in the induction motor and system. This kind of tool is useful to rapidly assess all power losses in an induction motor based on different imposed factors including voltage, load, stator frequency and ambient temperature.**

Keywords— *AC motors, electromagnetic model, efficiency, induction motor, iron losses, copper losses*

I. INTRODUCTION

The number of converter-driven electrical machines is increasing rapidly. Traditionally, an electrical machine was designed with a fixed stator voltage and frequency with only minimal variation tolerance and the performance and energy efficiency were compared at a single operation point with the grid supply. Converter supply creates new demands for the design procedure; however, it also gives more degrees of freedom to fulfil the application requirements. Motor treatment estimation can nowadays be assessed by simulation using Finite Element Analysis (FEA) tools. A major problem with these tools, however, is the time-consuming nature of the calculation processes for plotting losses and efficiency rates at various loads and inputs. Another problem related to commercial motors is that the exact construction of the machine cannot be known without full reverse engineering. If a calculation method with a programing basis such as MATLAB could be substituted with FEA tools, the computation could be resolved in seconds instead of requiring many hours. Thus far, however, studies using analytical calculation methods to model and predict losses have been unable to address the many factors (such voltage, stator frequency, and temperature) involved in operating at a wide range of loads and have not provided efficient methods. Recent work examining analytical methods for loss calculation in induction machines proposes utilization of an equivalent circuit to calculate losses and efficiency rates [1]. In the work, Nasir [1] endeavored to make a tradeoff between various prior analytical methods. [2] offers a comprehensive induction motor (IM) model fed by an inverter including losses such as copper, iron losses along with switching and

conduction losses. However, none of the methods included a preemptive and conducive model to predict the losses based on motor design and operation conditions mentioned earlier. In other words, their methods could not plot motor losses and motor efficiency rates based on expeditiously changing operating parameters such input voltage, frequency, ambient temperature, and load conditions. Among literatures regarding analytical loss calculation model, although Hannu [3] has succeeded to plot different loss figures based on 16 operating load points, his method is the way different from the model utilized in this paper.

This paper introduces an electromagnetic model, which is programmed with MATLAB, for power loss predication and efficiency computation along a wide range of operating load points of the motor. The approach focuses on development of a method for power losses calculation for a 5-kW IM in DOL supply with IEC standard loss segregation introduced earlier by the same research group [4]-[7]. The geometry of the machine and an analytical motor design-based approach are presented in [4], where the material parameters are utilized to define the loss elements. A fundamental 2D FEA for the 5-kW induction motor is given in [5] and the results of the analysis are compared with results measured with the IEC 60034-2-1 segregation of losses method at each supply frequency. In [6], power losses for this motor are evaluated and the results of commercial FEA software and open-source FEA software are compared. [7] addresses the loss evaluation sensitivity of the motor and induced harmonic content in FEA tools when applying different calculation time steps. Harmonic losses are omitted in the model discussed in this paper due to their complexity. An arbitrary loss increase to account for harmonic losses can be added to the sinusoidal supply losses when the results of the model are compared with the converter supply results.

The electromagnetic model can form a basis for multidisciplinary modelling. In this work, the goal of the multidisciplinary modelling is to combine different computational models to generate a comprehensive model that includes electromagnetic, mechanical, thermal and fluid dynamics models as presented in Fig. 1. This paper focuses on the electromechanical model, which is the fundamental basis of the comprehensive model. The idea of the complete model states "there is an undisputed relationship among these already mentioned models in Fig. 1". In other words, reference [8] shows the coupling of the electromagnetic model and mechanical model. The link between the thermal model and the other models is clarified in [9] and [10]. References [11] and [12] refer to the relationship of the fluid dynamics model with other models.

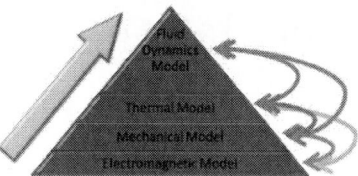

Fig. 1. Priorities of combined model for a motor

The rest of the paper is organized as follows. Section II overviews the contributing factors affecting the electromagnetic model directly or indirectly. Section III explains the approach used to compute each loss based on the contributing factors introduced in section II. Section IV compares analytical results with real measurements. In view of the adequate accuracy noted in section IV, section V discusses total losses with a crude fixed rate torque in analytical, measured, and other FEA tool results presented in previous study of this IM [6]. Section V presents concluding remarks and suggests areas for further study.

II. MODEL

This chapter summarizes the methods used in the loss calculation model. The aim is to give an overview of the model and the models are not presented in detail, since many of the methods are well-known. Nine factors affecting the model are presented in this section.

A. Primary Inputs

This part introduces the primary factors that are first inputted into the model before it is run:

1. Input voltage (U)
2. Stator frequency (f)
3. Load condition (torque) (T)
4. Core and winding temperature (θ)
5. Steel material applied ($B - H$ curve).

The "1. Input voltage", "2. Stator frequency" and "3. Load condition (torque)" were entered into the model in the form of IEC standard loss segregation [5]. The "4. Core and winding temperature" in the electromagnetic model, which gives the disposition of the core temperature factor, was disregarded, since it requires convoluted computations that cannot be included in the electromagnetic model. Popova [13] suggests a plan of action for combining the electromagnetic model and thermal model for a Totally Enclosed Fan Cooled (TEFC) induction motor of the same geometry type as in this work. Since temperature calculation is not done in this paper, the temperature rise in the stator windings measured at the laboratory was referred to the stator windings of the analytical model. To account for the rise in temperature in the rotor windings, an estimated value of 10 °C was added to the measured stator winding temperature rise. More information about the relationship between temperature and stator winding resistance can be found in [4] and for the rotor winding resistance in [5]. The model input for "5. Steel material applied" is the $B - H$ curve of steel

material M800-65A, which is utilized for peak flux density calculations.

B. Secondary inputs

The secondary inputs are consequentially conveyed based on primary input values:

6. Peak flux density ($\hat{B}$)
7. Air gap saturation factor (k_{sat})
8. Slip (s)
9. Skin effect

"6. Peak flux densities" in different parts of the induction motor are derived from the primary inputs. The maximum flux densities of the stator yoke $\hat{B}_{ys}$ and rotor yoke $\hat{B}_{yr}$ are calculated in the model, and then the maximum field strengths $\hat{H}_{yr}$ and $\hat{H}_{ys}$ are approachable through a mathematical interpolation process in the electromagnetic model using the $B - H$ curve of the steel material M800-65A. Integration of maximum flux densities with their respective correction factors enables the magnetic voltages of the stator yoke $\hat{U}_{mys}$ and rotor yoke $\hat{U}_{myr}$ to be acquired. For the magnetic voltage of stator tooth $\hat{U}_{mds}$ and rotor tooth $\hat{U}_{mdr}$, the stator tooth and rotor tooth flux densities for this IM are applied and then, using the $B - H$ curve, the respective field strengths of $\hat{H}_{ds}$ and $\hat{H}_{dr}$ are established.

Carter's principle is used to obtain the magnetic voltage of the airgap $\hat{U}_{m\delta e}$. An air gap equivalent δ_e is determined using Carter's principle. Then, we can calculate the magnetic voltage of the airgap $\hat{U}_{m\delta e}$ on account of δ_e. μ_0 is vacuum permeability and $\hat{B}_{air-gap}$ is peak flux density of the airgap calculated according to the variation of the input voltage and input stator frequency:

$$\hat{U}_{m\delta e} = \frac{\hat{B}_{air-gap}}{\mu_0} \delta_e \tag{1}$$

To acquire the magnetizing inductance L_m, δ_e has to be replaced with the effective air-gap length δ_{ef} to take the characteristics of the iron into account:

$$\delta_{ef} = \frac{\hat{U}_{m\delta e} + \hat{U}_{mds} + \hat{U}_{mdr} + \frac{\hat{U}_{mys}}{2} + \frac{\hat{U}_{myr}}{2}}{\hat{U}_{m\delta e}} \delta_e \tag{2}$$

In the following equation, m, $\acute{l}$, p, N_s, τ_p and k_{w1} are phase number, stator stack length, pole pair number, number of winding turns in one phase, stator pole pitch, and winding factor for the fundamental component:

$$L_m = \frac{m}{2} \frac{2}{\pi} \mu_0 \acute{l} \frac{4}{2p\pi} \frac{\tau_p}{\delta_{ef}} (k_{w1} N_s)^2 \tag{3}$$

L_m can slightly affect the stator inductance. The following equation shows the elements of stator leakage inductance used in calculation of the stator inductance L_s:

$$L_{s\sigma} = L_{\delta s} + L_{us} + L_{\sigma ds} + L_{ws} + L_{sq} \tag{4}$$

Where L_{us} is slot leakage inductance, $L_{\sigma ds}$ is tooth tip leakage inductance, L_{ws} is end winding leakage inductance, and L_{sq} is skew leakage inductance. The inductances are based on physical conditions such as permeance factor λ,

978-1-6654-6787-2/22 $31.00 © 2022 IEEE

stator slot Q_s etc. as shown in [14]. Air-gap leakage inductance $L_{\delta s}$ can be calculated with information of the harmonic winding factors as explained in the following. In the model, the air-gap leakage $\sigma_{\delta s}$ is computed in two parts of 600 harmonics in the form of $K_{\delta 1}$ and $K_{\delta 2}$:

$$K_{\delta 1} = \sum_{k=1}^{300} \left[\frac{\sin\left((1+2km)W_{\tau p}\frac{\pi}{2}\right) \frac{\sin\left((1+2km)q\frac{\alpha_{us}}{2}\right)}{\sin\left((1+2km)\frac{\alpha_{us}}{2}\right)}}{(1+2km)k_{w1}} \right]^2 \quad (5)$$

In Equation (5), $\alpha_{us} = \frac{p2\pi}{Q_s}$ and q is the number of stator slots per pole and phase and $W_{\tau p}$ is the winding pitch.

$$K_{\delta 2} = \sum_{k=-1}^{-300} \left[\frac{\sin\left((1+2km)W_{\tau p}\frac{\pi}{2}\right) \frac{\sin\left((1+2km)q\frac{\alpha_{us}}{2}\right)}{\sin\left((1+2km)\frac{\alpha_{us}}{2}\right)}}{(1+2km)k_{w1}} \right]^2 \quad (6)$$

Therefore, $\sigma_{\delta s} = K_{\delta 1} + K_{\delta 2}$. Since in asynchronous machines with a cage winding, the cage damps the harmonics, the air-gap inductance appears somewhat less important. This was estimated empirically by multiplying with a damping factor with a magnitude of 0.8. Hence, the stator air-gap leakage inductance is:

$$L_{\delta s} = 0.8\sigma_{\delta s}L_m \quad (7)$$

The magnetic voltage and inductance calculation elements can be seen in Fig. 2 in the following section. Another factor in the loss estimation model is the "7. air-gap saturation factor, k_{sat}". The air gap saturation factor emerges in the stator and rotor teeth, where the peak value of the flux density is considerably higher than a sinusoidal fundamental:

$$k_{sat} = \frac{\hat{U}_{mds} + \hat{U}_{mdr}}{\hat{U}_{m\delta e}} \quad (8)$$

The air gap saturation factor k_{sat} mainly varies based on $\hat{B}_{air-gap}$ and it is affected primarily by the input voltage and frequency. The more the machine teeth are saturated, the higher the induced air-gap voltage emerges. Upon the value of k_{sat}, the value of α is determined for the induced air-gap voltage E_{m1} in the form $\hat{B}_{air-gap} = \frac{\sqrt{3}E_{m1}}{N_s\omega k_{w1}\alpha\tau_p l}$, where ω is stator angular frequency.

The next factor is "8. Slip (s)" that in mechanical loss in section III, the path it is computed, will be described. Rotor frequency is slip times stator frequency and slip varies depending on rotor speed.

"9. Skin effect", in this model, influences the resistance and inductance of the rotor that accompanies a non-integral treatment of the rotor that equations for this computation can be found in [14].

III. Loss MODELS

The motor losses are divided into five major parts as given in IEC standard loss segregation: 1. Iron losses, 2. Stator winding losses, 3. Rotor winding losses, 4. Mechanical losses, and 5. Additional losses.

Fig. 2 shows the structure of the electromagnetic model. The model starts with the inputs given in black. Inputs are entered to run the script introduced in section II part A. The inputs in purple are secondary factors presented in section II part B. Magnetic voltages of different parts of the IM were colored in yellow. The red shapes are elements for the air gap calculation through Carter factors [14]-[15]. It can be seen in Fig. 2 that particular emphasis is given to the magnetic voltage of the airgap, which plays a significant role in the effective airgap computation, unlike other magnetic voltages, which are considered constant. The light blue shapes in Fig. 2 are resistances, inductances and impedances and they are

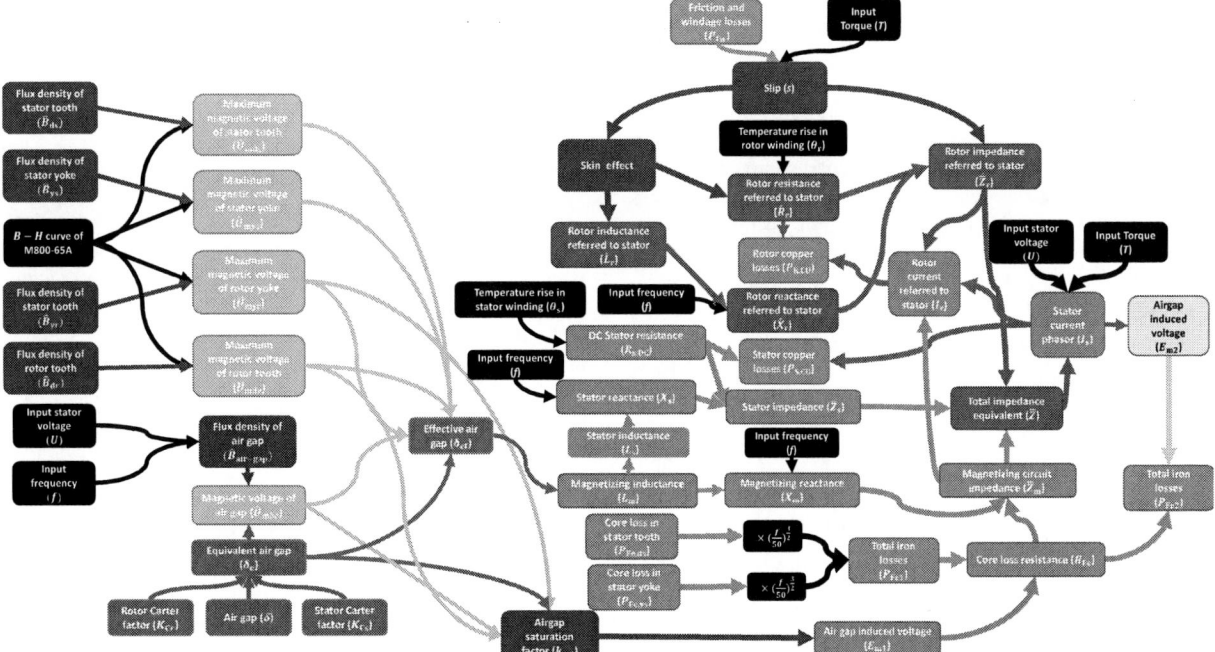

Fig. 2. Overview of the loss calculation model. Black: Primary contributing factors, Purple: Secondary contributing factors, Yellow: Maximum magnetic voltages of different parts of IM, Red: Airgap calculation elements, Light blue: Resistance, reactance, inductance, and impedance independent of rotor slip, Dark blue: Resistance, reactance, inductance, and impedance upon rotor slip, Orange: Airgap induced voltage independent of rotor slip, Dark green: Electrical current, Light green: Airgap induced voltage upon rotor slip, Gray: Different types of power losses

affected by temperature and frequency (without consideration the role of slip) and the dark blue shapes react to slip and skin effect. The various losses, grey boxes in Fig. 2, are described in the following sections.

A. Iron losses

The majority of the iron losses are generated in the stator teeth and yoke and only some in the rotor. In this 5-kW prototype induction motor, the rotor losses can be neglected because of the low rotor frequency. It should be noted, however, that these losses cannot be neglected in high speed machines. The laminated steel material used in the machine is M800-65A and it is assumed that high iron losses occur in the stator [4]-[5]. Iron losses of the stator, gray in Fig. 2, are computed with two steps. In the first step, iron losses are not affected by load; however, in the second step, the magnitude of load affects the iron losses. In the first step, a conventional iron loss method based on the motor steel lamination feature was employed as in [4] for the stator yoke $P_{\text{Fe,ys}}$ and stator teeth $P_{\text{Fe,ds}}$. Even if the load is not connected to the motor in analytical analysis, the iron losses can be computed. Since this method is mostly used for 50 Hz stator frequency, both equations are multiplied by $(\frac{f}{50})^{\frac{3}{2}}$ to make them flexible [5] for various input stator frequency f. P_{15} is specific iron loss per mass at 1.5 T and 50 Hz. Peak flux densities $\hat{B}_{\text{ys}}$ and $\hat{B}_{\text{ds}}$ along with correction coefficients k_{Fe} found in [4] were inserted into the equations. Summation of $P_{\text{Fe,ds}}$ and $P_{\text{Fe,ys}}$ returns total iron losses P_{Fe1}. Now, we attempt to acquire the core loss resistance R_{Fe}. The following equations are used.

$$P_{\text{Fe,ys}} = k_{\text{Fe,yoke}} P_{15} \left(\frac{\hat{B}_{\text{ys}}}{1.5 \text{ T}} \right)^2 m_{\text{ys}} \qquad (9)$$

$$P_{\text{Fe,ds}} = k_{\text{Fe,tooth}} P_{15} \left(\frac{\hat{B}_{\text{ds}}}{1.5 \text{ T}} \right)^2 m_{\text{ds}} \qquad (10)$$

$$R_{\text{Fe}} = \frac{m \times (E_{\text{m1}})^2}{P_{\text{Fe}}} \qquad (11)$$

E_{m1} in Equation (11), in orange in Fig. 2, the airgap induced voltage is independent of slip. Consequently, the iron losses cannot react with slip while the load changes. In order to reconcile iron losses with slip and rotor speed, we progress to the second step of iron loss calculation. In this step, we require airgap induced voltage dependent on slip, E_{m2}, in light green in Fig. 2. Based on the calculated total impedance equivalent $\bar{Z}$, which is affected by slip, shown in dark blue, and the input torque and voltage, the phasor stator current $\bar{I}_s$, in green in Fig. 2, can be calculated. $\bar{I}_s$ is computed by a simple division of $U_{\text{phase}}/\bar{Z}$, where U_{phase} is the phasor voltage. Based on the calculated values, the airgap induced voltage upon slip variation can be obtained with a simple equation of $\bar{E}_{\text{m2}} = U_{\text{Phase}} - \bar{I}_s \bar{Z}_S$. Thus, accurate iron losses can be evaluated based on slip with Equation (11) in the form $P_{\text{Fe2}} = \frac{m \times (|\bar{E}_{\text{m2}}|)^2}{R_{\text{Fe}}}$.

B. Stator copper losses

In the stator slots, the conductor windings are the source of the Joule losses.

$$P_{\text{S,Cu}} = m R_{\text{s,DC}} \bar{I}_s^2 \qquad (12)$$

Equation (12) was used to obtain the stator copper losses in the three-phase induction motor. $R_{\text{s,DC}}$ is the DC resistance value of the stator measured in the laboratory at 20 °C ambient temperature. The resistance varies with stator winding temperature rise. The difference between AC and DC stator copper losses is determined in the additional losses.

C. Rotor copper losses

Rotor copper loss $P_{\text{R,CU}}$ is mainly affected by torque, and to a minor extent by the slip, skin effect, and temperature rise in the rotor winding. $P_{\text{R,CU}}$ is obtained from Equation (13):

$$P_{\text{R,CU}} = m \acute{R}_r \acute{I}_r^2 \qquad (13)$$

The rotor current referred to stator $\acute{I}_r$ can be concluded as:

$$\acute{I}_r = \bar{I}_s \frac{\bar{Z}_m}{\bar{Z}_m + \acute{\bar{Z}}_r} \qquad (14)$$

D. Friction and windage losses

The mechanical losses P_{fw} occur due to friction in the rotor bearings and the forced air ventilation in the motor structure. Bearing losses are dependent on bearing type, lubricant features, shaft speed and torque. Windage losses are primarily connected with shaft speed and friction between environmental air and rotating motor parts. Equation (15) is based on synchronous speed and the protection degree of the motor (IP 55) proposed by [14]. k_p, D_r, l, v_r are (TEFC) experimental factors for windage and bearing losses, rotor diameter, core length in the machine, and velocity, respectively.

$$P_{\text{fw}} = k_p D_r (l + 0.6\tau_p) v_r^2 \qquad (15)$$

Through a mathematical interpolation process among input torque given before running the MATLAB script and a wide range of calculated torques T via equation (16), slip s, which a single-row matrix from 0 to 1, is found. With the obtained slip, rotor speed and other losses based on the acquired rotor speed can be calculated.

$$T = \frac{m(|\acute{I}_r|)^2 \frac{\acute{R}_r}{s} - P_{\text{fw}}}{2\pi \frac{f}{p}} \qquad (16)$$

E. Additional load losses

Additional losses P_{add} can appear next to the air-gap region when field distribution in the form of non-sinusoidal waveforms exists. Additional losses can occur even in stator slots and conductors as a result of the induced skin effect. In our model, additional losses cover losses such as harmonic losses, rotor iron losses, and the ignored losses from the difference between DC resistance and AC resistance in the stator windings. Equation (17) [4], where P_{in} and P_{out} are input and shaft power, respectively, is utilized to calculate losses of this loss type:

$$P_{\text{add}} = P_{\text{in}} \times [0.025 - 0.005 \log_{10}^{(\frac{P_{\text{out}}}{1 \text{ kW}})}] \qquad (17)$$

IV. COMPARISON TO MEASURED VALUES

The figures in this part of the paper show the analytical results obtained and the discrepancy between the analytical and measured results. To compare the analytically calculated loss components, four IEC segregation of losses test procedures were performed with different input frequency values. Voltage was adjusted to keep the flux density about constant, i.e., the *U/f* ratio is constant. The input frequencies used, and the line-to-line voltages are: 50 Hz –400V, 37.5 Hz – 300V, etc. [5]. The PMW induced losses are not considered, but typically they are relatively constant over the whole operation area. All the analytical figures were plotted at the same ambient temperature as used when ascertaining the experimental values. The aim of this comparison is to evaluate the discrepancy between the analytical results achieved with the electromagnetic model and the experimental results.

A. Iron losses

Fig. 3.a presents analytical results for iron losses at different loads. In [16], it is argued that iron losses can be influenced by load condition, since the losses drop as the load increases. Consequently, mild curvy contours appear in Fig. 3.a showing the analytical iron losses. Nevertheless, since the laboratory measurement of the iron losses is carried out at no-load condition, the result of measured loss remained unaffected by load variations. Fig. 3.b shows the margin between the experimental results and analytical results. A minor deviation in iron losses can be seen when reviewing the difference between the results.

In Fig. 3.b, the loss tolerance is inclined more to positive values than negative values because the rotor iron losses were ignored in the analytical values. The major share of deviation in iron losses is found at stator frequency of 25 Hz with 125% of rated torque. This deviation approaches 12 W. At this point, iron losses are 13.11% of the total losses in analytical results.

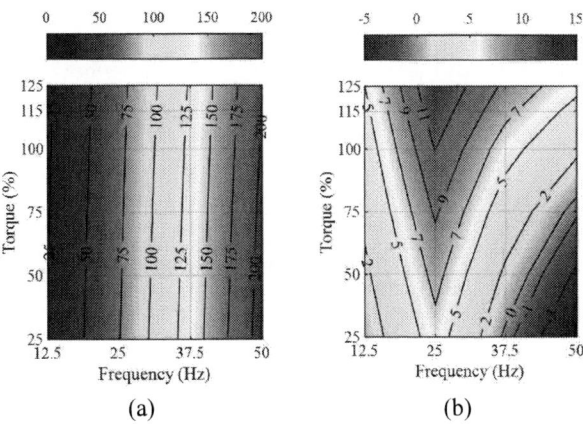

Fig. 3. a) Analytical iron losses (W) b) Deviation rates (W)

B. Stator copper losses

In Fig. 4.a, stator copper losses found with the analytical approach are lower at low stator frequency. This follows from the fact that the temperature rise in the stator winding is lower at lower frequency than at the rated frequency. In the analytical approach, the same ambient temperature was applied that appeared in laboratory measurement. Fig. 4.b shows that the discrepancy between the measured values and analytical results is trivial. This positive minor difference indicates the existence of DC resistance applied in this electromagnetic model since DC resistance measurement is lower than AC resistance that this loss deviation is noted in
Fig. 4. a) Analytical stator copper losses (W) b) Deviation rates (W)

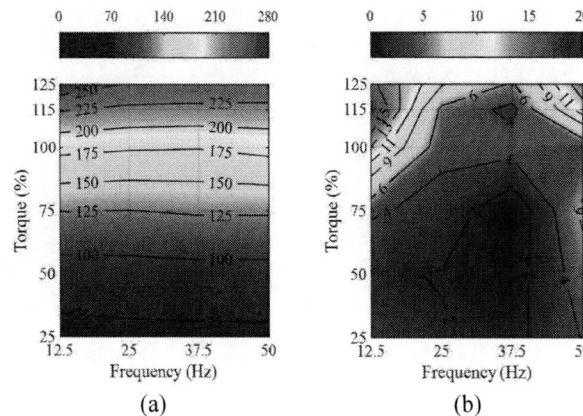

additional losses. The greatest discrepancy is found at 12.5 Hz at 115% of rated torque, where the discrepancy is 18 W, or in other words 7.36 %, whereas for other operation points this discrepancy is less than ±5%.

C. Rotor copper losses

The decreasing slope in Fig. 5.a with frequency decline is similar to the gradient found for stator copper losses due to lower temperature at lower frequency rates. The effect of torque on rotor and stator copper losses can be observed in both Figs 4 and 5. At rated torque and below, the analytical and measured results show a promising match (below 6% deviation) as a result of the AC resistance used in the model for rotor resistance. In overload conditions, particularly at low stator frequency of 12.5 Hz, Fig. 5.b displays indiscipline. At 12.5 Hz at rated torque (Fig. 5), the deviation can be clearly seen, and the values differ by up to 15.82%.

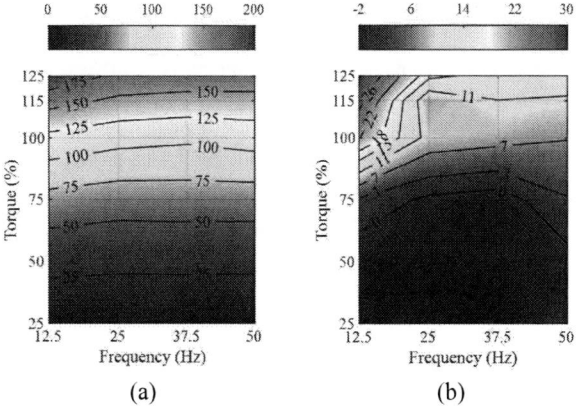

Fig. 5. a) Analytical rotor copper losses (W) b) Deviation rates (W)

D. Efficiency

Fig. 6.a shows the efficiency based on analytical calculation. Although the previous figures indicate that the analytical losses are somewhat lower than the measured losses, the efficiency deviation tolerance in Fig. 6.b is leading in negative side of efficiency deviation. This finding is due to higher values for additional losses and mechanical losses in the analytical approach than found with measured losses. At 37.5 Hz and 25% of rated torque, 3% non-conformity is found. This highest efficiency discrepancy between analytical and measured losses is 10.5 Watt in total losses estimation.

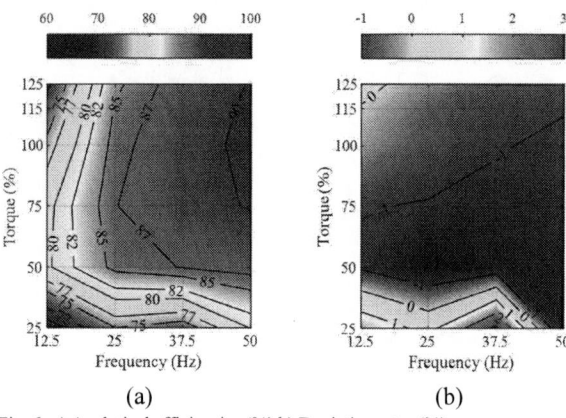

Fig. 6. a) Analytical efficiencies (%) b) Deviation rates (%)

V. DISCUSSION

An earlier study [6] considered loss calculation with commercial and open-source FEA tools to this motor employing IEC 60034-2-3 loss segregation. The results in [6] were obtained with time stepping FEA calculation, which is very time-consuming. Even when using a multicore supercomputer, calculation time is over 8 hours at a single operation point, if a 4 × 4 matrix of operating points is desired. On the other hand, the use of the analytical approach, applied in this paper, will enable these results to be computed in less than 1 minute. The harmonic losses induced by the PWM supply are relatively constant over the whole operation area [6] and can be easily added onto the fundamental losses. It should be kept in mind that FEA cannot model these high frequency losses, but additional loss models tuned to the specific case are used in a post-processing estimate of the losses. The losses found with the various approaches are shown in Fig. 7, where FEA values are taken from [6] and the experimental results and electromagnetic model results are from this study. The similarity between the measured values and the values found with the electromagnetic model indicate the validity of the electromagnetic model. Fig. 7 includes only the iron and copper losses.

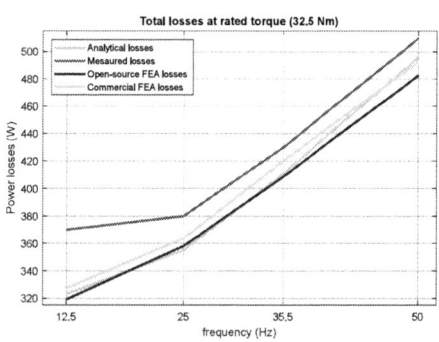

Fig. 7. Comparison of total losses of FEA values and measured values with analytical values acquired in this paper (unit in W).

VI. CONCLUSIONS

This paper presents an analytical loss model programmed with MATLAB as an alternative to FEA tools, whose use is time-consuming. The accuracy of the calculated values in the electromagnetic model was assessed with experimental values measured in laboratory. The proposed analytical model can be used to evaluate the energy efficiency of initial motor designs. For the 5-kW induction motor, the calculated efficiency deviates less than 2% over the whole operating area of the motor. The model does not consider frequency converter induced harmonic losses. Further study should thus investigate modeling of the harmonic losses of the frequency converter for this model.

REFERENCES

[1] B. Nasir, "An Accurate Iron Core Loss Model in Equivalent Circuit of Induction Machines", *Journal of Energy*, vol. 2020, pp. 1-10, 2020. Available: 10.1155/2020/7613737.

[2] Y. Liu and A. M. Bazzi, "A comprehensive analytical power loss model of an induction motor drive system with loss minimization control," *2015 IEEE International Electric Machines & Drives Conference (IEMDC)*, 2015, pp. 1638-1643, doi: 10.1109/IEMDC.2015.7409283.

[3] H. Kärkkäinen, L. Aarniovuori, M. Niemelä and J. Pyrhönen, "Converter-fed induction motor losses in different operating points," *2016 18th European Conference on Power Electronics and Applications (EPE'16 ECCE Europe)*, 2016, pp. 1-8, doi: 10.1109/EPE.2016.7695508.

[4] L. Aarniovuori, P. Rasilo, M. Niemelä and J. J. Pyrhönen, "Analysis of 37-kW Converter-Fed Induction Motor Losses," in *IEEE Transactions on Industrial Electronics*, vol. 63, no. 9, pp. 5357-5365, Sept. 2016, doi: 10.1109/TIE.2016.2555278.

[5] P. Lindh, L. Aarniovuori, H. Kärkkäinen, M. Niemela and J. Pyrhonen, "IM Loss Evaluation Using FEA and Measurements," *2018 XIII International Conference on Electrical Machines (ICEM)*, 2018, pp. 1220-1226, doi: 10.1109/ICELMACH.2018.8507154.

[6] M. Zaheer, P. Lindh, L. Aarniovuori and J. Pyrhönen, "Assessment of 5 kW Induction Motor Finite element computations with a Commercial and an Open-source software," *2019 International Aegean Conference on Electrical Machines and Power Electronics (ACEMP) & 2019 International Conference on Optimization of Electrical and Electronic Equipment (OPTIM)*, 2019, pp. 114-119, doi: 10.1109/ACEMP-OPTIM44294.2019.9007150.

[7] M. Zaheer, P. Lindh, L. Aarniovuori and J. Pyrhönen, "Converter-Fed Induction Motor Finite Element Analysis With Different Time Steps," *2020 XI International Conference on Electrical Power Drive Systems (ICEPDS)*, 2020, pp. 1-7, doi: 10.1109/ICEPDS47235.2020.9249341..

[8] Aroua. FOURATI, "An electromagnetic-mechanical modeling of an induction motor and rotor shaft coupled to angular approach", 22 eme Congres Francais de Mecanique, 24, 28, 2015.

[9] Schützhold, J. and Hofmann, W., 2013. Analysis of the Temperature Dependence of Losses in Electrical Machines. *IEEE*,.

[10] Xue, S., Feng, J., Guo, S. et al. (3 more authors) (2018) A New Iron Loss Model for Temperature Dependencies of Hysteresis and Eddy Current Losses in Electrical Machines.IEEE Transactions on Magnetics, 54 (1). 8100310. ISSN 0018-9464

[11] Pechanek, R., Kindl, V. and Skala, B., 2015. TRANSIENT THERMAL ANALYSIS OF SMALL SQUIRREL CAGE MOTOR THROUGH COUPLED FEA. MM Science Journal, 2015(01), pp.560-563.

[12] Xie, Y., Guo, J., Chen, P. and Li, Z., 2018. Coupled Fluid-Thermal Analysis for Induction Motors with Broken Bars Operating under the Rated Load. *Energies*, 11(8), p.2024.

[13] L. Popova, "COMBINED ELECTROMAGNETIC AND THERMAL DESIGN PLATFORM FOR TOTALLY ENCLOSED INDUCTION MACHINES", Master's program, Lappeenranta University of Technology, 2010.

[14] Pyrhönen, J., Jokinen, T. and Hrabovcová, V., 2014. Design Of Rotating Electrical Machines. Chichester: Wiley, Second Edition.

[15] W. Calixto, E. Marra, L. da Cunha Brito and B. de Alvarenga, "A new methodology to calculate Carter factor using genetic algorithms", *International Journal of Numerical Modelling: Electronic Networks, Devices and Fields*, vol. 24, no. 4, pp. 387-399, 2010. Available: 10.1002/jnm.785.

[16] H. Kärkkäinen, "CONVERTER-FED INDUCTION MOTOR LOSSES: DETERMINATION WITH IEC METHODS", Master's thesis, Lappeenranta University of Technology, 2015.

An Experimental Test Bench for Studying Sucker Rod Pump

Samuel Isaac Tecle
Ural Federal University
Yekateringburg, Russia
samuel47tecle@gmail.com

Nakataev Anton Andreevich
Ural Federal University
Yekateringburg, Russia
a.a.nakataev@urfu.ru

Anatolii Ziuzev
Ural Federal University
Yekateringburg, Russia
a.m.zyuzev@urfu.ru

Abstract—This paper presents a strategy to emulate a reduced scale of sucker rod pump using laboratory test bench. The strategy is based on combination of software and hardware system. Torque reference is generated using the proposed strategy based on real time model of sucker rod pump that runs in LabVIEW. The output of encoder and torque sensor are integrated through NI PCI- 6221 data acquisition card to the real time platform. The performance of the experimental test bench is evaluated using current, voltage, speed, and torque measurements and power calculations. The experimental results are compared to simulation results. The results indicate the capability of this test bench to serve as a platform for conducting experimental studies on sucker rod pump.

Keywords— Load emulation, test bench, sucker rod pump, electric drive.

I. INTRODUCTION

The promotion from experimental phase to production phase is a long and expensive process, especially for complex electromechanical system. First, the basic notions are validated using software simulations (like MATLAB, PSCAD), then proceeds with more and more deepened representation of circuit details and control functions [1]. Software simulations provide convenience in restructuring and assessing the actual processes taking place in the system, but thorough testing of the hardware under different conditions is required.

Nowadays, emulation techniques offer convenience for conducting various tests, especially for those systems hardly available in laboratory facility or impractical to keep them in laboratory. The use of test bench, Hardware-In-Loop (HIL) with its extension Power Hardware-In-Loop (PHIL) and their combination are common emulation techniques which have eased the experimentation phase for different applications. In the test bench setup, two machines (driving machine and load machine) are mechanically coupled on a common shaft. With this configuration, the load profile of wind turbine [2], electric vehicle [3], lift [4], etc. have been successfully emulated. HIL/PHIL is advanced design/testing method, which is capable to accurately emulate complex dynamic systems in safe, repeatable, and cost-effective way.

Sucker rod pumps are considered a well-established technology in the oil and gas industry [5]. Their installation is easy, and they have low capital and operational costs. However, still there is demand for improved development in its operational performance. Any solution must pass certain tests before it is applied, though it is not practical to keep sucker rod pump in laboratory. In addition, most installations are situated on remote site, possibly installed in severe environmental condition. Therefore, development of an experimental platform is essential.

To emulate mechanical load through test bench, the load machine is controlled in such a way that the characteristics of load it exerts on the driving machine is like the desired mechanical load would do. The control strategy adapted by the load machine depends on the operating condition and nature of the application to be imitated or imitating system. In applications, where the dynamic performance is not critical, static emulation [6] is sufficient. In this case the load developed by the load machine is predetermined or calculated from measured speed. For faster dynamics, this approach was upgraded in [7] by emulating simulated load under open loop conditions. However, for some applications, dynamic load emulation under closed loop condition is desirable because most problems are related to the dynamic behavior of mechanical loads. The principle of inverse mechanical dynamics was the foundation for most early research on dynamic load emulation [8]-[9]. The report in [8] briefly analyzes and discusses the issues related to stability and accuracy of this strategy on emulating dynamic load. To overcome the limitations imposed by the inverse mechanical dynamics, [8] [10] focused on implementation of improved load emulation strategy using feedback controller in combination with compensation. With this strategy, the desired load model dynamics is preserved while providing opportunity to improve the performance of the equipment through changes in software. This method allows to emulate time varying inertias and highly non-linear dynamics with satisfactory results. However, this method relays on the estimated mechanical parameters of the test bench. Parameter variation and certain unmodeled dynamics reduce the accuracy of emulation. To investigate the non-linear dynamics of the test bench and consider certain unmodeled dynamics, advanced methods based on non-linear control approaches were developed in [10].

In this paper the main objective is to reproduce the sucker rod pump system in small scale using test bench and provide platform for off-site test and experiment. This platform can help to verify diagnostic models and to conduct experiments related to optimal balance of counterweight and optimal efficiency operation.

II. SUCKER ROD PUMP MODELING

Fig. 1 presents block diagram representation for sucker rod pump dynamic simulation model. The overall model of sucker rod pump includes surface model, sucker rod string model, pump model, and reservoir model.

A. Surface Model

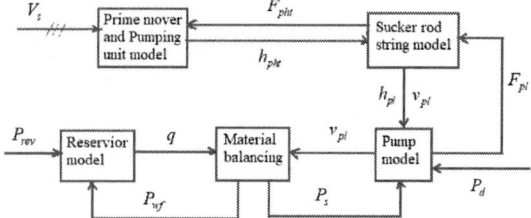

Fig.1. Dynamic simulation model.

The prime mover, gear reducer and the pumping unit are the main elements of the sucker rod pump in the surface. The prime mover is the source of motive power. Induction motors are commonly used as prime movers in sucker rod pump installations. In synchronous rotating reference frame, the state space model of induction motor [11]:

$$
\begin{cases}
\dfrac{di_{sd}}{dt} = -di_{sd} + \omega_s i_{sq} + ci_{md} + c\omega T_r i_{mq} + au_{sd} \\[2mm]
\dfrac{di_{sq}}{dt} = -\omega_s i_{sd} - di_{sq} - c\omega T_r i_{md} + ci_{mq} + au_{sq} \\[2mm]
\dfrac{di_{md}}{dt} = \dfrac{1}{T_r}\left(i_{sd} - i_{md}\right) + \left(\omega_s - \omega\right)i_{mq} \\[2mm]
\dfrac{di_{mq}}{dt} = \dfrac{1}{T_r}\left(i_{sq} - i_{mq}\right) - \left(\omega_s - \omega\right)i_{md} \\[2mm]
M_e = 1.5z_p\left(1-\sigma\right)L_s\left(i_{md}i_{sq} - i_{mq}i_{sd}\right) \\[2mm]
\dfrac{J_T}{z_p}\dfrac{d\omega}{dt} = M_e - M_s
\end{cases}
\tag{1}
$$

Where: (i_{sd}, i_{sq}), (i_{md}, i_{mq}), (u_{sd}, u_{sq}) are respective stator current, magnetizing current, and stator voltage in dq coordinate; ω_s, ω, z_p, J_T, M_s are synchronous speed, rotor speed, pair of poles, total reduced inertia, and reduced moment to the rotor respectively; and

$$
T_s = \frac{L_s}{R_s}; \quad T_r = \frac{L_r}{R_r}; \quad \sigma = 1 - \frac{L_m^2}{L_s L_r}
$$

$$
a = \frac{1}{\sigma L_s}; \quad b = \frac{1}{\sigma T_s}; \quad c = \frac{1-\sigma}{\sigma T_r}; \quad d = b + c
$$

$$
M_s = \frac{1}{u_r}\left(M_c\left(\varphi_{cr}\right) + \frac{\omega_{cr}^2}{2}\frac{dJ_\Sigma\left(\varphi_{cr}\right)}{d\varphi_{cr}}\right)
$$

Where: ω_{cr} is the speed of the crank, φ_{cr} is the crank angle, u_r is transformation ratio of belt-gear reducer system, J_Σ is total inertia about the crank, M_c is the total moment on the gearbox; R_s, L_s, R_r, L_r are respective stator resistance, stator inductance, rotor resistance, rotor inductance.

The function of the gear reducer is to reduce the high rotational speed generated by the prime mover to the required pumping speed. The relation between angular displacement of the crank and the rotor speed of the prime mover can be given by:

$$
\varphi_{cr} = \frac{1}{z_p u_r}\int \omega \, dt
\tag{2}
$$

The rotary motion of the crank is converted into reciprocating movement by pumping unit. To calculate power and torque requirements, knowledge of the polished rod's movement is essential. Fig.2 shows the schematic diagram of sucker rod pump. Let, AB=r, BC=k, CD=l_1, DE=l_2, OD=n_y and OA=m_x.

If we define:

$$
\alpha = 4((n_y - y_B)^2 + x_B^2); \quad \beta = 4(d(n_y - y_B) - 2x_B^2 n_y)
$$

$$
\gamma = d^2 - (2x_B l_1)^2 + (2x_B n_y)^2; \quad d = x_B^2 - k^2 + l_1^2 - n_y^2 + y_B^2
$$

Then, at any instant, the coordinate position of points A, B, C, D, and E can be calculated by expressions presented in Table I:

The angle subtended by a line connecting point C and point D with respect to the horizontal can be expressed by:

$$
aC = \tan^{-1}\left(\frac{y_c - n_y}{x_c}\right)
\tag{3}
$$

And the angular velocity of point E can also be obtained from (3):

$$
\omega_E = -\frac{\Delta aC}{\Delta t}
\tag{4}
$$

Where: Δt is the time interval.

Then the torque factor at the crank angle φ_{cr} can be calculated from (4):

$$
TF\left(\varphi_{cr}\right) = \frac{\omega_E z_p u_r}{\omega}
\tag{5}
$$

Therefore, the moment due to polished rod load and

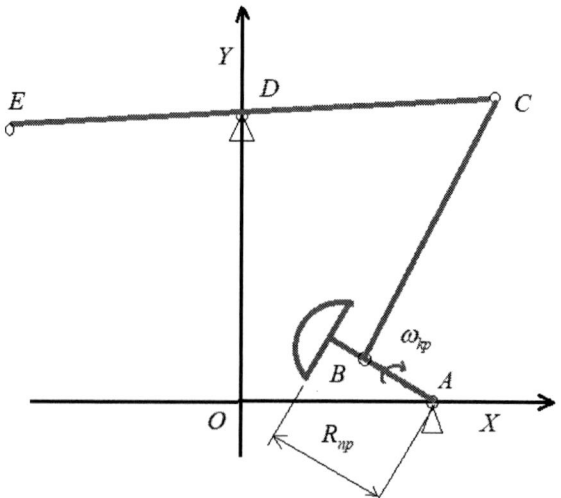

Fig.2. Schematic diagram.

counterweight about the crank can be expressed in (6) and (7) respectively.

$$M_{pht} = F_{pht} l_2 TF(\varphi_{cr}) \tag{6}$$

$$M_{cb} = m_1 g R_{np} \cos \varphi_{cr} \tag{7}$$

Where: F_{pht} is the polished rod load, m_1 is mass of the counterweight; g is gravity, R_{np} is the distance to center of mass of the counterweight.

Finally, the total moment on the gearbox (M_c) is given:

$$M_c = M_{pht} + M_{cb} \tag{8}$$

The total inertia about the crank due to the beam, horse head, equalizer, bearing and pitman can be calculated by (8):

$$J_\Sigma = \sum_{j=1}^{n} m_j \left(\frac{v_j z_p u_r}{\omega} \right)^2 \tag{9}$$

Where: m_j, v_j are the mass and velocity of the jth element respectively.

B. Sucker Rod String Model

The sucker rod string, which extends thousands of meter, stretches and recoils depending on the load on the plunger [12]. In [13]-[15], the sucker rod string model is described as a spring and damper system, dispersed into spring-mass-damper system. Equation (10), which is a boundary problem, expresses the mathematical model of the sucker rod string. The solution to this problem is the foundation in the analysis of the sucker rod pump.

$$\begin{cases} m_1 \ddot{u}_1 = k_o (u_o - u_1) - k_1 (u_2 - u_1) + c_o (\dot{u}_o - \dot{u}_1) - c_1 (\dot{u}_2 - \dot{u}_1) \\ m_2 \ddot{u}_1 = k_1 (u_1 - u_2) - k_2 (u_3 - u_2) + c_1 (\dot{u}_1 - \dot{u}_2) - c_2 (\dot{u}_3 - \dot{u}_2) \\ \vdots \\ m_{n-1} \ddot{u}_{n-1} = k_{n-2} (u_{n-2} - u_{n-1}) - k_{n-1} (u_n - u_{n-1}) \\ \qquad + c_{n-2} (\dot{u}_{n-2} - \dot{u}_{n-1}) - c_{n-1} (\dot{u}_n - \dot{u}_{n-1}) \\ m_n \ddot{u}_n = k_{n-1} (u_{n-1} - u_n) + c_{n-1} (\dot{u}_{n-1} - \dot{u}_n) - F_{pl}(t) \end{cases} \tag{10}$$

Where: m_i is the mass of the ith element, u_i is the displacement of the ith element, k_i is the stiffness coefficient

TABLE I.   CORDINATE POSITION OF MAIN POINTS

Coordinate point	x	y
A	m_x	0
B	$x_B = m_x + r \cdot \cos(\varphi_{cr})$	$y_B = r \cdot \sin(\varphi_{cr})$
C	$x_c = \sqrt{l_1^2 - (y_c - n_y)^2}$	$y_c = \dfrac{-\beta \pm \sqrt{\beta^2 - 4\alpha\gamma}}{2\alpha}$
D	0	n_y
E	$x_E = -x_C \dfrac{l_2}{l_1}$	$y_E = n_y - (y_C - n_y)\dfrac{l_2}{l_1}$

of the ith element, C_i is the damping coefficient of the ith element, F_{pl} is the load on the plunger.

C. Pump Model

The subsurface boundary condition, which is given by (11), is mainly influenced by the pump pressure.

$$F_{pl} = A_{pl}(P_d - P) + A_r P_d \tag{11}$$

Where: A_{pl} is cross-sectional area of plunger; A_r is the cross-sectional area of the rod; P is pump pressure; P_d is discharge pressure.

The pump pressure depends on the state of valves and movement of the plunger. During normal working state, there are four phases in one cycle of pump operation. Expansion phase occurs at the beginning of the upstroke movement before the standing valve opens. After the standing valve opens, the pumping phase precedes. Based on the transient variation rule of liquid pressure in the pump, the rate of liquid pressure [15] is described by (12).

$$\frac{dp}{dt} = \begin{cases} \dfrac{\left[v_n A_{pl} - \dfrac{\pi D_{pl} \delta^3}{12\mu L_p} (P_d - P) \right] P_d}{V_{og}(-1/n)(P/P_d)^{-\frac{1}{n}-1}}; & P > P_s \\ 0; & P = P_s \end{cases} \tag{12}$$

Where: D_{pl} is diameter of plunger; v_n is the velocity of the plunger; V_{og} is the volume of residual gas of the clearance volume; δ is the gap between the pump plunger and barrel; μ is kinematic viscosity of liquid; n is power law coefficient;

The compression phase occurs at the beginning of down stroke movement before the travelling valve opens. After the travelling valve opens, the discharging phase follows. Equation (13) describes the rate of liquid pressure in the pump [15].

$$\frac{dp}{dt} = \begin{cases} \dfrac{\left[v_n A_{pl} - \dfrac{\pi D_{pl} \delta^3}{12\mu L_p} (P_d - P) \right] P_s}{V_g(-1/n)(P/P_s)^{-\frac{1}{n}-1}}; & P < P_d \\ 0; & P = P_d \end{cases} \tag{13}$$

Where: V_g is the volume of free gas; P_s is suction pressure.

During pumping phase, the pump pressure is nearly equal to suction pressure. However, during the discharge phase, the pump pressure nearly equals the discharge pressure.

D. Reservoir Model

In the analysis of sucker rod pump, accurate knowledge of well's inflow performance relationship is vital. Several factors such as reservoir properties (rock permeability, pay thickness, etc.), fluid properties (viscosity, density, etc.), and well completion effects (perforation, well damage) affect the fluid inflow into the well [16]. However, all the quantitative analysis about oil reservoirs are based on Darcy's law, even though it assumes a steady state, linear flow of a single phase in a homogeneous porous media saturated with the same fluid, described by (14).

$$\frac{q}{A} = -\frac{K}{\mu}\frac{dp}{dl} \qquad (14)$$

Where: q is liquid rate; A is cross-sectional area; K is permeability; μ is liquid viscosity; dp/dl is pressure gradient.

For considerable length of time, several factors such as effective permeability, pay thickness, liquid viscosity, drainage radius, reservoir pressure can be assumed constant. This assumption results the simplest approach given by (15). With this assumption only the bottom hole pressure can be controlled.

$$q = PI\left(P_{rev} - P_{wf}\right) \qquad (15)$$

Where: PI is productivity index; P_{rev} is reservoir pressure; P_{wf} is bottom hole pressure.

If the pressure in the vicinity of well bore is lower than the bubble point pressure, free gas can be liberated. This effect reduces the liquid rate compared to single phase conditions. Such cases are better described by Vogel's inflow performance relationship given by the expression (16):

$$\frac{q}{q_{max}} = 1 - 0.2\frac{P_{wf}}{P_{rev}} - 0.8\left(\frac{P_{wf}}{P_{rev}}\right)^2 \qquad (16)$$

Where: q_{max} is the maximum liquid rate

III. AN APPROACH FOR EMULATING SUCKER ROD PUMP DYNAMICS

Electric drives are used to drive, translation, stroke, positioning etc. [17]. The relation between the input (electrical torque) and output (speed) can be linear or non-linear. In load emulation, the objective is to develop a load machine control structure, which when simplified equals the open loop transfer function of the desired system.

Consider the sucker rod pump is the desired mechanical load to be emulated. The mechanical equation in (1) is rewritten as:

$$J_{em}(\varphi_{cr})\frac{d\omega_{em}}{dt} = M_{e1} - M_s \qquad (17)$$

Where: J_{em}, ω_{em}, M_{e1}, M_s, ω_{cr}, φ_{cr} are the total reduced inertia, desired mechanical speed, torque developed by the driving machine in the desired mechanism, total moment on gear reducer reduced to the rotor of the prime mover as expressed in (1), angular speed of the crank, and angular position of the crank respectively.

On the other hand, the total dynamics of the test bench can also be described by:

$$M_m - M_e = J\frac{d\omega_m}{dt} + B\omega_m \qquad (18)$$

$$J = J_m + J_L; \quad B = B_m + B_L$$

Where: ω_m, M_m, M_e, J_m, J_L, B_m, B_L are mechanical speed, moment developed by the driving machine in the test bench, moment developed by load machine, mass inertia of driving machine, mass inertia of load machine, coefficient of friction of driving machine, coefficient of friction of the load machine respectively.

A scaling factor can be deduced from the nominal torque of the drive machine driving the real mechanism and the nominal torque of the drive machine in the test bench. Therefore, the scaling factor (f_c) can be defined as:

$$f_c = \frac{M_{n1}}{M_{n2}} \qquad (19)$$

Where: M_{n1} is nominal torque of the driving machine in the desired mechanism (sucker rod pump) and M_{n2} is the nominal torque of the driving machine in the test bench.

Now, (17) can be written as:

$$J_{em}\frac{d\omega_{em}}{dt} = f_c M_m - M_s \qquad (20)$$

The objective is to find a proper reference torque for the load machine so that the dynamic response of the imitated system and the imitating system are the same. When this condition is satisfied:

$$\begin{cases} \omega_m = \omega_{em} \\ \dfrac{d\omega_m}{dt} = \dfrac{d\omega_{em}}{dt} \end{cases} \qquad (21)$$

Using (18), (20) and (21), the reference torque of the load machine can be obtained by:

$$M_e = \left(1 - \frac{Jf_c}{J_{em}}\right)M_m + \frac{J}{J_{em}}M_s - B\omega_m \qquad (22)$$

On the bases of developed control law (22) and neglecting coefficient of friction, the control strategy shown in Fig.3 is proposed. F_T and F_L are the torque control loops for the driving machine and load machine respectively. The torque loops of real experimental system have bandwidth much higher than speed loops and thus the equivalent transfer functions can be considered as a unity gain [18].

This strategy relies on the estimated parameters of the test bench and bandwidth of the torque control loop. The impact of parameter variation and choice of feedback controllers can be analyzed based on the scheme in Fig.3. If the parameters of the feedback controllers are selected such that the torque loops can be approximated with a unity gain, then the relationship between the actual speed and desired speed is given by:

$$\omega_m = \left(\frac{\hat{J}}{J}\right)\frac{1}{J_{em}s}\left(f_c M_m - M_s\right) \qquad (23)$$

978-1-6654-6787-2/22 $31.00 © 2022 IEEE

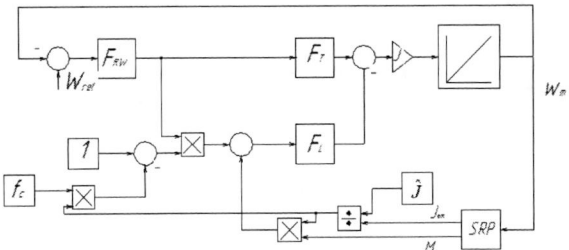

Fig.3. Control strategy

$$\omega_m = \left(\frac{\hat{J}}{J}\right)\omega_{em}$$

Where: $\hat{J}$ is the estimated inertia and s is Laplace operator.

Thus, the emulation is independent of torque and current control techniques. However, error in combined inertia estimation significantly affects the emulation. Similarly, in the case when the effect of torque control loops is considered, the following expression can be used in the analysis:

$$\omega_m = \left[\left(\frac{F_T}{F_L} - \left(1 - \frac{\hat{J}f_c}{J_{em}}\right)\right)M_m + \frac{\hat{J}}{J_{em}}M_s\right]\frac{1}{Js} \qquad (24)$$

The operation is limited to certain bandwidth. Thus, the capability to tracking the desired speed is affected by torque control loops.

IV. OVERVIEW OF THE EXPERIMENTAL TEST BENCH

To implement the proposed control strategy, an experimental setup is designed as shown in Fig.4 with the block diagram representation. The experimental system comprises two 1.5KW induction motors, Altivar 630, ABB ACS580, breaking resistor, standard measurement devices (voltage, current, speed and torque), and a PC (with NI PCI-6221 data acquisition card and LabVIEW installed). ABB ACS580 have several control macros, which are set of default parameter values suitable for a certain control

configuration [19]. For example, by selecting suitable macro, the control principle can be switched from scalar to vector or vise-versa. When vector control is selected, the ABB ACS580 provides an external torque reference as an alternative to the reference torque generated by the speed controller. ABB ACS580 has faster torque response because it is based on direct torque control (DTC) [20].

The two induction machines are mechanically coupled on a common shaft. Altivar 630 supplies the required power to the driving machine and ABB ACS580 supplies the required power for the load machine. Altivar 630 is configured to regulate the speed of the driving machine while ABB ACS580 controls the developed torque by the load machine. The reference torque is calculated based on real time model of the sucker rod pump. E40S6-1000-3-24 pulse encoder provides angular information of the rotor of the driving machine that is sampled at 1000 pulses per revolution. Using this angular information, angular speed of the driving machine is calculated. ZMDN (a high-precision rotary torque sensor) is installed on the shaft to measure the shaft torque of the driving machine. A PC with LabVIEW installed runs real time model of sucker rod pump. NI PCI- 6221 provides both analog I/O and digital I/O. Through NI PCI- 6221, the PC sends analog reference speed and torque signals used by the frequency converters while, at the same time it receives measured digital angular information and analog shaft torque signals form encoder and torque sensor.

For larger part of the pumping cycle, the load machine operates in generating mode while there is excursion into motoring mode in certain part of the cycle. To enable load machine operation in regenerative breaking mode, breaking resistor are installed.

V. MEASUREMENTS

To evaluate the performance of the experimental test bench current, voltage, speed, and torque measurements are crucial. Using these measurements, mechanical power, the active power at the input terminals of the Altivar 630 and motor are calculated.

Fig.5 shows the curves of simulated speed, measured speed, and reference speed. Due to the variable load, the measured speed experiences larger speed changes compared to simulated speed. Fig.6 gives comparative values of the scaled down simulated torque, reference torque and measured torque. The measured torque nearly follows the

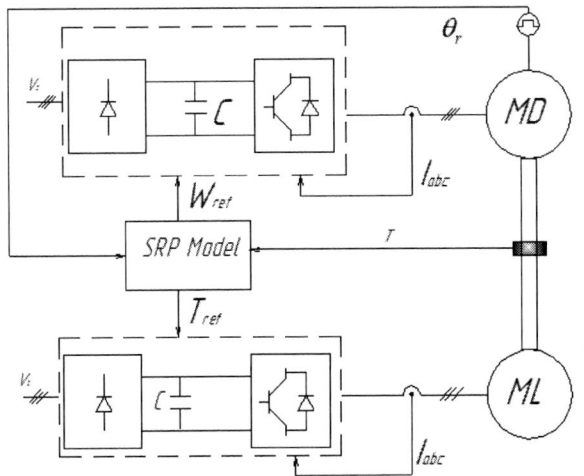

Fig.4. Experimental setup

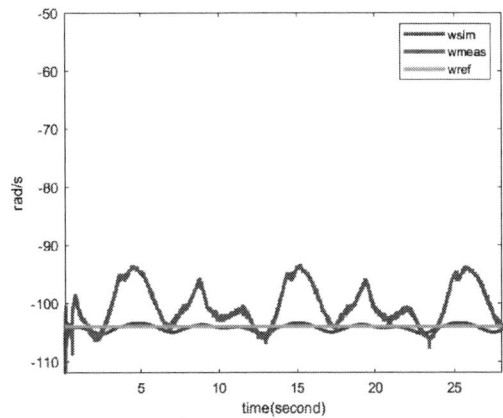

Fig.5. Simulated speed, measured speed and reference speed

978-1-6654-6787-2/22 $31.00 © 2022 IEEE

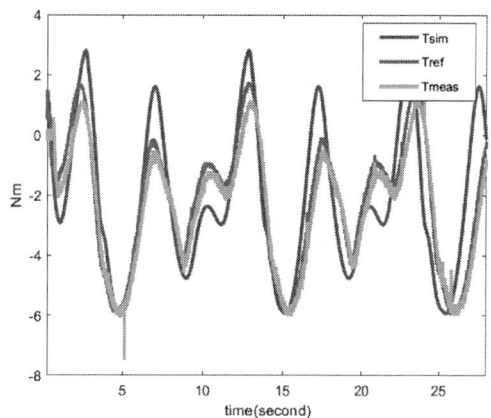

Fig.6. Simulated scaled torque, reference torque and measured torque.

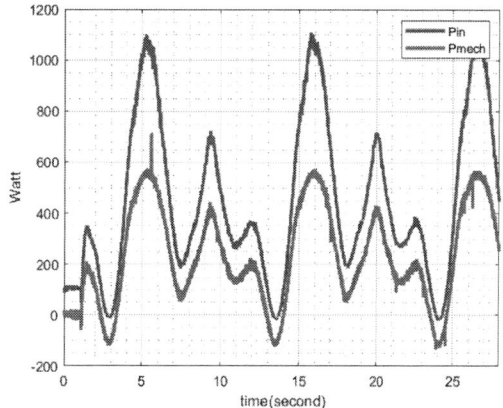

Fig.7. Active power input and mechanical power.

reference torque. The measured torque of experimental test bench is comparable to scaled down simulated torque. This validates the capability of the proposed emulation strategy. Fig.7 shows the active power at the input terminals of Altivar 630 and mechanical power. The curves have similar shape but the active power at the input terminals of Altivar 630 is larger than the mechanical power, especially at larger loads, for reasons associated with reduction in efficiency at reduced speed.

VI. CONCLUSION

In this work a control strategy for dynamic emulation of sucker rod pump using test bench is presented. This strategy is capable to emulate sucker rod pump although it neglects test bench parameter variation. The comparable results of the experiment and simulation validate the proposed strategy. The experimental setup can be modified to serve an effective platform for generating a training set (motor power curves for different sucker rod pump working states). Moreover, studies on optimal counterbalance and optimal energy efficiency operation could be conducted using this test bench.

REFERENCES

[1] L. G. B. Rolim, U. A. Miranda, B. W. França, R. M. Fernandes, and M. Aredes, "Hardware-in-the-loop evaluation of DSP-controlled converters," *2009 Brazilian Power Electron. Conf. COBEP2009*, 2009, pp. 805–809.

[2] S. Kouadria, S. Belfedhal, Y. Meslem and E. M. Berkouk, "Development of real time Wind Turbine Emulator based on DC Motor controlled by hysteresis regulator," 2013 International Renewable and Sustainable Energy Conference (IRSEC), 2013, pp. 246-250.

[3] A. Rassõlkin and V. Vodovozov, "A test bench to study propulsion drives of electric vehicles," 2013 International Conference-Workshop Compatibility And Power Electronics, 2013, pp. 275-279, doi: 10.1109/CPE.2013.6601169.

[4] S. Saarakkala, A. Alahäivälä, M. Hinkkanen and J. Luomi, "Dynamic emulation of multi-mass mechanical loads in electric drives," Proceedings of the 2011 14th European Conference on Power Electronics and Applications, 2011, pp. 1-10.

[5] S. Fakher, A. Khlaifat, M.E. Hossain, *et al.* "A comprehensive review of sucker rod pumps' components, diagnostics, mathematical models, and common failures and mitigations," *Petrol Explor Prod Technol* 11, 2021, pp.3815–3839.

[6] E. R. Collins and Y. Huang, "A programmable dynamometer for testing rotating machinery using a three-phase induction machine," in IEEE Transactions on Energy Conversion, vol. 9, no. 3, pp. 521-527, Sept. 1994, doi: 10.1109/60.326471..

[7] P. Sandholdt, E. Ritchie, J. K. Pedersen and R. E. Betz, "A dynamometer performing dynamical emulation of loads with nonlinear friction," Proceedings of IEEE International Symposium on Industrial Electronics, vol.2, 1996, pp. 873-878.

[8] Z. Hakan Akpolat, G. M. Asher and J. C. Clare, "Dynamic emulation of mechanical loads using a vector-controlled induction motor-generator set," in IEEE Transactions on Industrial Electronics, vol. 46, no. 2, pp. 370-379, April 1999, doi: 10.1109/41.753776.

[9] Z.H. Akpolat, G.M. Asher, J.C. Clare, "Experimental dynamometer emulation of non-linear mechanical loads", IEEE Trans. on Induisthy Applications, vol. 35, no. 6, 1999, pp. 1367-1273.

[10] M. Rodic, K. Jezernik and M. Trlep, "A feedforward approach to the dynamic emulation of mechanical loads," 2004 IEEE 35th Annual Power Electronics Specialists Conference (IEEE Cat. No.04CH37551), 2004, pp. 4595-4601 Vol.6, doi: 10.1109/PESC.2004.1354812.

[11] Quang N. P and Dittrich J. A. *Vector control of three-phase AC machines - System development in the practice*, 2nd edition, Springer, 2015.

[12] S. I. Tecle and A. Ziuzev, "Analysis of Motor Power Curve for Detecting Fault Conditions in Sucker Rod Pump," *2020 27th International Workshop on Electric Drives: MPEI Department of Electric Drives 90th Anniversary (IWED)*, Moscow, Russia, 2020, pp. 1-6.

[13] W. Li, S. Dong, and X. Sun, "An improved sucker rod pumping system model and swabbing parameters optimized design," *Math. Probl. Eng.*, vol. 2018, 2018.

[14] M. Xing and S. Dong, "An improved longitudinal vibration model and dynamic characteristic of sucker rod string," *Journal of Vibroengineering* 16 (2014), pp. 3432-3448.

[15] M. Xing, "Response analysis of longitudinal vibration of sucker rod string considering rod buckling," *Adv. Eng. Softw.*, vol. 99, pp. 49–58, 2016.

[16] Takács Gábor, Sucker rod pumping handbook. Gulf Professional Publishing, 2015, pp. 41–52.

[17] M. Zalman and R. Macko, "Design and realization of programmable emulator of mechanical loads," *IFAC Proceedings Volumes* 38 (2005), pp. 246-250.

[18] J. Arellano-Padilla, G.M. Asher, M. Sumner, "Control of an AC Dynamometer for Dynamic Emulation of Mechanical Loads With Stiff and Flexible Shafts," Industrial Electronics, IEEE Transactions on, vol.53, no.4, pp.1250-1260, June 2006.

[19] ABB general purpose drives, Acs580 drive standard control program: firmware manual, October 30, 2020. Accessed on: Dec. 15, 2021. [online]. Available: https://new.abb.com/products/3AUA0000130331/acs580-01.

[20] ABB, DTC: A motor control technique for all seasons , ABB white paper, Accessed on: Dec. 11, 2021. Available: https://new.abb.com/drives/dtc.

Optimized Field-Weakening Strategy for Control of PM Synchronous Motors

Anton Dianov
System lab
DaeYoung R&D center
Yongin, Republic of Korea
anton.dianov@gmail.com

Abstract—Field-weakening mode of operation is an important feature of modern permanent magnet (PM) synchronous electrical drives, which overcomes speed limitation due to the lack of supplying voltage and increases their speed operational range. Therefore, field-weakening (FW) algorithm is an inalienable part of the overwhelming majority of control systems for PM synchronous motors. At the same time, seamless integration of the filed-weakening techniques into the conventional control systems is not considered well and some reported solutions may decrease system reaction time and cause stability issues. This paper proposes an improved field-weakening strategy, which includes optimized FW controller and novel current limitation strategy in the field-weakening mode. These features simplify structure of the resulting control scheme, decrease computational burden on the microcontrollers and decreases system reaction time, caused by wind-up of the speed controller operating in parallel with the conventional FW algorithms.

Keywords—*Variable speed drives, Permanent magnet motors, Torque control, Energy efficiency.*

I. INTRODUCTION

With the world trends on prioritizing green energy and decrease of CO_2 emissions, efficiency of electrical equipment became an important factor, which significantly impacts design of electrical and electromechanical systems [1], [2]. For this reason, Permanent Magnet Synchronous Motors (PMSM) attract more attention in a number of low- and middle-power applications, where their compactness, efficiency and torque (power) density are prioritized [3], [4]. Since these motor drives focus on efficiency, their control schemes include loss minimization [5] or Maximum Torque Per Ampere (MTPA) algorithms [6], [7], which increase motor utilization.

The point of the maximum drive efficiency is typically selected to correspond the most frequent operational conditions, which increases overall efficiency of motor drive. At the same time, there is a need to extend working range of PM motor drives and provide possibility of operation at higher speeds [8], which is required in many applications: spindles, traction drives, washing machines, compressors, etc. The conventional control schemes may not run PMSMs in high-speed region, where back-EMF is higher than the applied stator voltage, therefore field-weakening algorithms are involved. They modify direct component of the stator current in order to decrease rotor's field and induced back-EMF, making rotation at higher speeds possible [9]. This strategy increases stator current, which in turn, decreases motor drive efficiency [10]. However, system efficiency is typically evaluated over operational cycle, which includes various operational modes [11], therefore, taking into account that working time in the FW mode is typically quite short, the overall motor drive efficiency remains high.

Field weakening algorithms typically enhance conventional control schemes, therefore they are designed as extensions for them. They were designed for Field Oriented Control (FOC) and Direct Torque Control (DTC) schemes [12], however this study considers only FOC as the most popular solution.

Since FW techniques are implemented as extensions for the conventional FOC schemes, the interaction between them significantly impacts the performance of motor drive. Incorrect design of the field-weakening algorithm may significantly worsen the system reaction time and stability. Furthermore, PMSM operation in the field-weakening mode is described with the complicated equations, which are hard for calculations, especially in the low-cost systems [13]. As a result, various simplification techniques, which can decrease burden to the microcontroller, became critical for the development.

In order to solve the abovementioned problems, this paper propose an optimized field weakening strategy, which improves integration of the field-weakening algorithm into conventional FOC system. It eliminates wind-up of the speed controller inherent to the conventional techniques and decreases system reaction time. Furthermore, the author suggests modifications, which simplify the field-weakening controller and significantly accelerate calculations. They exclude square root calculations and significantly decrease load to MCU, which is critical for low-cost systems.

II. FIELD-WEAKENING STRATEGIES

All field-weakening techniques modify direct current in order to weak the field and decrease rotor back-EMF, however do it in other ways.

A. Field-weakening approaches

The authors of [14] provide a brief review and classification of the FW algorithms. They divide approaches to weaken the rotor field into feedforward, feedback and mixed approaches. The feedforward techniques generally use motor models, thus these algorithms significantly depend on the motor parameters [15]. Furthermore, since analytical solutions are complicated, different simplifying hypothesis are linearization techniques [16] and fitting functions are used [17].

Feedback techniques typically compare the required (commanded) voltage with the maximum available, which depends on the DC-link voltage. If the commanded voltage is higher than the maximum available voltage, they increase field weakening component of stator current by the means of

978-1-6654-6787-2/22 $31.00 © 2022 IEEE

controller. These techniques do not depend on motor parameters, thus they are used in the overwhelming majority of motor drives [18].

Mixed approaches involve motor model and improve reaction time comparing to feedback strategies. Furthermore, they may compensate motor parameters variation or mismatching. However, these techniques require additional computational power and need complicated tuning, thus they are not suitable for many applications [19].

The feedback techniques by the principle of operation can be divided into stator current phase modifying methods and stator current direct component modifying algorithms [14]. The first group of methods indirectly varies direct current by rotating the stator current vector [20]. They are simpler in implementation and significantly easier in the limitation of stator current, however usage of these algorithms results in non-linear control in the field-weakening mode and more difficult analysis and tuning of the control systems. The second group of methods straightly modifies the direct component of stator current, which makes this approach easier for understanding and tuning [21]. However, the main drawback of these techniques is difficulties with the proper current limitations, which are typically performed in two stages: before and after field weakening. It results in possible mistakes in tuning and wind-up of the speed controller [22]. Taking into account that field-weakening algorithms with straight control of direct current are more popular, this paper considers only them.

B. Conventional control scheme

It was explained above, that modern control schemes of PMSMs include MTPA algorithms, which are typically implemented as shown in Fig. 1 in black, where ω_{ref} and ω are reference and actual motor speeds, i_{dc} and i_{qc} are commanded values of direct and quadrature currents, i_d and i_q are actual values of direct and quadrature currents, u_{dc} and u_{qc} are commanded values of direct and quadrature voltages, $\Delta\omega$, Δi_d and Δi_q are speed, direct and quadrature current errors, respectively. The speed controller outputs commanded value of stator current, which is decomposed as:

$$\begin{cases} i_d(I_s) = I_s \cdot \sin\gamma(I_s) \\ i_q(I_s) = I_s \cdot \cos\gamma(I_s) \\ \gamma(I_s) = \arccos\left(\dfrac{-\Psi_m + \sqrt{\Psi_m^2 + 8(L_d - L_q)^2 I_s^2}}{4(L_d - L_q)I_s} \right) \end{cases}, \quad (1)$$

where L_d and L_q are direct and quadrature inductances, Ψ_m is rotor flux linkage and γ is MTPA angle.

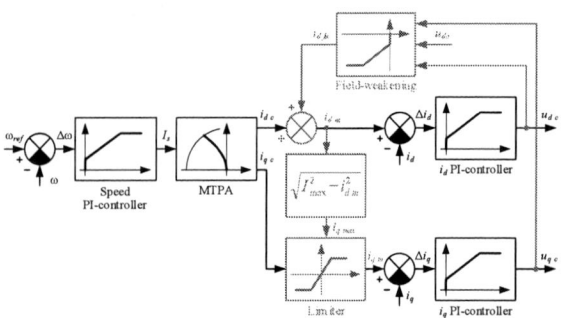

Fig. 1. Vector control scheme with MTPA.

C. Integration with conventional control scheme

When a field-weakening algorithm is added to the conventional control scheme, it is typically modified as shown in Fig. 1 in red [23]. It inputs commanded voltages and DC-link voltage u_{dc}, which are used for calculation of field-weakening component i_{dfw}. This component is added to the commanded direct current, producing modified command for direct current i_{dm}. Since direct commanded current was increased, it is necessary to limit the quadrature commanded current in order to keep stator current below maximum value I_{max}, which is typically implemented by limiter block.

This conventional algorithm is simple for understanding, however has several drawbacks, which decrease performance of the motor drive. As it can be seen from Fig. 1, the commanded current is initially calculated by the speed controller, which limits output at I_{max}. However, after that, the current components may be rearranged by the field-weakening algorithms, resulting in increase of direct current and decrease of quadrature currents. Since speed controller has no any information about these modifications it causes wind-up and decrease reaction time.

Furthermore, conventional field-weakening approach needs calculation of two square roots, which significantly load microcontrollers. As a result, usage of the conventional FW techniques in limited in many low-cost applications. In order to solve the abovementioned problem, this manuscript suggests new field-weakening strategy.

III. PROPOSED FIELD-WEAKENING STRATEGY

Despite the differences in control methods, all the considered techniques have similar part – filed-weakening controller, which is considered below.

A. Field-weakening controller

The conventional field-weakening controller, Fig. 2a, inputs commanded stator voltage magnitude $|\mathbf{u}^*|$ or its components in stationary $\alpha\beta$ or rotational dq reference frames: u_α^*, u_β^* and u_d^*, u_q^*, respectively. Then, if voltage components were provided, a magnitude of the voltage vector is calculated and compared with the maximum available phase voltage. If the control system, uses Space Vector PWM (SVPWM) and does not involve overmodulation strategies, the maximum available phase voltage is $u_{dc}/\sqrt{3}$. After that, the difference between maximum available and commanded voltages Δu is sent to the proportional-integral (PI) type controller, which outputs FW component of direct current i_{dfw}, or stator current phase angle γ, depending on the exact technique. In the first case, i_{dfw} has to be added to the previously calculated (by the MTPA block) direct current commanded value and send to the current controller. The specific of the PI-controller is its output, which can be only negative for motors, where quadrate inductance is higher than the direct inductance, and only positive for motors with $L_q < L_d$.

The main drawback of the conventional FW controller is its low reaction time and stability problem. These issues arise from the fact that current controller produces non-zero output, when the commanded voltage exceeds the existing voltage. Thus, the control system needs increase of i_{dfw} at the current calculation step, while the new value will be calculated one calculation step after and half step will be

978-1-6654-6787-2/22 $31.00 © 2022 IEEE

taken to make the command effective. As a result, the dynamic worsens and stability decreases.

In order to solve this problem, the FW algorithm is designed to have a voltage gap (typically 3~5%), Fig. 2b. The maximum available phase voltage is multiplied with a gap coefficient k_{gap}, which typical value is 0.95~0.97%. As a result, the field-weakening algorithm activates before all existing voltage is utilized, which decreases reaction time and increases system stability. The exact value of the gap coefficients depends on the requirements to the motor drive dynamic and typically does not exceed 10% due to the efficiency decrease issues.

One more disadvantage of the considered FW controllers is the necessity of square root calculation. This operation significantly loads microcontroller and not desired for low-cost systems. In order to exclude square roots from the FW algorithms, a schematic demonstrated in Fig. 2c is recommended. It operates with squared voltages, therefore square root calculation is excluded, however this control

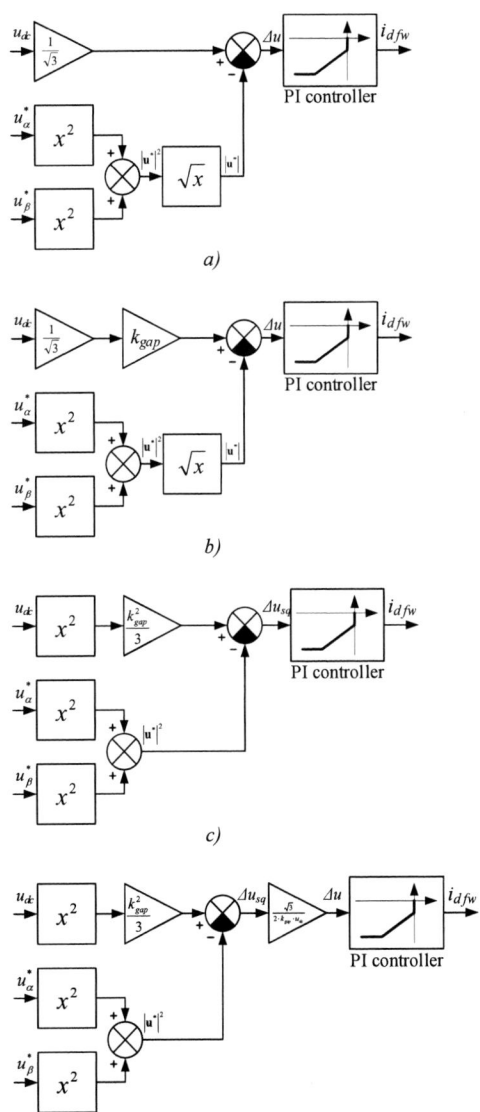

Fig. 2. Field-weakening controllers. a) Conventional controller; b) Controller with a voltage gap; c) Optimized controller; d) Linearized optimized controller.

scheme becomes non-linear and perfectly fits only slight field-weakening below 30~40%.

In order to improve the previous idea and decrease non-linearity, the following considerations may be involved. The error of squared voltages Δu_{sq} can be expressed as:

$$\Delta u_{sq} = \frac{u_{dc}^2 \cdot k_{gap}^2}{3} - \left|\mathbf{u}^*\right|^2 =$$
$$\left(\frac{u_{dc} \cdot k_{gap}}{\sqrt{3}} - \left|\mathbf{u}^*\right|\right)\left(\frac{u_{dc} \cdot k_{gap}}{\sqrt{3}} + \left|\mathbf{u}^*\right|\right) = \Delta u \cdot \left(\frac{u_{dc} \cdot k_{gap}}{\sqrt{3}} + \left|\mathbf{u}^*\right|\right), \quad (2)$$

Assuming that the commanded voltage value $\left|\mathbf{u}^*\right|$ is close to the maximum available phase voltage $u_{dc} \cdot k_{gap}/\sqrt{3}$, equation (2) could be written as:

$$\Delta u_{sq} \approx \Delta u \cdot \frac{2 \cdot u_{dc} \cdot k_{gap}}{\sqrt{3}}. \quad (3)$$

Therefore, schematic demonstrated in Fig. 2d, allows elimination of square root from the field-weakening algorithm and exclude non-linearities caused by operation with squared values. Its implementation needs additionally only multiplication with a constant and division by the value of DC-link voltage. This division may also increase computational burden, since the most of low-cost processors do not have H/W acceleration of divisions. However, the value of inversed DC-link voltage is necessary for proper operation of SVPWM, thus it is already calculated. As a result, the proposed linearization needs only two additional multiplications. Furthermore, multiplication with a constant $\sqrt{3}/2 \cdot k_{gap}$ may be excluded, if the gains of PI controller are corrected to consider this term.

B. Limitation technique

In order to simplify calculations and exclude speed controller wind-up, it is desired to have only one stator current limitation, which is implemented in the speed controller. Thus, it is necessary to obtain dependence of the maximum value of the commanded stator current I_{sc} on the commanded field-weakening current i_{dfw}, as shown in Fig. 3. This vector diagram can be described with the following system of equations:

$$\begin{cases} \left(i_{dfw} + i_{dc}\right)^2 + i_{qc}^2 = I_{max}^2 \\ i_{dc} = -\dfrac{\Psi_m + \sqrt{\Psi_m^2 + 8\left(L_d - L_q\right)^2 I_{sc}^2}}{4\left(L_d - L_q\right)} \\ i_{qc}^2 = I_{sc}^2 - i_{dc}^2 \end{cases} \quad (4)$$

Combining all expressions of (4) into one equation, results:

$$I_{sc}^4 - I_{sc}^2 \left(2I_{max}^2 + \frac{i_{dfw}\Psi_m}{\left(L_d - L_q\right)}\right) + \\ \left(\left(I_{max}^2 - i_{dfw}^2\right) + \frac{i_{dfw}\Psi_m}{\left(L_d - L_q\right)}\right)\left(I_{max}^2 - i_{dfw}^2\right) = 0 \quad (5)$$

In turn, solution of (5) is:

$$I_{sc}(i_{dfw}) =$$

$$= \sqrt{\begin{array}{c} I_{max}^2 + \dfrac{i_{dfw}\Psi_m}{2(L_d - L_q)} + \\ + i_{dfw}\sqrt{2(I_{max}^2 - i_{dfw}^2) + \left(i_{dfw} + \dfrac{\Psi_m}{2(L_d - L_q)}\right)^2} \end{array}}, \quad (6)$$

which defines desired limitation.

Expression (6) is extremely hard for calculations, especially in the low-cost systems, however it can be easily approximated with polynomials. The dependency of the maximum commanded stator current on the field-weakening current is demonstrated in Fig. 4a, which illustrates that this function is smooth and its derivative is also smooth and slowly varies outside the neighborhood of point $[-I_{smax}, 0]$. It means that equation (6) can be easily approximated with polynomials, if the neighborhood of point $[-I_{smax}, 0]$ is not considered. Taking into account that field-weakening component of a current is typically limited at 80~90% of the maximum stator current I_{max}, expression (6) can be easily

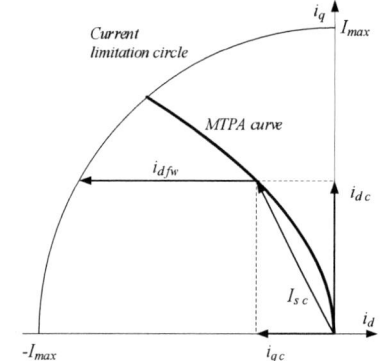

Fig. 3. Stator currents vector diagram

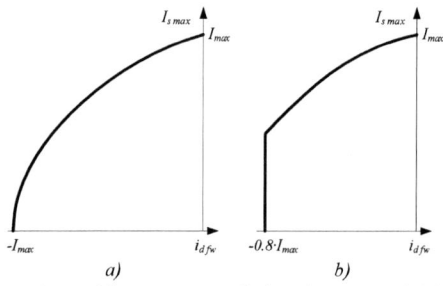

Fig. 4. Dependence of the stator current limit on the commanded field-weakening current. a) Theoretical curve; b) Second order polynomial approximation

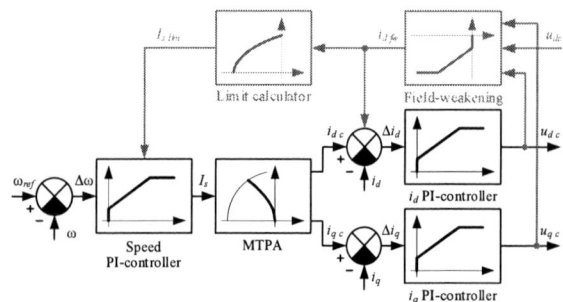

Fig. 5. Integration of the proposed strategy into conventional control scheme

approximated with low-order polynomial. It is demonstrated in Fig. 4b, which illustrates approximation with a second order polynomial at the interval of $[-0.8I_{max}..0]$.

Integration of the proposed FW strategy into conventional control scheme is demonstrated in Fig. 5. As it clearly seen, the proposed method requires less changes and is free of square root calculations. Furthermore, it directly modifies limit values of the speed controller, which excludes wind-up.

IV. EXPERIMENTAL SETUP

Testing setup used for experimental verification of the proposed algorithm is demonstrated in Fig. 6. It includes commercial type IPMSM, used in reciprocating compressors of refrigerators and air conditioners. The motor was optimized to decrease its cost and simplify manufacturing using approach considered in [24, 25]. Motor parameters are shown in Table I and almost do not depend on operational conditions (inductance deviation is less than 5%). Inverter unit is also a mass productive device, designed to drive series of PM motors including the test one. It is based on three-phase intelligent power module from "ST Microelectronics" - STGIPN3H60 (3 A/600 V), which contains 6 MOSFETs and corresponding gating circuits. The inverter was designed for operation in 220 V (50 – 60) Hz mains and contains corresponding filters and fuse. The control system is based on a 60 MIPS Cortex-M3 microcontroller, which operates inverter switches at 6 kHz PWM. The information on currents and voltages is obtained by means of a DC-link voltage sensor and two current sensors [26], which signals are converted by a built into the microcontroller 12-bit ADC.

The control code used for the experiments was based on the commercial version and was enhanced for experiments and debugging. The drive operates in position sensorless mode, where information on rotor speed and position is provided by the back-EMF based estimator [27]. The estimator performance was tested using incremental

TABLE I. EXPERIMENTAL MOTOR PARAMETERS

Parameter	Value
Pole pairs, p	3
Stator resistance, R_s [Ω]	5.1
Direct axis inductance, L_d [mH]	68
Quadrature axis inductance, L_q [mH]	97
Flux linkage, ψ [V/rad/s]	0.135
Rated power, P_{rated} [W]	250

Fig. 6. Experimental set-up

encoders included into the test rig, and signal processing algorithms discussed in [28]–[30]. According to these experiments, the estimation algorithm works stably at 10 Hz and higher, providing estimation error less than five electrical degrees.

The starting is performed in the open-loop with immediate seamless closing as considered in [31], [32]. After that, the motor drive follows the received speed commands, which are processed by the S-shaped limiter. It limits maximum rotor acceleration with the maximum allowed values for the target compressor. In order to provide better dynamic, current PI controllers are designed according to [33], [34]. Calculation of square roots are accelerated using techniques explained in [35], which decreases MCU burden and adapts control system for usage in low-cost systems. The overvoltage protections are implemented as described in [36]–[38], while phase loss detection involves technique from [39].

V. EXPERIMENTAL RESULTS

In order to evaluate the proposed strategy and compare it with the conventional approach, a series of experiments has been conducted. In these experiments, the test motor was started and accelerated until 2000 rpm at load torque of 0.3 N·m. After that, the motor operated about two minutes under these conditions in order to be sure that the system is stable and transients are over. After that, the speed command was changed to 7000 rpm and the motor started acceleration. This experiment was conducted two times: for the control system with the conventional field-weakening algorithm shown in Fig. 1 and for the control system with the proposed field-weakening algorithm illustrated by Fig. 5.

The signals of the test motor drive are demonstrated in Fig. 7–9, where Fig 7 illustrates and compares commanded stator currents, obtained at the output of the speed controllers; Fig 8 demonstrates sensed motor torques under operation of the proposed a) and conventional b) FW algorithms; Fig. 9 shows motor phase currents under operation of the proposed a) and conventional b) FW algorithms. In these figures motor operates at 2000 rpm in zone I, accelerates without field-weakening in zone II, accelerates with field-weakening, but without stator current limitation, in zone III, accelerates with field-weakening and with stator current limitation, in zone IV and operates at the target speed of 7000 rpm in zone V.

It is clearly seen from the demonstrates figures that behaviors of the conventional and proposed methods are similar in zones I-III, where the motor operates without limitation of stator current. However, when the stator current is limited, the speed controller, operating together with the conventional FW algorithm, fails and its output raises to the maximum value causing wind-up. At the same time, the proposed FW algorithm provides normal operation of the speed controller. Furthermore, when the acceleration is over, motor torque decreases and motor currents may be less than the maximum value, the proposed FW algorithm guarantees normal reaction of the speed controller. Simultaneously, the conventional FW technique causes wind-up of the speed controller and significant corresponding delay in current control.

FUTURE WORK

As it was stated above, this manuscript considers only PM motors with constant parameters, which do not vary depending on the motor load and environmental conditions. At the same time, parameters of a substantial part of producing motors significantly deviate from their rated

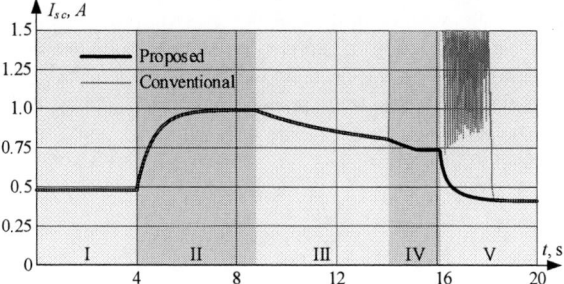

Fig. 7. Commanded stator current (speed controller output)

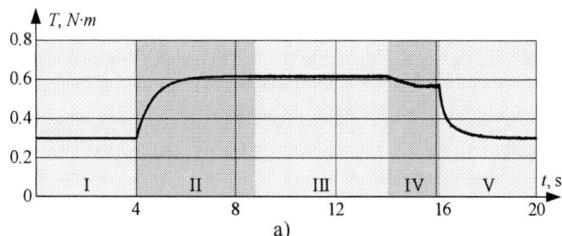

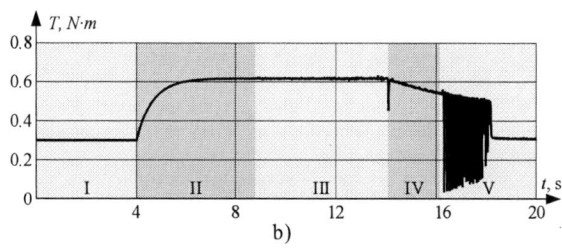

Fig. 8. Motor torque: a) Proposed field-weakening algorithm, b) Conventional field-weakening

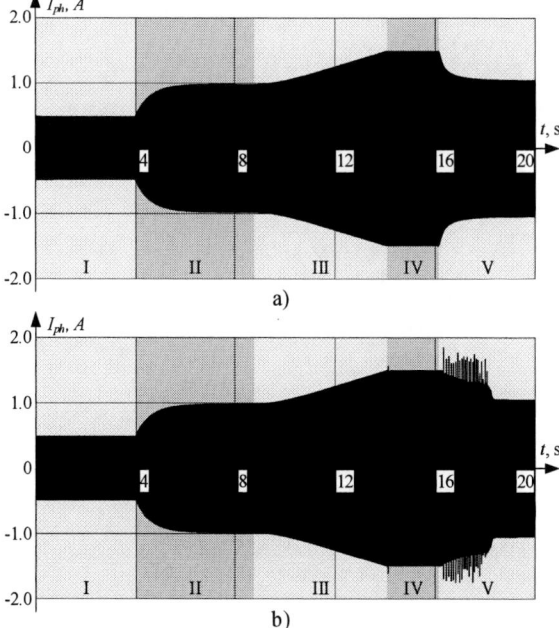

Fig. 9. Phase currents: a) Proposed field-weakening algorithm, b) Conventional field-weakening

values, thus these changes must be taken into account. In the future research the author plans to adapt the proposed idea to the control system of motors, which parameters vary depending on different conditions.

CONCLUSION

This paper presents an optimized field-weakening strategy, which improves interaction of a FW controller with the conventional control scheme. The proposed solution simplifies calculation of current limitations, excludes speed controller wind-up and decreases reaction time. Furthermore, the author proposes an approach to exclude square roots from the calculations, which decreases burden to microcontroller and makes implementation easier for low-cost systems.

REFERENCES

[1] V. Goman, V. Prakht, V. Kazakbaev and V. Dmitrievskii, "Comparative Study of Induction Motors of IE2, IE3 and IE4 Efficiency Classes in Pump Applications Taking into Account CO2 Emission Intensity". Appl. Sci., 10, 8536, pp. 1–10, Nov. 2020.

[2] V. Goman, V. Prakht, V. Kazakbaev and V. Dmitrievskii, "Comparative Study of Energy Consumption and CO2 Emissions of Variable-Speed Electric Drives with Induction and Synchronous Reluctance Motors in Pump Units". Mathematics, 9, 2679, pp. 1–16, Oct. 2021.

[3] X. Zhang, et al., "A Robust Field-Weakening Approach for Direct Torque and Flux Controlled Reluctance Synchronous Motors With Extended Constant Power Speed Region," IEEE Transactions on Industrial Electronics, vol. 67, no 3, pp. 1813-1823, 2020.

[4] N. Zhao, N. Schofield, "Field-Weakening Capability of Interior Permanent-Magnet Machines With Salient Pole Shoe Rotors," IEEE Transactions on Magnetics, vol. 53, no 11, pp. 1-7, 2017.

[5] J. Lee, K. Nam, S. Choi and S. Kwon, "Loss-minimizing control of PMSM with the use of polynomial approximations," IEEE Trans. Power Electron, vol. 24, no. 4, pp. 1071-1082, Apr. 2009.

[6] A. Dianov and A. Anuchin, "Adaptive maximum torque per ampere control of sensorless permanent magnet motor drives," Energies, 13, 5071, pp. 1–13, Sep 2020.

[7] A. Dianov and A. Anuchin, "Adaptive maximum torque per ampere control for IPMSM drives with load varying over mechanical revolution," IEEE J. of Emerg. Sel. Topics in Power Electron., doi: 10.1109/JESTPE.2020.3029647.

[8] Y.D. Yoon, W.J. Lee and S.K. Sul, "New flux weakening control for high saliency interior permanent magnet synchronous machine without any tables," in European Conf. Power Electron. Appl., 2007, pp. 1-7.

[9] M. Preindl, S. Bolognani, "Optimal State Reference Computation With Constrained MTPA Criterion for PM Motor Drives," IEEE Trans. Power Electronics, vol. 30, no 8, pp. 4524–4535, Aug. 2015.

[10] M.S. Islam, et al., "Field Weakening Operation of Slotless Permanent Magnet Machines using Stator Embedded Inductor," IEEE Trans.Ind. Appl. vol. 57, no. 3, pp. 2387-2397, May/Jun. 2021.

[11] V. Goman, S. Oshurbekov, V. Kazakbaev, V. Prakht and V. Dmitrievskii, "Energy Efficiency Analysis of Fixed-Speed Pump Drives with Various Types of Motors". Appl. Sci., 9, 529, pp. 1–12, Dec. 2019.

[12] Y. Inoue, S. Morimoto and M. Sanada, "Comparative Study of PMSM Drive Systems Based on Current Control and Direct Torque Control in Flux-Weakening Control Region," IEEE Trans. Ind. Appl., vol. 48, no. 6, pp. 2382-2389, Nov./Dec. 2012.

[13] K. Chen, et al., "Interior Permanent Magnet Synchronous Motor Linear Field-Weakening Control," IEEE Transactions on Energy Conversion, vol. 31, no 1, pp. 159-164, 2016.

[14] S. Bolognani, S. Calligaro and R. Petrella, "Adaptive Flux-Weakening Controller for Interior Permanent Magnet Synchronous Motor Drives," in IEEE J. Emerg. Sel. Topics in Power Electron., vol. 2, no. 2, pp. 236-248, June 2014.

[15] R. Ancuti, I. Boldea, and G.-D. Andreescu, "Sensorless V/f control of high-speed surface permanent magnet synchronous motor drives with two novel stabilizing loops for fast dynamics and robustness," IET Electr. Power Appl., vol. 4, no. 3, pp. 149–157, Mar. 2010.

[16] K. Zhou, M. Ai, D. Sun, N. Jin and X. Wu, "Field Weakening Operation Control Strategies of PMSM Based on Feedback Linearization," Energies, 12, 4526, pp. 1–18, Nov. 2019.

[17] S. Huang, Z. Chen, K. Huang and Jian Gao, "Maximum torque per ampere and flux-weakening control for PMSM based on curve fitting," in IEEE Veh. Power Prop. Conf., 2010, pp. 1-5.

[18] J. Su, et al., "Model Predictive Control Based Field-Weakening Strategy for Traction EV Used Induction Motor," IEEE Transactions on Industry Applications, vol. 54, no 3, pp. 2295-2305, 2018.

[19] S. Bolognani, S. Calligaro, R. Petrella and F. Pogni, "Flux-weakening in IPM motor drives: Comparison of state-of-art algorithms and a novel proposal for controller design," in 14th European Conf. Power Electron. Applications, 2011, pp. 1-11.

[20] D. Lu and N.C. Kar, "A review of flux-weakening control in permanent magnet synchronous machines," IEEE Vehicle Power and Propulsion Conference, 2010, pp. 1-6.

[21] V. Manzolini, et al., "An effective voltage control loop for a deep flux-weakening in IPM synchronous motor drives," In Proc. 2017 IEEE Energy Conversion Congress and Exposition, pp. 3979-3985, 2017.

[22] W. Xu, et al., "Parameter Optimization of Adaptive Flux-Weakening Strategy for Permanent-Magnet Synchronous Motor Drives Based on Particle Swarm Algorithm," IEEE Trans. Power Electron., vol. 34, no. 12, pp. 12128-12140, Dec. 2019.

[23] D. Sun, et al., "A Hybrid PWM-Based Field Weakening Strategy for a Hybrid-Inverter-Driven Open-Winding PMSM System," IEEE Transactions on Energy Conversion, vol. 32, no 3, pp. 857-865, 2017.

[24] V. Dmitrievskii, V. Prakht and V. Kazakbaev, "IE5 Energy-Efficiency Class Synchronous Reluctance Motor With Fractional Slot Winding," in IEEE Trans. Ind. Appl., vol. 55, no. 5, pp. 4676-4684, Oct. 2019.

[25] V. Dmitrievskii, V. Prakht and V. Kazakbaev, "Ultra Premium Efficiency (IE5 Energy-Efficiency Class) Synchronous Reluctance Motor with Fractional Slot Winding," in XIII Int. Conf. on Electrical Machines (ICEM), 2018, pp. 1015-1020.

[26] A. Anuchin, et al. "Accuracy analysis of shunt current sensing by means of delta-sigma modulation in electric drives," In Proc. 17th Int. Ural Conf. on AC Electric Drives (ACED), pp. 1–5, 2018.

[27] A. Dianov, N.S. Kim and S.M. Lim, "Sensorless starting of horizontal axis washing machines with direct drive," in Int. Conf. on Elect. Mach. and Syst., 2013, pp. 1–6.

[28] A. Anuchin, et al., "Speed estimation algorithm with specified bandwidth for incremental position encoder," In Proc. 2016 17th Int. Conf. on Mechatronics - Mechatronika (ME), pp. 1–6, 2016.

[29] A. Anuchin, et al., "Synchronous constant elapsed time speed estimation using incremental encoders," IEEE/ASME Trans. Mechatronics, vol. 24, no 4, pp. 1893–1901, Jul. 2019.

[30] A. Anuchin, et al., "Optimized method for speed estimation using incremental encoder," In Proc. International Symposium on Power Electronics, pp. 1–5, 2017.

[31] A. Dianov, "Instant and seamless closing of control system of IPMSM after open-loop starting," in XVIII Int. Scientific Tech. Conf. Alternating Current Electric Drives, 2021, pp. 1–6.

[32] A. Dianov, N.S. Kim and S.M. Lim, "Sensorless starting of direct drive horizontal axis washing machines," J. of Int. Conf. on Elect. Mach. and Syst., vol. 3, no 2, pp. 148–154, 2014.

[33] A. Anuchin, et al., "Adaptive Efficient Control for Switch-Reluctance Drives with DCDC-regulator for Inverter Supply," In Proc. EPE Power Electronics and Motion Control Conference, pp. 1–5, 2004.

[34] A. Anuchin, V. Kozachenko, "Current loop dead-beat control with the digital PI-controller," In. Eur. Conf. Power Electron. Appl., pp. 1–8, 2014.

[35] A. Dianov, A. Anuchin, "Review of Fast Square Root Calculation Methods for Fixed Point Microcontroller-based control systems of Power Electronics," Int. J. of Power Electron. and Drive System, vol. 3, no. 11, pp. 1153-1164, 2020.

[36] H.D. Do, et al., "Overvoltage protection for interior permanent magnet synchronous motor test bench," In Proc. Int. Workshop on Electric Drives: Optimization in Control of Electric Drives, pp. 1–5, 2017.

[37] A. Anuchin, et al., "Insulation monitoring system for electric drives in TN networks," In Proc. International Conference on Modern Power Systems, pp. 1–4, 2017.

[38] M. Yakovenko, et al., "Implementation of a protected low-cost voltage-source inverter," In Proc. X International Conference on Electrical Power Drive Systems, pp. 1–4, 2014.

[39] A. Dianov, A. Anuchin, "Phase Loss Detection Using Current Signals: A Review," IEEE Access vol. 9, pp. 114727-114740, Aug. 2021.

978-1-6654-6787-2/22 $31.00 © 2022 IEEE

Field-Oriented Control of the Induction Motor as Part of the Shunting Locomotive Powertrain Considering Core Losses and Magnetic Saturation

Igor Zhurov
Departament of complete traction electric drives
"Engineering center "Ruselprom" Ltd.
Moscow, Russia
igor.zhurov.work@gmail.com

Sergey Bayda
Departament of complete traction electric drives
"Engineering center "Ruselprom" Ltd.
Moscow, Russia
sbai@mail.ru

Stanislav Florentsev
Departament of complete traction electric drives
"Engineering center "Ruselprom" Ltd.
Moscow, Russia
florentsev@ruselprom.ru

Abstract—The analysis of the magnetic saturation and core losses effects influence on a field-oriented controlled induction motor current and torque control quality is given. The magnetization circuit active resistance makes it possible to take core losses into account in induction motor electromagnetic processes equations. The rotor flux linkage vector magnitude and rotation angle estimation unit synthesis based on the rotor circuit equations is made considering core losses. The mutual inductance and the core losses circuit resistance estimating technique in a no-load motor mode are presented.

Keywords—induction motors, traction motors, parameter estimation, magnetic flux, machine vector control, control system analysis, optimal control, current control, motor drives, torque control, vehicles

I. INTRODUCTION

Today, the field-oriented control (FOC) is one of the most commonly used methods for the induction motor (IM) torque and rotation angular speed controlling, which allow achieving high quality indicators in static and dynamic operation modes. This also applies to IM as railway traction electric drive part, also called electromechanical powertrains (EMP), where there is a wide range of torque and an angular frequency variation [1], [2], [3].

However, creating a high-quality and efficient EMP requires some key design steps. The greatest contribution to EMP efficiency makes at the design stage of traction motor and frequency converter. A high-quality manufacturing of these units provides highly efficiency of entire EMP [3].

Further EMP quality improvement becomes possible at the control system optimal strategies synthesis stage. With regard to an IM-based electric drive, different control strategies are known today, the most used among which are "Maximum Torque per Ampere" and "Maximum Efficiency" (i.e. "Total Losses Minimization") [4], [5].

It is known for the FOC system the most important unit from the torque quality control point of view is the rotor flux linkage observer (FLO) which allows estimating its vector magnitude and rotation angle, since determination accuracy of these parameters is directly related to the correct stator current components feedbacks formation.

There are different approaches for the FLO synthesis, but it is necessary to use motor electromagnetic processes equations in one way or another. One variant of the FLO structure is based on the rotor circuit equations in d,q reference frame [6], [7].

At the same time, it is important to consider the influence of core losses, which are quite pronounced at high frequency values. In [8] and [9] the sliding-mode flux linkage observer structure were presented, however, there is no analysis of the core losses influence on the estimation quality. In [10] and [11] an observer with the core losses compensation is proposed. However, a common drawback of sliding-mode observers is the unified technique lack for the correction gains matrix synthesizing.

Additionally it is necessary to consider motor magnetic saturation effect, which manifests itself as a change of the mutual inductance according to dependence on the magnetic flux linkage value. The saturation characteristic should be taken into account in FLO unit as well as in other units of the control system structure.

Therefore, taking into account both magnetic saturation and core losses influence, the FOC system for the IM electric drive synthesis becomes relevant. To achieve this goal, a method based on a no-load motor mode is proposed for a preliminary mutual inductance and core losses circuit resistance estimation, which are subsequently put into the control system structure.

The obtained results of the study are considered within the framework of the shunting diesel locomotive "TEM-6m" traction electric drive modernization project on the basis of the traction induction motor. This motor has six pole pairs, high torque and low speed, which allows realizing a gearless traction electric drive for the shunting locomotive.

II. INDUCTION MOTOR EQUIVALENT CIRCUIT

The literature analysis shows that IM core losses can be taken into account in its equivalent circuit in two ways, as shown in work [12]. This is most often done by connecting an active resistance element R_c in parallel with the magnetizing circuit, as shown on Fig. 1.

In this case, the stator and rotor circuits equations in d,q reference frame are significantly complicated due to a seemingly insignificant change of an equivalent circuit. This also leads to the complexity of the FOC system synthesis.

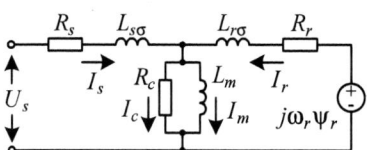

Fig. 1. Induction motor equivalent circuit in a stator reference frame, taking core losses into account using parallel connection of the resistance R_c relative to the magnetizing circuit

This work was supported by the "Ruselprom" Ltd.

978-1-6654-6787-2/22 $31.00 © 2022 IEEE

In this context, from the core losses influence point of view, the most sensitive element is the FLO unit, which structure is based on the rotor circuit equations. So, if both rotor flux linkage vector parameters calculate incorrectly, the values of the stator current components I_d and I_q as well as the torque will be also determined incorrectly. This fact necessitates taking into account the core losses circuit resistance both in the structure of the FLO and in other units of the control system structure.

One of the first attempts to design the FOC system for IM drive is described in [13], where an equivalent circuit shown on Fig. 1 is taken as a basis. However, this work analysis shows complexity of this approach since it leads to a synthesis complication because of equations, underlying the control system structure, are very complicated. Therefore it requires converting of IM equivalent circuit to a form suitable for simplified synthesis, as well as for the relative simplicity and clarity of the control system structure.

One of the ways to solve this problem is to bring an equivalent circuit with parallel connection of R_c (Fig. 1) to a circuit with a series connection of an equivalent element R_m (Fig. 2) so that both circuits are identical. Such an equivalent circuit is considered, for example, in works [12], [14].

For simplicity, let introduce notation: the circuit (Fig. 1) will be called "p-circuit" (p - from "parallel"), and the circuit (Fig. 2) will be called "s-circuit" (s - from "series").

The transformation of p-scheme to s-scheme performs taking into account the relationships between the elements of these schemes, which are quite simple to obtain [15]:

$$
\begin{cases}
R_m = \dfrac{R_c \left(\omega L_{m,p}\right)^2}{R_c^2 + \left(\omega L_{m,p}\right)^2}; \\
L_{m,s} = \dfrac{R_c^2 L_{m,p}}{R_c^2 + \left(\omega L_{m,p}\right)^2},
\end{cases}
\tag{1}
$$

where R_c, $L_{m,p}$ are the core loss circuit resistance and the mutual inductance of the p- scheme, R_m, $L_{m,s}$ are the core loss circuit resistance and the equivalent mutual inductance of the s- scheme, ω is an angular frequency.

Nevertheless, when using the s-scheme, the control system synthesis is still rather difficult due to its parameter expressions (1) relative complexity, which requires further transformations and simplifications.

Note, as a rule, the R_c core loss resistance in p- scheme is measured in hundreds of Ohms, and the mutual inductance for high-power machines is measured in a thousandths of Henry. Thus, in the equation (1), the square of the core loss circuit resistance in the denominator is much larger than the square of the inductive resistance, which means that the latter can be neglected [16].

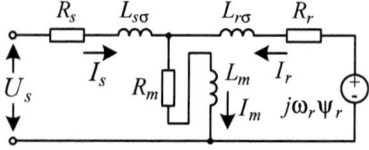

Fig. 2. Induction motor equivalent circuit in a stator reference frame, taking core losses into account using series connection of the resistance R_c in the magnetizing circuit

Then, system (1) is transformed to the form (2), where the expression for the parameter R_m is greatly simplified, and the mutual inductance for both circuits becomes identical and can be denoted as L_m:

$$
\begin{cases}
R_m = \dfrac{X_m^2}{R_c}; \\
L_{m,s} = L_{m,p} = L_m.
\end{cases}
\tag{2}
$$

III. INDUCTION MOTOR MATHEMATICAL DESCRIPTION AND FLUX LINKAGE ESTIMATION UNIT STRUCTURE SYNTHESIS

The differential equations of IM stator and rotor circuits' electrical balance are supplemented by a term that causes the voltage drop across an element R_m. In this case, Park equations in form of generalized vectors looks like (3):

$$
\begin{cases}
\mathbf{U_s} = R_s \mathbf{I_s} + R_m \mathbf{I_m} + \dot{\boldsymbol{\psi}}_\mathbf{s} + j\omega_c \boldsymbol{\psi}_\mathbf{s}; \\
0 = R_r \mathbf{I_r} + R_m \mathbf{I_m} + \dot{\boldsymbol{\psi}}_\mathbf{r} + j\left(\omega_c - \omega_r\right)\boldsymbol{\psi}_\mathbf{r}.
\end{cases}
\tag{3}
$$

For the stator and rotor flux linkages equations are in the form (4), since core losses don't affect them:

$$
\begin{cases}
\boldsymbol{\psi}_\mathbf{s} = L_{s\sigma} \mathbf{I_s} + L_m \mathbf{I_m} = L_s \mathbf{I_s} + L_m \mathbf{I_r}; \\
\boldsymbol{\psi}_\mathbf{r} = L_{r\sigma} \mathbf{I_r} + L_m \mathbf{I_m} = L_m \mathbf{I_s} + L_r \mathbf{I_r}.
\end{cases}
\tag{4}
$$

In the above equations, $\mathbf{U_s}$, $\mathbf{U_r}$, $\mathbf{I_s}$, $\mathbf{I_r}$ are the stator and rotor voltages and currents respectively, R_s, R_r are the resistances of the stator and rotor circuits, $L_{s\sigma}$, L_s are the leakage inductance and the total stator inductance, $L_{r\sigma}$, L_r are the leakage inductance and the total rotor inductance, $\omega_r = z_p \omega$ is the rotor electrical rotational angular frequency, z_p is the pole pairs number, ω_c is d,q reference frame rotational angular frequency.

The magnetizing circuit current is the sum of the stator and rotor currents (5):

$$
\mathbf{I_m} = \mathbf{I_s} + \mathbf{I_r}.
\tag{5}
$$

To exclude the rotor current and stator flux linkage from the equations (3), they should be expressed from equations (4) which lead to the system (6):

$$
\begin{cases}
\mathbf{I_r} = \dfrac{1}{L_r}\boldsymbol{\psi}_\mathbf{r} - \dfrac{L_m}{L_r}\mathbf{I_s}; \\
\boldsymbol{\psi}_\mathbf{s} = \left(L_s - \dfrac{L_m^2}{L_r}\right)\mathbf{I_s} + \dfrac{L_m}{L_r}\boldsymbol{\psi}_\mathbf{r} = \sigma L_s \mathbf{I_s} + \dfrac{L_m}{L_r}\boldsymbol{\psi}_\mathbf{r}.
\end{cases}
\tag{6}
$$

Here σ is the leakage coefficient. Substituting the first equation of system (6) into equation (5) leads to (7):

$$
\mathbf{I_m} = \dfrac{1}{L_r}\boldsymbol{\psi}_\mathbf{r} + \left(1 - \dfrac{L_m}{L_r}\right)\mathbf{I_s}.
\tag{7}
$$

Basing on equations (4), (6) and (7), and introducing following designations: $K_r = L_m/L_r$, $\overline{K}_r = 1 - K_r$, represent the equation system (3) in the form (8):

$$
\begin{cases}
\mathbf{U_s} = R_s \mathbf{I_s} + \sigma L_s \dot{\mathbf{I}}_s + K_r \dot{\boldsymbol{\psi}}_\mathbf{r} + R_m \left(\overline{K}_r \mathbf{I_s} + \dfrac{\boldsymbol{\psi_r}}{L_r} \right) + \\
\qquad + j\omega_c \left(\sigma L_s \mathbf{I_s} + K_r \boldsymbol{\psi_r} \right); \\
0 = \dfrac{\boldsymbol{\psi_r}}{T_r} - K_r \mathbf{I_s} + \dot{\boldsymbol{\psi}}_\mathbf{r} + R_m \left(\overline{K}_r \mathbf{I_s} + \dfrac{\boldsymbol{\psi_r}}{L_r} \right) + \\
\qquad + j(\omega_c - \omega_r) \boldsymbol{\psi_r}.
\end{cases} \tag{8}
$$

Expanding equation system (8) along d and q axes after performing some transformations leads to system (9) for stator circuit and (10) for rotor circuit. The rotor flux linkage component ψ_d is further denoted as ψ and the component ψ_q is equal to zero.

$$
\begin{cases}
U_d = R_s I_d + \sigma L_s \dot{I}_d + K_r \dot{\psi} - \\
\qquad - \omega_c \sigma L_s I_q + R_m \overline{K}_r I_d + \dfrac{R_m}{L_r} \psi; \\
U_q = R_s I_q + \sigma L_s \dot{I}_q + R_m \overline{K}_r I_q + \\
\qquad + \omega_c (\sigma L_s I_d + K_r \psi);
\end{cases} \tag{9}
$$

$$
\begin{cases}
0 = \dfrac{\psi}{T_r} - K_r R_r I_d + \dot{\psi} + R_m \overline{K}_r I_d + \dfrac{R_m}{L_r} \psi; \\
0 = -K_r R_r I_q + (\omega_c - \omega_r) \psi + R_m \overline{K}_r I_q.
\end{cases} \tag{10}
$$

A similar induction motor mathematical description is given in [17] and [18] with the difference that the core losses are separated to losses in the stator and losses in the rotor. It was also shown that the rotor core losses can be neglected due to their small value relative to the stator core losses.

Equation system (10) is taken as a basis for the FLO unit structure (Fig. 3) and additional terms of system (9) that include R_m parameter, might be taken into account in the cross-link compensation unit structure to improve dynamic performance of the electric drive.

The FLO unit uses a stator current components I_d, I_q as input signals as well as a rotor angular electric speed ω_r. An equivalent circuit parameters L_m, $L_{r\sigma}$, R_r, R_m The observer structure is detailed, in which equivalent circuit parameters are separated from each other due to the fact they can be variable values. The rotor flux linkage vector magnitude and rotation angle are formed at the output of FLO unit as estimated signals.

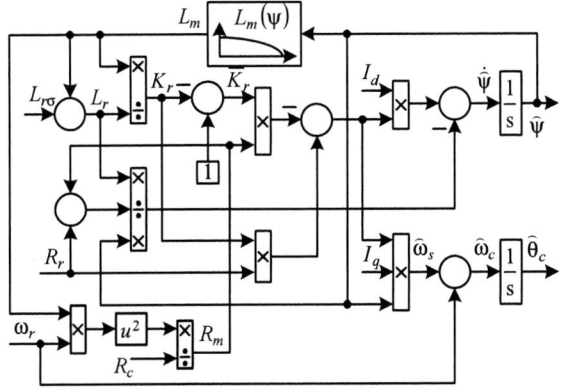

Fig. 3. Flux linkage observer (FLO) unit structure with core losses and magnetic saturation taking into account

IV. A PROPOSED TECHNIQUE FOR MAGNETIZING CIRCUIT PARAMETES ESIMATION

Equations (9) and (10) must be transformed to the steady-state form, which leads to equations (11) and (12).

$$
\begin{cases}
U_d = \left(R_s + K_r^2 R_r + \overline{K}_r^2 R_m \right) I_d - \\
\qquad - \sigma L_s \omega_c I_q - \dfrac{K_r R_r - \overline{K}_r R_m}{L_r} \psi; \\
U_q = \left(R_s + K_r^2 R_r + \overline{K}_r^2 R_m \right) I_q + \\
\qquad + \sigma L_s \omega_c I_d - K_r \omega_r \psi;
\end{cases} \tag{11}
$$

$$
\begin{cases}
\psi = \dfrac{L_r \left(K_r R_r - \overline{K}_r R_m \right)}{R_r + R_m} I_d; \\
\omega_c = \omega_s + \omega_r = \dfrac{R_r + R_m}{L_r} \dfrac{I_q}{I_d} + \omega_r.
\end{cases} \tag{12}
$$

The first parameter need to be estimated is the mutual inductance L_m. An analysis of the equation (2) shows that at low frequency values the core losses are manifested rather weakly and the parameter R_m can be neglected.

When idling, a torque-forming current I_q is equal to zero and the d,q-basis angular frequency ω_c is equal to the rotor electric angular frequency ω_r. Additionally, for the steady state mode, neglecting parameter R_m, the rotor flux linkage magnitude is proportional to a flux-forming current $\psi_r = L_m I_d$. Therefore, rewrite system (11) in the form (13):

$$
\begin{cases}
U_d = R_s I_d; \\
U_q = \omega_r L_s I_d.
\end{cases} \tag{13}
$$

The phase voltage and phase current amplitudes squares should be calculated by the equation (14):

$$
\begin{cases}
U_{\text{ph.m.}}^2 = U_d^2 + U_q^2; \\
I_{ph.m.}^2 = I_d^2 + I_q^2.
\end{cases} \tag{14}
$$

To derive the relationship between current, voltage and a mutual inductance equations (13) and (14):

$$
U_{ph.m} = I_{ph.m.} \sqrt{R_s^2 + \left(\omega_r L_s \right)^2}. \tag{15}
$$

Introduce a conventional notation:

$$
R_e = U_{ph.m.} \big/ I_{ph.m.}. \tag{16}
$$

From equation (15) express the mutual inductance L_m:

$$
\widehat{L}_m = \sqrt{R_e^2 - R_s^2} \big/ \omega_r - L_{s\sigma}. \tag{17}
$$

The estimating procedure for the mutual inductance is performed at a fixed low frequency using an auxiliary machine to unwind the shaft of the motor under study. By varying the magnetic flux of the machine using the flux-forming current I_d and measuring the stator phase voltage, the parameter L_m is calculated according to equation (17).

978-1-6654-6787-2/22 $31.00 © 2022 IEEE

The next step is the resistance R_m estimation procedure, which is performed in a similar way. For this goal, equations (11) and (12) are also reduced to the steady state form (18) taking into account parameter R_m:

$$\begin{cases} U_d = \left(\begin{array}{c} R_s + K_r^2 R_r + \overline{K}_r^2 R_m - \\ - \dfrac{\left(K_r R_r - \overline{K}_r R_m\right)^2}{R_r + R_m} \end{array} \right) I_d; \\ U_q = \left(\sigma L_s + \dfrac{L_m\left(K_r R_r - \overline{K}_r R_m\right)}{R_r + R_m} \right) \omega_r I_d. \end{cases} \quad (18)$$

Some transformation of system (18) leads to system (19):

$$\begin{cases} U_d = \left(R_s + \dfrac{R_r R_m}{R_r + R_m} \right) I_d; \\ U_q = \left(L_s - \dfrac{L_m R_m}{R_r + R_m} \right) \omega_r I_d. \end{cases} \quad (19)$$

The squares of the left and right sides of (19) using (14) and (16) must then be added, transferring the current square to the left side, as a result of which equation (20) can be obtain:

$$R_e^2 = \left(R_s + \frac{R_r R_m}{R_r + R_m} \right)^2 + \omega_r^2 \left(L_s - \frac{L_m R_m}{R_r + R_m} \right)^2. \quad (20)$$

In equation (20) introduce designation (21):

$$K_{rm} = \frac{R_m}{R_r + R_m}. \quad (21)$$

Considering (21) perform transformation (20) to the form (22), which is a quadratic equation with respect to parameter K_{rm}:

$$\left(R_r^2 + (L_m \omega_r)^2 \right) K_{rm}^2 + 2\left(R_s R_r - L_s L_m \omega_r^2 \right) K_{rm} + \\ + \left(R_s^2 - R_e^2 + (L_s \omega_r)^2 \right) = 0. \quad (22)$$

After solving equation (22) one should choose one of two roots (23):

$$K_{rm} = \frac{L_s L_m \omega_r^2 - R_s R_r - \sqrt{D}}{R_r^2 + (L_m \omega_r)^2}. \quad (23)$$

Here D is the quadratic equation (22) discriminant (24) which is:

$$D = \left(R_s R_r - L_s L_m \omega_r^2 \right)^2 - \\ - \left(R_r^2 + (L_m \omega_r)^2 \right)\left(R_s^2 - R_e^2 + (L_s \omega_r)^2 \right). \quad (24)$$

Expressing of R_m through R_r and K_{rm} leads to (25):

$$\hat{R}_m = \frac{R_r K_{rm}}{1 - K_{rm}}. \quad (25)$$

V. SIMULATION RESULTS

The developed methods analysis for the rotor flux linkage and the magnetization circuit parameters estimating is done using simulation with *Matlab/Simulink* program package. All calculations are made for the traction induction motor TAD-320-12-U2 parameters, that given in Table 1.

TABLE I. NOMINAL PAPMETERS OF THE TRACTION INDUCTION MOTOR TAD-320-12-U2

Parameter	Symbol	Unit	Value
Nominal Power	P_n	kW	320
Nominal Angular Speed	N_{nom}	rpm	204
Nominal Torque	$T_{e,nom}$	kNm	15
Pole pairs	z_p	-	6
Stator Resistance	R_s	mOhm	865
Rotor Resistance	R_r	mOhm	688
Stator Inductance	$L_{s\sigma}$	mH	0.301
Rotor Inductance	$L_{r\sigma}$	mH	0.306

An incorrect determining of the magnetization circuit parameters leads to a significant error in torque processing. Fig. 4 shows the results of the torque step response of 15 kNm, which is considered traction motor nominal value. A 10% determining error in the mutual inductance with a deviation both up and down leads to an almost 5% torque processing error (Fig.4 (a)). On the other hand, a 20% determining error in the core losses circuit resistance results in a torque error of approximately 2.5% (Fig.4 b).

First, investigate the core losses influence on the mutual inductance estimation accuracy according to equation (17) at different rotation angular frequency. In the first case, assume that there are no core losses (i.e. $R_m = 0$). In the second case, let the parameter $R_c = 250$ Ohm into the IM model, and the value of R_m calculate using the first equation of system (2). Also the saturation effect is take into account in the induction motor model structure as a static dependence $L_m(\psi)$ (Fig. 5).

Fig. 6 shows the reference current I_d trajectory, which value varies in range from 100 A to 500 A with a step of 100 A, as well as the angular rotation speed trajectory, which varies from 100 rpm to 500 rpm with a step of 100 rpm. The current I_d trajectory is used for the mutual inductance estimation procedure, and the angular speed trajectory is used for the core losses resistance estimation.

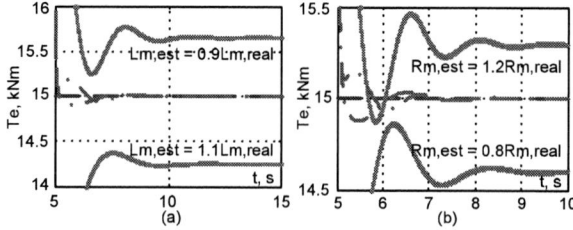

Fig. 4. The mutual inductance dependency on the flux linkage magnitude

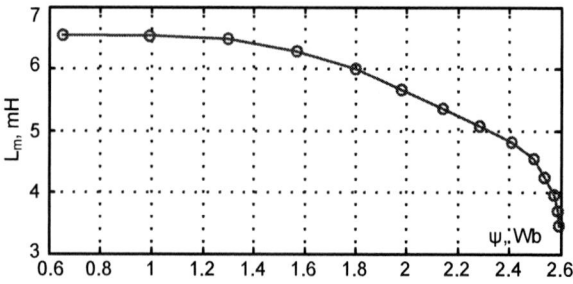

Fig. 5. The mutual inductance dependency on the flux linkage magnitude

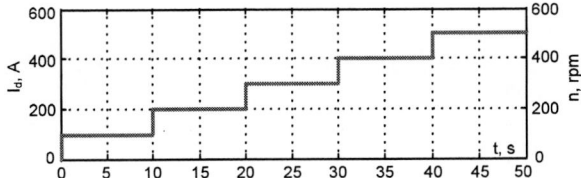

Fig. 6. The flux-forming current I_d trajectory for the mutual inductance estimation procedure and the angular rotation speed trajectory for the core losses circuit resistance estimation procedure

Fig. 7 show simulation results of the mutual inductance estimating procedure at angular speed values equal to 20 rpm and 200 rpm. There are estimated and true values of the mutual inductance curves, as well as the rotor flux linkage curve.

Fig. 7 (a) and (b) graphs correspond to the zero value of the resistance R_m, from which it can be seen that in this case there is no error for the mutual inductance L_m estimating in steady state mode, regardless the angular rotation frequency value.

For a nonzero R_m value the mutual inductance estimation error increases a lot with frequency increasing, as can be seen from Fig. 7, (d). In addition, the rotor flux linkage magnitude value changes, which is also due to the core losses influence in accordance with the first equation of (14). Hence, the mutual inductance estimation procedure should be carried out at low speed, since the core losses influence is minimal, as shown on Fig. 7 (c).

Second, consider the parameter R_m estimation features using proposed technique. Equation (20) contains almost all the parameters of the machine equivalent circuit, which requires an analysis of the sensitivity of the proposed method to their variation.

Fig. 8 shows the results of the parameter R_m estimation procedure simulation with the 20% deviation of the stator and rotor resistances, in both positive and negative directions of the real values. From the given graphs it can be seen that the indicated deviation of the resistance R_s (Fig.8, (a)) has almost no effect on the accuracy of the parameter R_m estimation.

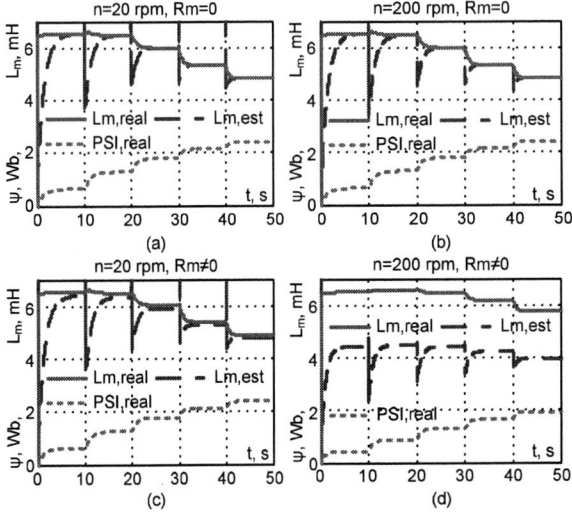

Fig. 7. Results of the parameter L_m estimating procedure for the rotational speed different values, with (a), (b) and without (c), (d) taking into account of core losses

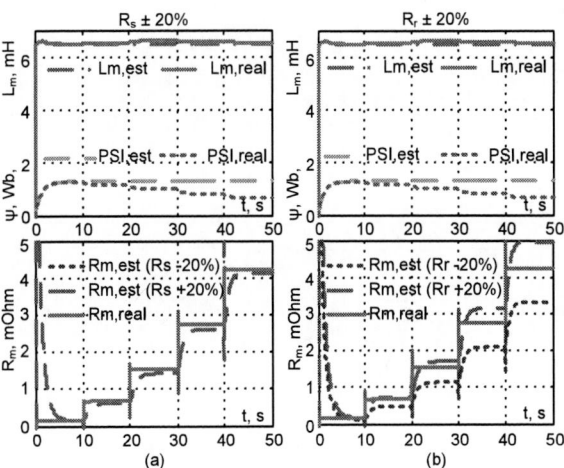

Fig. 8. The estimating procedure results of the R_m parameter at different rotation speed values with the positive and negative 20% deviation of R_s (a) and R_r (b) from the real values taking into account

A rather low estimation accuracy is observed when the rotor resistance R_r deviates in both positive and negative directions (Fig.8, (b)). This happens due to the much greater parameter R_r weight in equation (28) as well as due to its presence in the equation system (11), which is the basis for the FLO unit structure.

Note that resistances R_s and R_r change mainly due to the machine windings heating, and the mutual inductance L_m depends on the magnetic flux linkage magnitude observed by FLO unit. Consequently, when the parameter R_m estimating procedure carrying out, the signal of estimated parameter L_m value obtained from the dependence characteristic $L_m(\psi)$ will also be formed from the FLO unit.

Nevertheless, since initially unknown parameter R_m is included to FLO structure, the flux linkage will be estimated with some error. At the same time, the mutual inductance estimated value will not correspond to its true value since it depends on the calculated flux linkage value. This will lead to an estimation error for the parameter R_m to an extent.

Referring to the characteristic $L_m(\psi)$ (Fig. 5), it can be seen the flux linkage value is relatively small in unsaturated region, so the mutual inductance remains approximately the same. Additionally, assuming the core loss circuit resistance weakly depends on the magnetic flux linkage magnitude, means that for this parameter estimating one can set a certain fixed flux-forming current I_d value.

Therefore, to perform the estimating procedure of the resistance R_m, it is advisable to set a constant I_d value which corresponds to unsaturated region of the static dependency graph $L_m(\psi)$.

Now analyze the influence of the flux-forming current magnitude to the estimation accuracy of the parameter R_m. The Fig. 9 graphs show the results for two different values of current I_d. The rotation speed trajectory for this case corresponds to that on Fig. 6.

From the analysis of Fig. 8 graphs it follows that at a low speed the core losses are minimal, therefore, the resistance value R_m is sufficiently small so the flux linkage and the mutual inductance are estimated with a high accuracy at any I_d current value.

978-1-6654-6787-2/22 $31.00 © 2022 IEEE

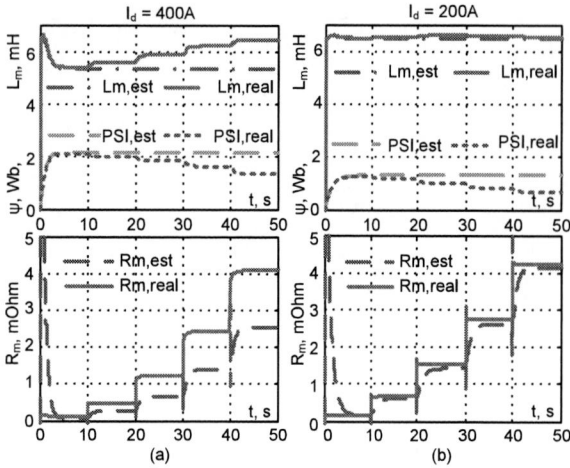

Fig. 9. The estimating procedure results of the parameter Rm with the rotation frequency varying, as well as with different values of the flux-forming current: $I_d = 400$ A (a) and $I_d = 200$ A (b)

With the speed increasing, the core losses are growing as well as the R_m resistance is. When $I_d = 400$A (Fig. 9 (a)) there is an incorrect estimation of the rotor flux linkage magnitude and the mutual inductance since a significant deviation of the flux estimate leads to a significant increase in the mutual inductance. As a result the resistance R_m estimating accuracy also turns out to be rather low. However, in case with $I_d = 200$A (Fig. 9 (b)) the estimation accuracy increases significantly due to relatively small deviations of the estimated L_m parameter value from the true value, despite the low observing accuracy of the flux linkage.

Additionally, the estimated value of flux linkage always decreases with the frequency and core losses increase which causes one more conclusion: one should choose the highest possible magnetizing current value on the unsaturated curve region. This is explained by the fact as the calculated flux shifts downward there is a slight change of L_m, which can be seen from the curve $L_m(\psi)$ to increase the accuracy of the resistance R_m estimation.

Using estimated L_m parameter for the given rotor flux linkage values and R_m parameter for given frequency values, it could be put into the FLO unit structure in a lookup table form or in a polynomial dependence form.

VI. CONCLUSIONS

The structure of the rotor flux linkage observer unit for the induction motor with FOC system is proposed, taking into account the magnetic saturation and core losses. A magnetizing circuit parameters estimation technique for the induction machine equivalent circuit is presented.

The research results can to be applied in both electromechanical transmissions and other types of electric drives based on an induction machine with a magnetic flux wide range variation and with high core losses.

The main disadvantage of the proposed technique is the dependence of the magnetizing circuit parameters estimation quality on other winding parameters. The most important among them is the rotor resistance, which cannot be measured directly and depends on the machine temperature state, and therefore also needs to be estimated in any pattern.

REFERENCES

[1] S. N. Florentsev and V. S. Polyukhovich, "Traction electric euipment set for industrial shunting locomotives," 2018 17th International Ural Conference on AC Electric Drives (ACED), 2018, pp. 1-8

[2] S. N. Florentsev and D. B. Izosimov, "Features of the design of traction induction motors," 2012 2nd International Electric Drives Production Conference (EDPC), 2012, pp. 1-6

[3] I. Zhurov, S. Bayda and S. Florentsev, "Modeling of a Diesel Locomotive Induction Motor Drive with the Field-oriented Control when Operating in a Limited Voltage and High Rotation Frequency Mode," 2021 28th International Workshop on Electric Drives: Improving Reliability of Electric Drives (IWED), 2021, pp. 1-5

[4] S. -. Kim, S. -. Sul and M. -. Park, "Maximum torque control of an induction machine in the field weakening region," Conference Record of the 1993 IEEE Industry Applications Conference Twenty-Eighth IAS Annual Meeting, 1993, pp. 401-407 vol.1

[5] Lu Xianliang and Wu Hanguang, "Maximum efficiency control strategy for induction machine," ICEMS'2001. Proceedings of the Fifth International Conference on Electrical Machines and Systems (IEEE Cat. No.01EX501), 2001, pp. 98-101 vol.1

[6] Sokolovskiy G. G. Elektroprivody peremennogo toka s chastotnym regulirovaniyem [Electrical Drives of Alternating Current with Frequency Controlling]. Moscow, ACADEMIA, 2006. 267 p.

[7] Vinogradov A. B. Vektornoe upravlenie elektroprivodami peremennogo toka [Vector Control of AC Electric Drives], Ivanovo: GOUVPO "Ivanovskii gosudarstvennyi energeticheskii universitet im. V. I. Lenina", 2008.

[8] H. Xie, F. Wang, W. Zhang, C. Garcia, J. Rodríguez and R. Kennel, "Sliding Mode Flux Observer Based Predictive Field Oriented Control for Induction Machine Drives," 2020 IEEE 9th International Power Electronics and Motion Control Conference (IPEMC2020-ECCE Asia), 2020, pp. 3021-3025

[9] A. Derdiyok, M. K. Guven, H. Rehman and Longya Xu, "A new approach to induction machine flux and speed observer with on-line rotor time constant estimation," IEMDC 2001. IEEE International Electric Machines and Drives Conference (Cat. No.01EX485), 2001, pp. 102-107

[10] K. Kubota and K. Matsuse, "Compensation for core loss of adaptive flux observer-based field-oriented induction motor drives" Proceedings of the 1992 International Conference on Industrial Electronics, Control, Instrumentation, and Automation, 1992, pp. 67-71 vol.1

[11] W. Sangtuntong and S. Sujitjorn, "An adaptive sliding-mode observer incorporating core loss," 2004 IEEE Region 10 Conference TENCON 2004., 2004, pp. 574-577 Vol. 4

[12] Wang, K.; Huai, R.; Yu, Z.; Zhang, X.; Li, F.; Zhang, L. Comparison Study of Induction Motor Models Considering Iron Loss for Electric Drives. Energies 2019

[13] E. Levi, "Impact of iron loss on behavior of vector controlled induction machines," in IEEE Transactions on Industry Applications, vol. 31, no. 6, pp. 1287-1296, Nov.-Dec. 1995

[14] M. Hasegawa, S. Doki and S. Okuma, "Compensation method of stator iron loss of vector controlled induction motor using robust flux observer," 6th International Workshop on Advanced Motion Control. Proceedings (Cat. No.00TH8494), 2000, pp. 287-292

[15] L. H. Altas and A. M. Sharaf, "Modified induction motor drive speed control strategies for loss reduction and efficiency enhancement," [1991 Proceedings] The Twenty-Third Southeastern Symposium on System Theory, 1991, pp. 39-44

[16] S. Moulahoum, O. Touhami, R. Ibtiouen and M. Fadel, "Induction Machine Modeling With Saturation and Series Iron Losses Resistance," 2007 IEEE International Electric Machines & Drives Conference, 2007, pp. 1067-1072

[17] Kheldoun, Aissa & Khodja, Djalal Eddine. (2009). Vector Control Using Series Iron Loss Model of Induction, Motors and Power Loss Minimization.

[18] Jinhwan Jung and Kwanghee Nam, "A vector control scheme for EV induction motors with a series iron loss model," in IEEE Transactions on Industrial Electronics, vol. 45, no. 4, pp. 617-624, Aug. 1998

2022 29th International Workshop on Electric Drives: Advances in Power Electronics for Electric Drives (IWED), Moscow, Russia. Jan 26 – 29, 2022

MPPT Algorithms for Wind Turbines: Review and Comparison

Lukin Aleksandr, Galina Demidova,
Omar Alassaf, Dmitry Lukichev
Faculty of Control System and Robotics
ITMO University
Saint-Petersburg, Russia
alevluukin@itmo.ru, demidova@itmo.ru

Anton Rassõlkin, Toomas Vaimann
Department of Electrical Power Engineering
and Mechatronics
Tallinn University of Technology
Tallinn, Estonia
anton.rassolkin@taltech.ee

Abstract— **Wind turbines play an important role in energy generation using renewable sources. The wind is very changeable, and such turbines require techniques for finding and tracking maximum power density. The presented work includes an overview of maximum power point tracking techniques applicable to small-size wind turbines based on the Magnus effect. An overview of existing algorithms is presented. Overviewed algorithms are assessed using a simulation in MATLAB with a model of the experimental Magnus effect wind turbine.**

Keywords—wind energy, wind power generation, automatic control systems, power system simulation

I. INTRODUCTION

Magnus effect (or Magnus force) is a force that appears when a rotating cylindrical or spherical body is submerged into a moving gas or fluid with a relative motion between the body and the medium. It can often be observed in sports in the form of the curved trajectory of a thrown ball. This physical phenomenon has attracted more attention during the last two decades thanks to its possible application as a sail for marine vessels [1]. Another application of the Magnus force is in the field of renewable energy generation. Magnus Wind Turbines (MWTs) present a novel approach and provide energy generation at a wider range of wind speeds than conventional wind turbines [2].

Studied MWT is controlled by varying the rotational speed of the cylinders, which, in turn, generate lift and create torque on the generator shaft. Cylinders are powered by DC motors, making the whole system very sensitive to the balance between generated power and power consumed by the motors. This leads to a problem of finding and tracking the maximum power point at which the speed of a motor can be minimized, and power generation can be maximized. The goal of this paper was to perform research to assess the applicability of modern maximum power point tracking (MPPT) techniques to MWTs and provide a review of currently existing work in the field of MPPT for MWTs.

II. STATE OF ART IN MPPT FOR WIND TURBINES

An extensive review of MPPT techniques is presented in [3]. Overviewed algorithms include Tip Speed Ratio (TSR), Optimal Torque (OT), Power Signal Feedback (PSF), conventional and modified Hill Climb Search (HCS), Incremental Conductance (INC), Optimal Relation Based (ORB), Hybrid, Fuzzy based (FLC), Neural Network (NN), adaptive and Multivariable Perturb and Observe. The authors separate algorithms into three distinct groups: Indirect Power Controllers (IPS) are focused on optimizing mechanical power; Direct Power Controllers (DPS) directly maximize the output electrical power of the generator; novel advanced algorithms which don't fit into the first two groups. The overviewed algorithms were also classified according to their other parameters, such as complexity, convergence speed, computational memory requirement, wind sensor requirement, performance under varying wind conditions, and prior training or knowledge.

A different approach to the classification of MPPT techniques with relation to the type of generator is proposed by Thongam and Ouhrouche in [4]. This paper classified MPPT algorithms for Permanent Magnet Synchronous Generator, Squirrel Cage Induction Generator, and Doubly-Fed Electric Generator.

Another review is given by Kazmi et al. in [5]. In this paper, the authors analyze literature in the field of MPPT control and provide critical analysis of advantages and drawbacks. According to this paper, the most effective control algorithms up to date are self-tuning sensorless adaptive HCS and adaptive TSR. The main advantage of these types of MPPT algorithms is self-tuning capability. The conventional algorithms require building a generation lookup table and additional wind speed sensors, while HCS provides an easier-to-implement sensorless approach.

Fig. 1. MPPT techniques with relation to wind turbine size according to [6].

978-1-6654-6787-2/22 $31.00 © 2022 IEEE

An analysis of existing wind turbine control algorithms and their applications is presented by Rathi and Sandhu in [6]. The authors used a simulation approach to determine which algorithm is more suitable for such an application according to the size and power rating of the wind turbine. Overviewed algorithms include TSR, non-adaptive HCS, and Fuzzy Logic (FL)-based MPPT, which were applied to models of three wind turbines: small-size Vestas (13.5 m, 225 kW), medium-size General Electric (50 m, 2 MW), and large Sea-Titan (95 m, 10 MW). MATLAB simulations were performed for each turbine, and the maximum amount of generated power was assessed. Authors conclude that TSR is mostly suitable for large-scale wind turbines, while HCS is suitable for all sizes. The scholars highlight the main disadvantage of non-adaptive fixed-step MPPT as slow response time. Meanwhile, the FL-based MPPT algorithm has shown higher convergence speed and overall better performance than all the previous ones. The results of this paper are summarized on a diagram in Fig. 1.

Another review of MPPT algorithms was presented by Mirecki et al. in [7]. In this work, four approaches to optimal control were studied: Torque Control, Speed control, Current controlled diode bridge, and FL MPPTs. The results were assessed by three different parameters. The analysis has shown the good performance of torque control and current control MPPT. FL control MPPT was shown to be less effective than other techniques, mostly due to high complexity and lower power extraction at certain parameters. However, the authors highlight that the ability to function without wind sensor data and wind turbine characteristics is a significant advantage of these algorithms.

More deep analysis of FL algorithmsshows significant advances in Fuzzy MPPT algorithms with multiple papers in the field published. Yaakoubi et al. in [8], a proposed FL

control algorithm is compared to conventional HCS MPPT. The results show that applying an advanced FL algorithm allows extraction of more power than the conventional techniques. Another application of FL-based control for MPPT is presented by Li et al. in [9]. In this paper, the effectiveness of the FL algorithm was estimated using four performance indicators: energy captured by the turbine, MPPT time during a slow change in wind speed, MPPT time during a fast change in wind speed, and power fluctuation magnitude during steady state. The authors have shown that adaptive algorithms in the form of FL control can significantly improve the performance of wind energy conversion systems.

The summary of reviewed papers is presented in Table I. The literature analysis shows that the most promising algorithms for MWT are HCS, FLC, and variations of adaptive and hybrid algorithms. Currently, the field of MPPT for MWT remains relatively unexplored, with only a few published works, due to the relative novelty of the technology and no commercial solutions available. Overview of algorithms for MWT control is presented in the next chapter.

III. MWT CONTROL ALGORITHMS

A. Fixed-step HCS

Research in the field of non-adaptive fixed-step HCS algorithms for MWT was carried out by Babaeva et al. [10].

In this paper, the authors present a simple non-adaptive HCS MPPT algorithm with a fixed step. The implementation of this technique allowed to perform automatic tracking of the maximum power point. At the same time, the authors point out the significant power consumption by motors at the initial stages of the transitional process and highlight the

TABLE I. REVIEW OF EXISTING MPPT ALGORITHMS

	Method	Description	Advantages	Disadvantages
Direct	**Tip Speed Ratio (TSR)**	Based on tracking of optimal relation between rotor speed and wind speed	• Fast • Easy to implement	• Requires wind speed sensor • Ineffective for small turbines
	Optimal Torque (OT)	Based on regulating mechanical torque to reach optimal value	• Fast • Easy to implement	• Requires rotor speed sensor • Requires known opt. torque
	Power Signal Feedback (PSF)	Tracking generated power and maintaining its optimal value	• Fast • Easy to implement • Power stability	• Requires wind speed and rotor speed sensors • Requires known power curve
Indirect	**Hill Climb Search (HCS)**	Measuring and evaluating system's response to variation in control parameter	• Easy to implement • Sensorless	• Fixed-step HCS leads to torque instability
	Incremental Conductance (INC)	Based on generated electric power measurement	• Easy to implement • Sensorless	• Low convergence speed
	Optimal relation Based (ORB)	Based on known functions of generated power and current	• Easy to implement • Sensorless	• Requires known power-current function
	Hybrid	A combination of different control methods	• High convergence speed • Good tracking	• Harder to implement than non-hybrid methods
Other	**Fuzzy Based (FLC)**	Based on Fuzzy Logic Controllers	• High convergence speed • Good performance under varying wind conditions	• Hard to implement • Requires developed rules • More complex controllers
	Neural Networks (NN)	Based on Neural Networks	• High convergence speed • Good performance under varying wind conditions	• Requires training • Requires complex NN controllers
	Adaptive	• Quick switching between previous algorithms • Adapting to environment	• High convergence speed • Good performance under varying wind conditions	• Requires training and knowledge • Requires complex NN controllers
	Multivariable Perturb and Observe (MP&O)	Using controlled perturbations and observing its reaction	• No training required • Good performance under varying wind conditions	• Low convergence speed • Large wind farms only

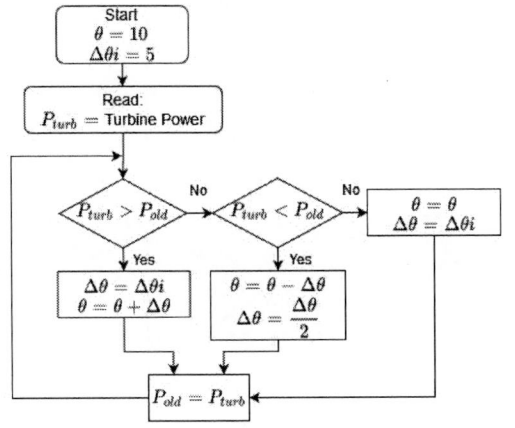

Fig. 4. Half-step HCS MPPT

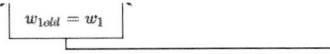

Fig. 3. Fixed step HCS MPPT

importance of further investigations in net power generation. The proposed algorithm is presented in Fig.2.

B. Adaptive step HCS

A comparison between fixed and adaptive step HCS algorithms is presented by Jinbo et al. [11]. This model uses an analytical wind power coefficient function applied to a wind turbine model. The control is based on the varying rotational speed of the cylinders. Two HCS algorithms were studied. The first was based on a fixed-step HCS, which is relatively easy to implement. The presented algorithm uses a constant voltage reference value to perform the MPPT. The increase in the size of this fixed step leads to a higher response rate. However, as the authors highlight, it also leads to higher controller instability around the maximum power point. To combat these problems, the authors propose an adaptive HCS algorithm, shown in Fig.3., which measures the size of the power increment and adapts the step size for the MPPC controller of the cylinders accordingly. The paper shows that both algorithms are fully applicable to MWTs, with adaptive HCS showing fast convergence and good stability around the maximum power point.

The same team presents a more advanced version of this research [12]. In this paper, the authors modify the previously developed system by adding the model of PMSG and boost converter. These modifications allow adding an additional MPPT controller for the boost converter, thus creating a system where both load and rotational speed can be controlled via HCS algorithms. The simulation results show that this approach presents a feasible solution to the problem of Magnus MPPT. The system automatically activated when the wind speed is high enough to ensure net power positive, which is around 4.5 m/s (an effective operational area for MWTs).

C. Half-step HCS

Another team working in the field of MPPT for MWTs was led by Correa [13]. This team presented a different approach to HCS algorithms by utilizing an adaptive control based on the divided step method. The method is similar to the simple non-adaptive HCS. However, each time the maximum power point (or maximum generation point, MGP) is reached, the increment of change in cylinder speed

is halved, as shown in Fig.4. Therefore, this control gradually reduces oscillations around the MGP, significantly reducing fluctuations of generated energy. The authors note that even though the presented method is more straightforward than the previously mentioned adaptive control, it still has a high impact on the efficiency of the MWT. In case of presented work, the algorithm was modified to save the step value after each increment.

The literature review has shown four works in the field of MPPT for MWTs. The analysis of the presented work highlighted multiple trends in control of MWTs. The first trend is the widespread use of HCS techniques. This fact aligns with the research results that showed that HCS algorithms show good performance in small wind-turbine systems, such as a vast majority of experimental MWTs.

All presented works are based on either simple or more complex adaptive HCS algorithms. This fact can be explained by the fact that the main advantage of HCS is that it doesn't require prior knowledge of the wind turbine. This advantage is very important in the field of MWTs, where simulation and modeling present a challenging task due to the complex nature of the Magnus effect. Another trend is the use of cylinder rotation speed as the main control parameter. This goes against the general trend in small-scale wind turbine control, where the main control parameter is the

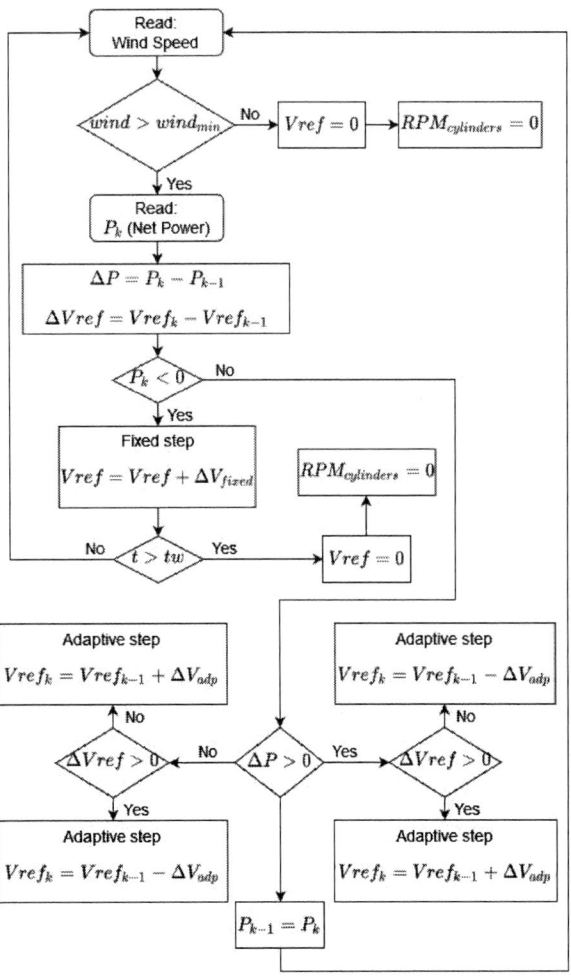

Fig. 2. Adaptive Step HCS MPPT

generator torque (since small-scale WTs with variable blade pitch are very rare due to mechanical complexity). Even though examples of torque-controlled MPPT for MWTs exist, they are coupled with speed-based controllers. This trend can be explained by the fact that in the case of MWTs, the cylinders' rotation speed plays a role similar to pitch control in the case of conventional horizontal axis wind turbines. But unlike pitch control, speed control is a fundamental feature of MWTs, and thus, it is reasonable to utilize this type of control to minimize the amount of energy consumed by the motor and increase the amount of energy generated by the turbine. The literature review has also shown that the FL-based control has demonstrated its effectiveness in MPPT for small wind turbines. Despite that, the FL algorithms were not yet studied with application to MWTs.

IV. WIND TURBINE DESCRIPTION

The experimental MWT operates with the help of two rotating cylinders. Each cylinder is powered with the help of a DC motor, which allows them to rotate with a speed of up to 6000 rpm, which is enough to cover the wide variety of wind speeds. The cylinders are located on a beam connected to the shaft via the hub. The speed of the turbine and cylinders was measured using [15], while the control of the generator was implemented using MTPA algorithm considered in [16], which was adapted for operation in generation mode. A Permanent Magnet Synchronous Generator (PMSG) with a gearbox is used to generate power, however, in the case of this work, the generator is not simulated, and only the mechanical power of the shaft is studied. A more detailed description of the generator is given in [14].

The MWT model was built using MATLAB Simulink. It consists of three distinct parts: wind model, aerodynamic model, and mechanical model, as shown in Fig. 5. For the purposes of this simulation, the moment of inertia of the motor and the cylinders is ignored. The wind model used in this work is a constant wind speed of 3 m/s (the speed at which the MWT is presumably more effective than conventional turbines).

The aerodynamic model is based on the Blade Element Momentum (BEM) theory. The BEM shows the amount of kinetic energy extracted by the windwheel of determined radius from the wind at a given speed as mechanical torque on the shaft of the windwheel, as shown in (1)

$$P = \frac{1}{2} \cdot A \cdot \rho \cdot V^3 \cdot C_p. \tag{1}$$

Here, P is the mechanical power, A is the area of the windwheel, p is the air density, V is the wind speed. These parameters show the maximum amount of kinetic energy contained in a moving mass of air. Meanwhile, C_p is a variable called the wind power coefficient. This coefficient shows the effectiveness of wind turbine and usually varies between 0 and 0.569 (so-called Betz limit, a theoretical limit on the amount of energy that can be extracted from the wind). C_p is a function that depends on multiple parameters, such as tip speed ratio, dimensions of the wind wheel, and constructive features of the wind turbine. Multiple models for C_p of Magnus wind turbines exist, and in the presented work, the analytical model was used [17]. The analytic function for C_p is presented in (2)

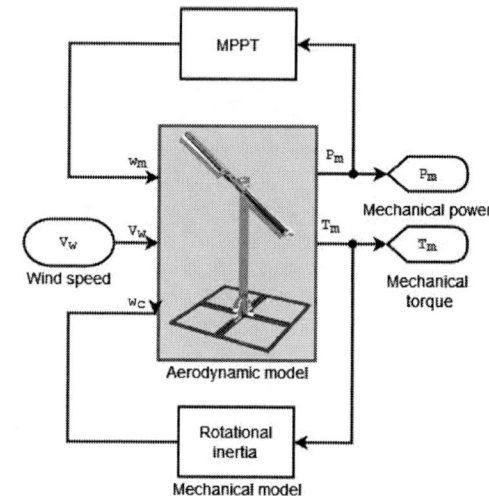

Fig. 5. Model of the wind turbine with MPPT

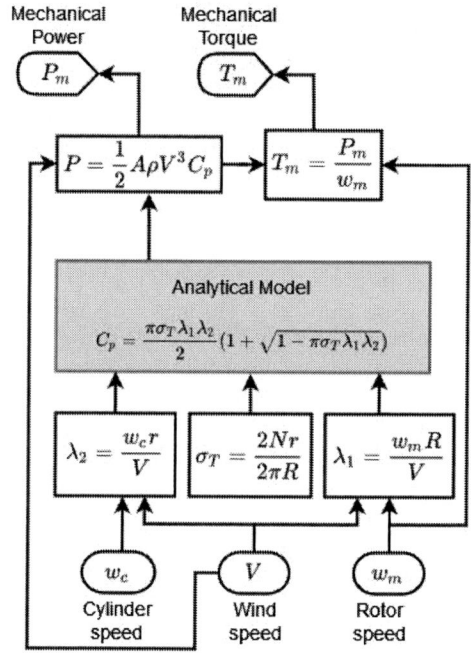

Fig. 6. Aerodynamic model of the wind turbine

$$C_p = \frac{\pi \sigma_T \lambda_1 \lambda_2}{2} \left(1 + \sqrt{1 - \pi \sigma_T \lambda_1 \lambda_2}\right). \tag{2}$$

In this equation, λ_1 is the tip speed ratio (3)

$$\lambda_1 = \frac{\omega_m R}{V}, \tag{3}$$

λ_2 is the relative speed of the cylinder's rotation (4)

$$\lambda_2 = \frac{\omega_c r}{V}, \tag{4}$$

and σ_T is a mechanical parameter - the rotor tip solidity (5)

$$\sigma_T = \frac{2Nr}{2\pi R}. \tag{5}$$

978-1-6654-6787-2/22 $31.00 © 2022 IEEE   82

Here, ω_m is the rotational speed of the rotor, ω_c is the rotational speed of the cylinder, R is the rotor radius, r is the cylinder radius, and N is the number of cylinders. The aerodynamic model is presented in Fig. 6.

The equation shows that an important parameter for the power coefficient is the speed of the rotor. To estimate the rotor speed, simple one-mass inertia was used. It takes into account the inertia of the rotor and cylinders and the viscous friction coefficient in the ball bearings; however, it does not take into account tortional effects and the load torque on the rotor shaft. The generator is also not included in the model. This inertia model (6) allows to estimate the speed of the windwheel shaft and perform dynamic simulations

$$T = J \cdot \frac{d\omega(t)}{dt} + B \cdot \omega(t) \qquad (6)$$

V. SIMULATION RESULTS

Three simulations were performed in the course of this work. The first simulation was running to study power generation of the MWT with cylinder running at maximum speed. Wind speed was set to 3 m/s since it's close to the theoretical value at which the MWT becomes effective in terms of net power generation. During the first simulation, the speed of the cylinder increased gradually from 0 to 614 rad/s, simulating acceleration from the standby mode.

The results of the simulation for mechanical power are presented in Fig. 7. The plot shows that studied wind speed, there is a point where cylinders rotate high enough to pass the maximum power point. This mode of operation is to be avoided since, in this mode, the cylinders consume extra power required to run the motor and at the same time generate less energy than physically possible at the given wind speed. In the case of the MWT, the maximum generated power at 3 m/s is 11.1 W at 337 rad/s. These values were used as a benchmark to further assess different MPPT techniques.

A. Fixed-step HCS

To combat the effect of passing maximum power point, MPPT algorithms must be utilized. The simulation results for generated mechanical power are presented in Fig.8. The cylinder speed variation is shown in Fig.9. Numerical results of the simulation are presented in Table II.

The first MPPT algorithm reviewed in the course of this work is a simple non-adaptive fixed-step HCS. For the presented algorithm, the step size increment was chosen to be 30 rad/s, and the MPPT controller is tuned to analyze the power coefficient with 1 second interval. The simulation shows that with a simple MPPT, the model was able to reach the MGP and tune the rotational speed of the cylinders so it would remain there, maximizing power generation. However, the plot also shows significant fluctuations of mechanical power and the maximum power point.

This effect can be explained by observing the MPPT signal for the cylinders. The plot shows the speed of the cylinders during the operation of the MWT. The chosen speed increment of 30 rad/s shows a good convergence time. However, it also leads to significant speed ripples at the maximum power point. This leads to fluctuations of mechanical torque and an overall decrease in the quality of produced energy. This ripple effect can be minimized by utilizing adaptive MPPT algorithms.

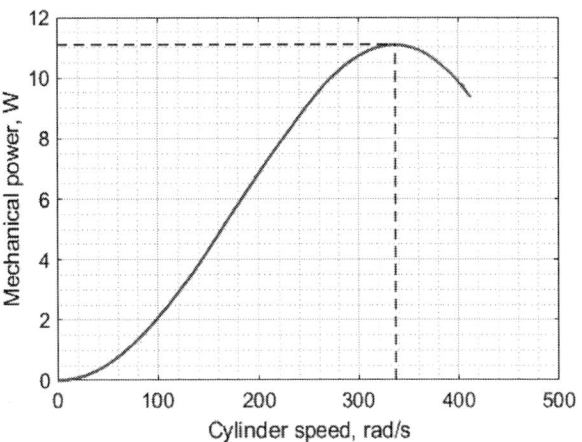

Fig. 7. Power curve for MWT at wind speed of 3 m/s

B. Adaptive Step HCS

The mechanical power plot shows that introducing the adaptive step algorithm allowed minimizing torque ripples while simultaneously achieving high convergence time. However, the simulation has shown that the studied algorithm reaches a steady-state with some error, which leads to increased power consumption by the cylinder drives and decreases power production simultaneously.

C. Half-step HCS

Half-step HCS showed the same convergence time as the conventional algorithm. However, as the system reached its maximum power point, the size of the step started to decrease and rapidly reached the target value with no torque ripples observed. The cylinder speed plot shows a steady decrease in rotational velocity with each incremental step.

VI. CONCLUSIONS

In the course of the presented work, state of art in the field of MPPT for wind turbines was presented. Initial analysis has shown that HCS and Fuzzy control techniques present interest due to their high response speed, simplicity, and ability to function without sensors. In [3] the authors show that HCS, INC, and ORB algorithms are easier to implement and provide reliable control. However, TSR, OT, and PFS show better response speed.

TABLE II. RESULTS OF THE SIMULATION

	HCS	Adaptive HCS	Half-Step HCS	Benchmark
Average power, W	10.96	10.99	11.08	11.08
St Dev of power	0.12	$1\cdot10^{-3}$	25e-4	N/A
Average speed, rad/s	332.89	353.92	337.65	336.72
St Dev of speed	20.19	$6.27\cdot10^{-13}$	0.92	N/A

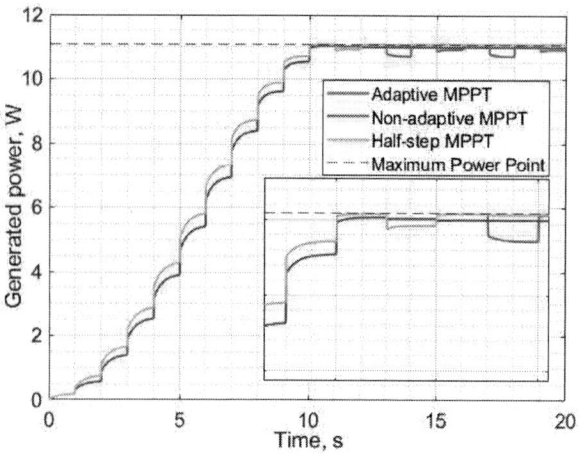

Fig. 8. Mechanical power, extracted with different MPPT techniques

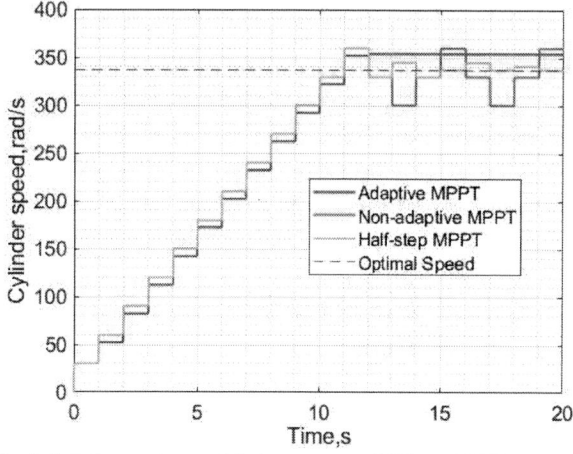

Fig. 9. Cylinder rotation speed during operation of different MPPT algorithms

Further analysis has shown that contemporary works in MPPT for MWTs focus on HCS techniques. Three algorithms were reviewed: conventional HCS, half-step HCS, and adaptive HCS. All these algorithms were assessed by integrating them into a Simulink model of the MWT. A step input for wind speed was later applied. A conventional HCS was used as a benchmark for comparison with advanced algorithms. The research has shown that both half-step and adaptive techniques show good convergence time, with adaptive HCS being significantly faster than half-step HCS. The downside of adaptive HCS is a steady-state error which leads to higher rotational speed (an, thus, energy consumption by the motor) and lowers energy generation.

In conclusion, it can be said that both adaptive and half-step HCS algorithms are fully applicable to the experimental MWT, while conventional HCS can be used as a benchmark to assess the effectiveness of MPPT techniques. The research has also shown that FL control is a promising and unexplored approach in the field of MWT MPPT, and further work on control of the experimental wind generator should focus on both adaptive MPPT methods and FL Control.

REFERENCES

[1] P. Nuttall and J. Kaitu'u, "The Magnus Effect and the Flettner Rotor: Potential Application for Future Oceanic Shipping," The Journal of Pacific Studies, vol. 36(2), pp. 161-182, 2016

[2] N.M. Bychkov, A. V. Dovgal, and V. V. Kozlov, "Magnus wind turbines as an alternative to the blade ones," Journal of Physics: Conference Series, vol. 75, No. 1, p. 012004, July 2007

[3] D. Kumar and K. Chatterjee, (2016). "A review of conventional and advanced MPPT algorithms for wind energy systems," Renewable and sustainable energy reviews, vol. 55, pp. 957-970, 2016

[4] J.S. Thongam and M. Ouhrouche, "MPPT control methods in wind energy conversion systems," Fundamental and advanced topics in wind power, vol. 15, pp. 339-36, 2011

[5] S. M. R. Kazmi, H. Goto, H. J. Guo and O. Ichinokura, "Review and critical analysis of the research papers published till date on maximum power point tracking in wind energy conversion system," 2010 IEEE Energy Conversion Congress and Exposition, pp. 4075-4082, September 2010

[6] R. Rathi and K.S. Sandhu, "Comparative analysis of MPPT algorithms using wind turbines with different dimensions & ratings," 2016 IEEE 1st International Conference on Power Electronics, Intelligent Control and Energy Systems (ICPEICES), pp. 1-4, July 2016

[7] A. Mirecki, X. Roboam and F. Richardeau, "Comparative study of maximum power strategy in wind turbines," 2004 IEEE International Symposium on Industrial Electronics, vol. 2, pp. 993-998, May 2004

[8] A. El Yaakoubi, A. Asselman and A. Djebli, "A MPPT strategy based on fuzzy control for a wind energy conversion system," Procedia Technology, vol. 22, pp. 697-704, 2016

[9] X. P. Li, W. L. Fu, Q. J. Shi, J. B. Xu and Q. Y. Jiang, "A fuzzy logical MPPT control strategy for PMSG wind generation systems," Journal of Electronic Science and Technology, vol. 11(1), pp. 72-77, 2013

[10] M. Babayeva, A. Abdullin, N. Polyakov and S. Aranovskiy, "MPPT Algorithms for Magnus Effect Wind Turbine Control System" 2020 XI International Conference on Electrical Power Drive Systems (ICEPDS), pp. 1-4, October 2020

[11] M. Jinbo, G. Cardoso, F. A. Farret, D. L. Hoss and M. C. Moreira, "Fixed and adaptive step HCC algorithms for MPPT of the cylinders of Magnus wind turbines," 3rd Renewable Power Generation Conference (RPG 2014), pp. 1-6, September 2014

[12] M. Jinbo, F. A. Farret, G. Cardoso Junior, D. Senter and M. Franklin Lorensetti, "MPPT of Magnus Wind System with DC Servo Drive for the Cylinders and Boost Converter," Journal of Wind Energy, 2015

[13] L. C. Corrêa, J. M. Lenz, C. G. Ribeiro, J. G. Trapp and F. A. Farret, "MPPT for Magnus wind turbines based on cylinders rotation speed," *IECON 2013-39th Annual Conference of the IEEE Industrial Electronics Society*, pp. 1718-1722, November 2013

[14] G. L. Demidova, A. Anuchin, A. Lukin, D. Lukichev, A. Rassolkin and A. Belahcen," Magnus wind turbine: Finite element analysis and control system," Proceedings of the 2020 International Symposium on Power Electronics, Electrical Drives, Automation and Motion, SPEEDAM 2020, pp. 59-64, 2020

[15] A. Anuchin, A. Dianov and F. Briz, "Synchronous Constant Elapsed Time Speed Estimation Using Incremental Encoders," in IEEE/ASME Transactions on Mechatronics, vol. 24, no. 4, pp. 1893-1901, Aug. 2019, doi: 10.1109/TMECH.2019.2928950.

[16] A. Dianov and A. Anuchin, "Adaptive Maximum Torque per Ampere Control of Sensorless Permanent Magnet Motor Drives," Energies, vol. 13, no. 19, p. 5071, Sep. 2020 [Online]. Available: http://dx.doi.org/10.3390/en13195071

[17] D. Luo, D. Huang and G. Wu, "Analytical solution on Magnus wind turbine power performance based on the blade element momentum theory," Journal of Renewable and Sustainable Energy, vol. 3(3), 2011

Implementation and Analysis of Rolling Bearing Faults Caused by Shaft Currents

Karolina Kudelina, Siarhei Autsou, Bilal Asad, Toomas Vaimann, Anton Rassõlkin, Ants Kallaste

Department of Electrical Power Engineering and Mechatronics
Tallinn University of Technology
Tallinn, Estonia
karolina.kudelina@taltech.ee

Abstract— **Bearing failures make the greatest proportion of damages affecting electrical machines. At the same time, there are various types of the bearing failures but the most common are related to damages caused by shaft currents in the machines leading to problems in industry. In this case, diagnostics and timely prediction of potential faults are important. The paper gives an overview of the problem describing main bearing damages. Particularly, different shaft currents related faults are discussed in detail. Implementation of these faults in laboratory environment is introduced. Analysis of rolling bearing faults caused by shaft currents is presented.**

Keywords—ball bearings, condition monitoring, electric motors, predictive maintenance, shafts.

I. INTRODUCTION

Bearing damages have been claimed as the most spread type of mechanical fault in any machines type by many researches. Authors in discuss the necessity of inverse problem theory in electric machine fault diagnosis [1]. In [2], there is presented a high-performance and energy-efficient fault diagnosis method with the usage of effective envelope analysis. Authors in [3] introduce fault detection for a permanent magnet synchronous motors. In [4], problem of bearing shaft currents is addressed. In [5], a large motor reliability survey of industrial installations is presented.

Nowadays, production is rapidly switching over predictive maintenance, which allows to minimize downtime during scheduled equipment controls and increase the system reliability [6]. To train the system to identify future faults, a huge amount of data in needed for algorithm trainings. The more input data is used, the more precise results are in the output. In this paper, there is presented a method for data gathering of bearing faults related to shaft currents. This data will be used for algorithm training and predictive system development.

Bearing damages can be caused by various environmental and operational factors, such as material fatigue, contamination of the bearing and lubricant, wrong emplacement, lubricant problems, and shaft currents in particular. Development of power electronics and its wide usage in electrical machines lead to serious issues in industry [7], [8]. There are many ways to identify bearing currents in electrical machine; by visual inspection or measurements with usage of different methods (Rogowski coil [9], currents transformer [10], vibration and ultrasound analysis [11], etc).

The research leading to these results received funding from the EEA/Norway Grants 2014–2021, Industrial Internet methods for electrical energy conversion systems monitoring and diagnostics.

Fig. 1. Material fatigue of rolling bearing.

II. BEARING FAULTS

By monitoring the operation of bearings, measuring temperature, noise, vibration, and periodically analyzing the quality of the lubricant, the risk of bearing damage can be significantly reduced. Due to the fact that the operating conditions of electrical machines can be different, the bearing can be affected by many fault types. The reasons for bearing failures can be different ambient and manufacturing factors.

A. Material Fatigue

Material fatigue is usually caused by cyclic and continuous loads [12]. These loads affect largely on bearing by cracking material and decreasing the material strength, which causes in turn its decomposition. Over time, cracking expands to the larger surface, and, eventually, bearing becomes incompatible for further operation. An example of material fatigue is shown in Fig. 1. The number of revolutions, which bearing makes before the first material fatigue signs become noticeable and appear on raceways as well as rolling elements, defines durability of the bearing.

B. Contamination

Corrosion is a process between metal and environment, which results in material consumption or dissolution [13]. Corroded bearing usually gives additional vibrations, which are clearly on frequency spectra. In this case, corrodible products interfere with the balls during the rotation of the shaft. The usage of the bearings in an aggressive environment leads to the more rapidly developing corrosion of the material. An example of corroded bearing is shown in Fig. 2.

Fig. 2. Corrosion of rolling bearing.

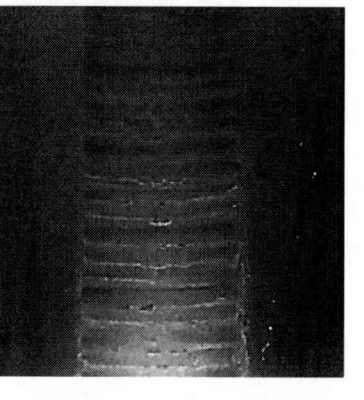  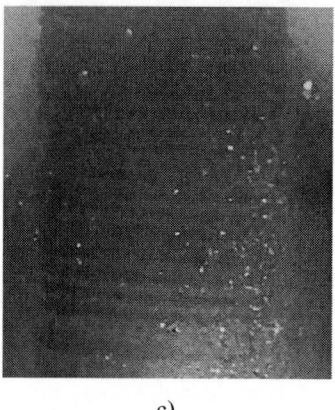

a)   b)   c)

Fig. 3. Bearing currents' faults: a) fluting, b) frosting, and c) pitting.

The lubricant can also become contaminated when water or some chemical substances enter the bearing. Since the lubricant properties are deteriorated, this in turn causes bearing corrosion. To prevent early corrosion sighs, the right gasket should be selected. Also, it is important to keep clean during assembly and not to use contaminated lubricants.

C. Shaft Currents

Bearings are frequently affected by shaft currents, especially in industry. This fault occurs when current goes through rolling elements from one bearing ring to another. Many parameters determine the fault development, such as rotational speed, current value, lubricant condition, and exposure time. In initial stages, faults caused by shaft currents are not visible and barely can be detected. By developing, increased noise and vibration indicate the fault presence. Defects that are to be inflicted due to shaft currents differ apparently from any other fault type. Depending on motor operating conditions, as shown in Fig 3, these faults can be related to one of the most common types of current faults: "fluting", "frosting" and "pitting".

D. Mechanical Damages

Many mechanical damages are to be occurred due to wrong placement of electrical machine parts. This can lead to damages of rings and cages; which example is shown in Fig. 4.

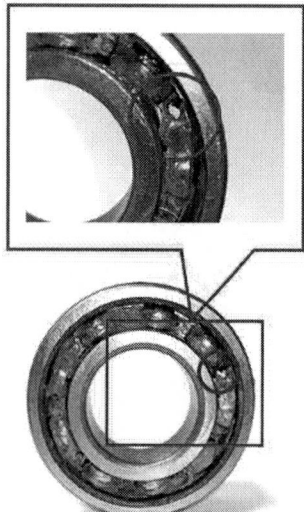

Fig. 4. Mechanical damages of bearing cage.

Before the mounting, the surfaces of the shafts have to be checked, completely cleaned and greased with a thin lubricant layer. Shaft alignment should be controlled, as well its compliance with the technical documentation and standards. Applying external forces during mounting is not allowed. The most suitable way to place the bearing to the shaft is to apply even pressure around the entire bearing ring.

Moreover, damages can be caused by manufacturing factors and be inflicted in production. Manufacturing faults are not usually considered during the design process and they tend to occur during the machines operation [14]. Before mounting, the bearing should be checked for manufacturing faults: compliance with the appearance, ease of rotation, and clearances to the requirements of technical documentation.

E. Lubricant problems

Frequently, lubricant condition can indicate a potential motor problem, especially in case of shaft currents. One of these indicators is darkening of the lubricant. An example of the darkened lubricant due to bearing shaft currents is shown in Fig. 5. In this case, lubricant darkens because of oxidizing caused by electrical discharges.

Fig. 5. Darkened lubricant.

Lubrication problems can also be related to either scarce or surplus lubrication. Scarce lubrication causes additional friction of motor moving parts that causes material cracking, while surplus lubrication leads to the excessive shaft slipping. Besides, wrongly selected lubricant can significantly contribute to lubricant contamination and bearing destruction.

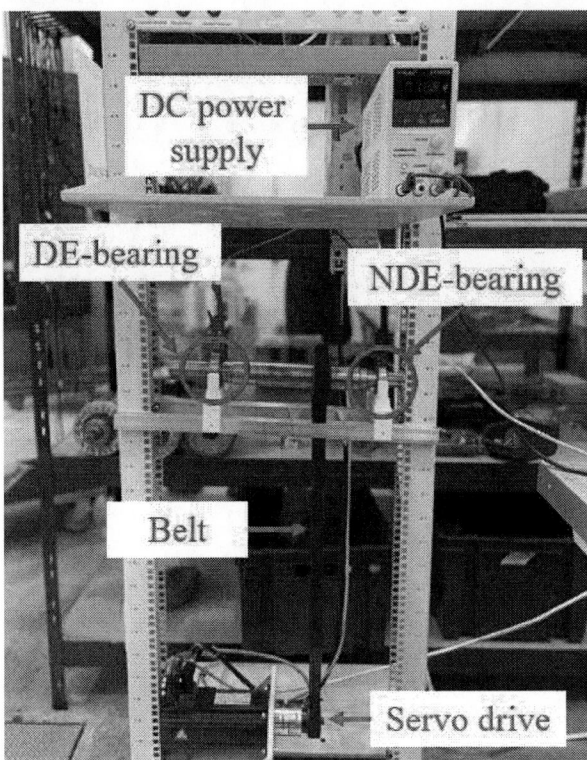

Fig. 6. Experimental test bench.

III. Experiments and Result Discussion

Fluting occurs in the combination of low voltage and constant rotational speed. When the motor runs at varying speeds, another well-detectable fault called frosting will appear. Pitting is caused by high voltage and low rotational speed. To fulfill these conditions and study different cases, an experimental test bench was built, as shown in Fig. 6.

a) Case 1 – 100 rpm and 10 A: As shown in Fig. 7, there was observed a darkening of the lubricant. DE-bearing as well as NDE-bearing have slightly darkened inner and outer raceways. Also, insignificant darkening of rolling elements was seen.

Fig. 7. DE-bearing at 100 rpm and 10 A.

Fig. 8. DE-bearing at 100 rpm and 20 A.

b) Case 2 – 100 rpm and 20 A: As shown in Fig. 8, this case revealed more serious case of lubricant darkening. Experiment was stopped due to the impossibility of the bearing to operate. In this case, DE-bearing has slight fluting in the inner raceway and darkened outer raceway. NDE-bearing has darkened inner as well as outer raceways, as shown in Fig. 9. Both have darkened rolling elements.

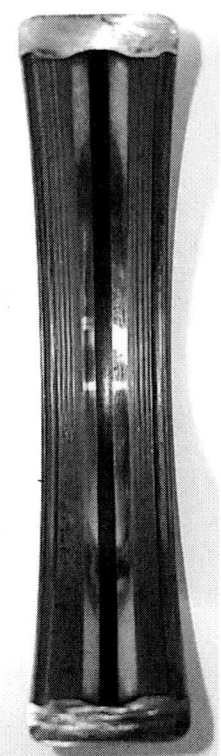

Fig. 9. DNE-bearing at 100 rpm and 20 A.

Fig. 12. DE-bearing at 800 rpm and 20 A.

e) *Case 5 – 800 rpm and 20 A:* As shown in Fig. 12, this case revealed a serious case of lubricant darkening. Experiment was urgently stopped due to the impossibility of the bearing to operate. As seen in Fig. 13, increase in rotational speed and current reveal an obvious case of pitting in combination with slight fluting of inner rceway and fluting on outer raceway in case of DE- bearing.

Fig. 10. DE-bearing at 500 rpm and 10 A.

c) *Case 3 – 500 rpm and 10 A:* This case revealed a slight darkening of the lubricant. Increase in rotational speed relealed an obvious case of fluting on inner raceway and darkened outer raceway in case of DE-bearing, as shown in Fig. 10. NDE-bearing has darkened inner and outer raceways with slight fluting trails. Both bearings have darkened rolling elements.

d) *Case 4 – 800 rpm and 10 A:* As shown in Fig. 11, lubricant changed color. DE-bearing has fluting phenomenon on inner raceway and darkened outer raceway with slight fluting. In case of NDE-bearing, darkeneing of raceways was observed.

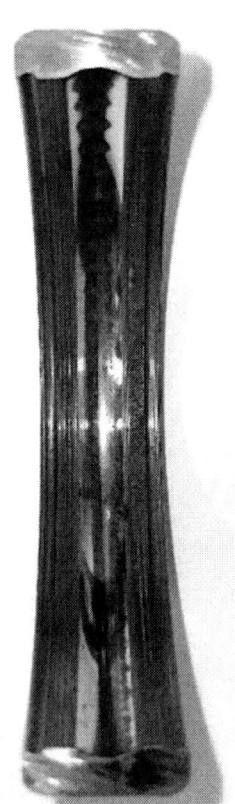

Fig. 11. DE-bearing at 800 rpm and 10 A.

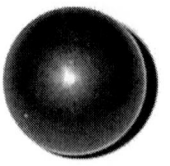

Fig. 13. DE-bearing at 800 rpm and 20 A.

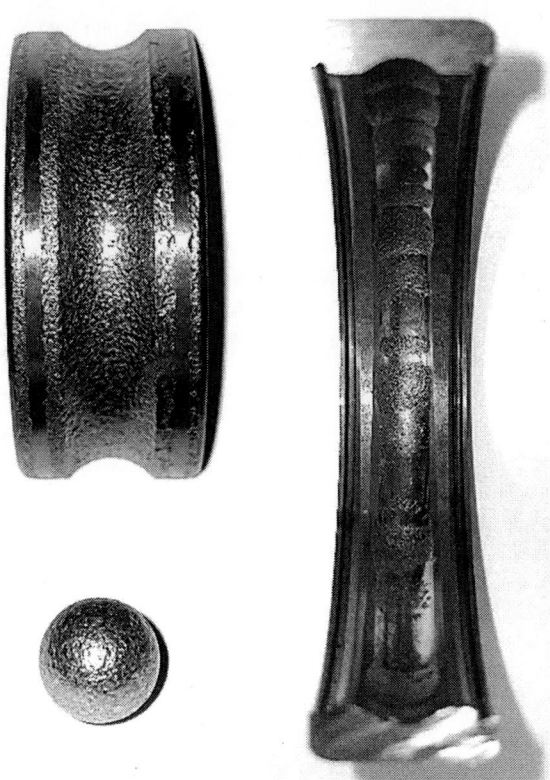

Fig. 14. DNE-bearing at 800 rpm and 20 A.

At the same time, as shown in Fig. 14, NDE-bearing presented a case of frosting on inner and outer raceways. Both have darkened rolling elements with pitting or frosting phemonema. In Table I, there is introduced a comparitive analysis of all studied cases with shaft current faults.

IV. CONCLUSION

Nowadays, bearing failures make the greatest proportion of damages affecting electrical machines and leading to fatal consequences in industry. These damages can be caused by various environmental and operational factors. The most spread bearings failures are related to damages caused by shaft currents. Therefore, timely diagnostics as well as prediction of such faults become a critical factor in production. Manufacturing is being switched rapidly over predictive maintenance. However, to train the system to identify precisely future faults, a huge amount of data in needed. The more input data is used, the more precise results are in the output. In this paper, there is presented a method for data gathering of bearing faults related to shaft currents.

In this paper, an overview of the problem describing main bearing damages, particularly, shaft currents related faults were discussed. Analysis of rolling bearing faults caused by shaft currents was presented. Implementation of these faults in laboratory environment was introduced. This data will be used for algorithm training and predictive system development, which is considered as future work in order to prevent timely machine breakdowns.

REFERENCES

[1] T. Vaimann, A. Belahcen, and A. Kallaste, "Necessity for implementation of inverse problem theory in electric machine fault diagnosis," *IEEE 10th Int. Symp. Diagnostics Electr. Mach. Power Electron. Drives*, pp. 380–385, 2015.

[2] M. Kang, J. Kim, and J. M. Kim, "High-performance and energy-efficient fault diagnosis using effective envelope analysis and denoising on a general-purpose graphics processing unit," *IEEE Trans. Power Electron.*, vol. 30, no. 5, pp. 2763–2776, 2015.

[3] J. A. Rosero, J. Cusido, A. Garcia, J. A. Ortega, and L. Romeral, "Broken bearings and eccentricity fault detection for a permanent magnet synchronous motor," *IECON Proc. (Industrial Electron. Conf.*, pp. 964–969, 2006.

TABLE I. COMPARITIVE ANALYSIS OF ALL STUDIED CASES

Test Conditions			Experimental Outcomes			
Rotational speed, rpm	Current, A	Duration, days	Bearing type	Inner raceway	Outer raceway	Rolling element
100	10	7	DE-bearing	Slightly darkened raceway	Darkened raceway	No changes
		7	NDE-bearing	Darkened raceway	Darkened raceway	Slightly darkened rolling elements
100	20	5 (experiment stopped)	DE-bearing	Slight fluting	Darkened raceway	Darkened rolling elements
		5 (experiment stopped)	NDE-bearing	Slightly darkened raceway	Darkened raceway	Darkened rolling elements
500	10	7	DE-bearing	Fluting	Darkened raceway	Darkened rolling elements
		7	NDE-bearing	Darkened raceway	Darkened raceway / slight fluting	Darkened rolling elements
800	10	7	DE-bearing	Fluting	Darkened raceway / slight fluting	Darkened rolling elements
		7	NDE-bearing	Slightly darkened raceway	Darkened raceway / slight fluting	Darkened rolling elements
800	20	4 days (experiment stopped)	DE-bearing	Fluting / pitting	Slight fluting	Darkened rolling elements / pitting
		4 days (experiment stopped)	NDE-bearing	Frosting	Frosting	Darkened rolling elements / frosting

[4] Tom Bishop, "Dealing with Shaft and Bearing Currents," 2017.

[5] P. O. Donnell, C. Heising, C. Singh, and S. J. Wells, "Report of Large Motor Reliability Survey of Industrial and Commercial Installations: Part 3," *IEEE Trans. Ind. Appl.*, vol. IA-23, no. 1, pp. 153–158, 1987.

[6] Y. Lei, N. Li, S. Gontarz, J. Lin, S. Radkowski, and J. Dybala, "A Model-Based Method for Remaining Useful Life Prediction of Machinery," *IEEE Trans. Reliab.*, vol. 65, no. 3, pp. 1314–1326, 2016.

[7] A. Muetze, A. Binder, H. Vogel, and J. Hering, "What can bearings bear?," *IEEE Ind. Appl. Mag.*, vol. 12, no. 6, pp. 57–64, 2006.

[8] Y. Le, J. Sun, and B. Han, "Modeling and Design of 3-DOF Magnetic Bearing for High-Speed Motor Including Eddy-Current Effects and Leakage Effects," *IEEE Trans. Ind. Electron.*, vol. 63, no. 6, pp. 3656–3665, 2016.

[9] S. Quabeck, L. Braun, N. Fritz, S. Klever, and R. W. De Doncker, "A Machine Integrated Rogowski Coil for Bearing Current Measurement," *13th Int. Symp. Diagnostics Electr. Mach. Power Electron. Drives*, pp. 17–23, 2021.

[10] L. Kütt, J. Järvik, T. Vaimann, M. Shafiq, M. Lehtonen, and J. Kilter, "High-frequency current sensor for power network on-line meaurements," *13th Int. Sci. Conf. Electr. Power Eng. 2012, EPE 2012*, vol. 1, no. May 2014, pp. 367–371, 2012.

[11] J. L. H. Silva and A. J. M. Cardoso, "Bearing failures diagnosis in three-phase induction motors by extended Park's Vector approach," *31st Annu. Conf. IEEE Ind. Electron. Soc.*, pp. 2591–2596, 2005.

[12] K. Kudelina, B. Asad, T. Vaimann, A. Rassõlkin, A. Kallaste, and D. V Lukichev, "Main Faults and Diagnostic Possibilities of BLDC Motors," *27th Int. Work. Electr. Drives MPEI Dep. Electr. Drives 90th Anniv.*, 2020.

[13] K. Kudelina, B. Asad, T. Vaimann, A. Rassolkin, and A. Kallaste, "Effect of Bearing Faults on Vibration Spectrum of BLDC Motor," *2020 IEEE Open Conf. Electr. Electron. Inf. Sci.*, pp. 1–6, 2020.

[14] K. Kudelina, B. Asad, T. Vaimann, A. Rassõlkin, and A. Kallaste, "Production Quality Related Propagating Faults of Induction Machines," *Int. Conf. Electr. Power Drive Syst.*, 2020.

Cylindrical Blades Magnus Wind Turbine Optimization and Control System

Omar Alassaf, Aleksandr Lukin, Galina Demidova, Gleb Kozlov, Andrey Volkhontsev, Nikolai Poliakov

Faculty of Control System and Robotics
ITMO University
Saint-Petersburg, Russia
alevlukin@itmo.ru, demidova@itmo.ru

Abstract— **Magnus wind turbine is a type of small wind turbine for private houses in the Energy 4.0 concept. Due to its novelty, there is no strong topology for such facilities. In this paper, the construction of the Magnus wind turbine is modified for the best energy efficiency. The cylindric blades as a main part of the Magnus wind turbine which are generally producing the Magnus Force was also modified. The control system for the cylindric blades based on brushless direct current motor was performed.**

Keywords—wind energy, wind power generation, rotating cylinder, finite element analysis, control system, brushless motors

I. Introduction

The concept of distributed energy systems [1] includes a new type of small wind turbine based on the Magnus effect [2]. The first version of the Magnus wind turbine is shown in Fig. 1a [3]. The design of the cylindric blades consists of a joist, on which two brackets are installed. Between these brackets, two cylinders are rotating by DC motors mounted on a beam. This construction of beam with the attached brackets had an extremely low streamlining and the open DC motors position has also worsened this characteristics. In addition, the DC motor used has several negative qualities, such as significant heating and low speed. BLDC motors (Fig. 1b) were proposed as a replacement for DC motors. The use of one long structural profile for two blades at once significantly reduces the modularity of the structure, increases the weight and makes the stand less portable. Furthermore, the cylinders were made of aluminium and had an extremely high mass and moment of inertia. The cylindric blades design also could be modified by using ridged cylinders [4]. According to [5] the smooth cylindrical blades topology for Magnus wind turbine has shown best aerodynamic characteristiccs. The numerical analysis of smooth cylinders was shown in [6]. The geometry, drag and lifting force was computed in ANSYS. Taking into account manuscripts above the topology of cylindrical blades remains unchanged. In several implementations, the Magnus Wind turbine has five cylindrical blades [7, 8]. But considering economic and flow analysis the topology has remained unchanged, too. The permanent synchronous generator (PMSG) was used as the main part of converting system of the Magnus wind turbine as shown in Fig.1. This research deals with the modification design of the Magnus wind turbine configuration, especially, with respect to cylindrical blades. The control system of cylindrical blade based on FOC control was also synthesized.

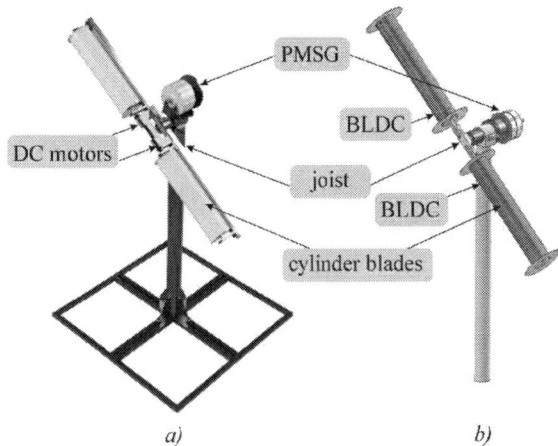

Fig. 1. Magnus wind turbine design: a) the first topology; b) modified version

II. Magnus Wind Turbine Design

The main research goal in the area Magnus wind turbine design deal with modify cylinders' shaft, choosing bearings and replacement motors.

A. Cylinder drag force calculation

The strength calculations deal with forces acting on the cylinder during the rotation. It is necessary to take into account the drag force with which the wind acts on the cylindrical blade.

$$F_D = \frac{c_x \cdot S \cdot v^2 \cdot \rho}{2},\qquad(1)$$

here S - the surface area; v - the wind speed; ρ - the air density; c_x - the drag coefficient. The cylinder data are shown in the Appendix.

The drag force was calculated on the maximum wind speed of 40 m/s and equal $F_D = 49.4$ N. Additionally, to provide safety the coefficient for drag force equal to two was used. In this case, the drag force equal $F_D = 100$ N and the force acting by the oncoming wind flow can be neglected.

978-1-6654-6787-2/22 $31.00 © 2022 IEEE

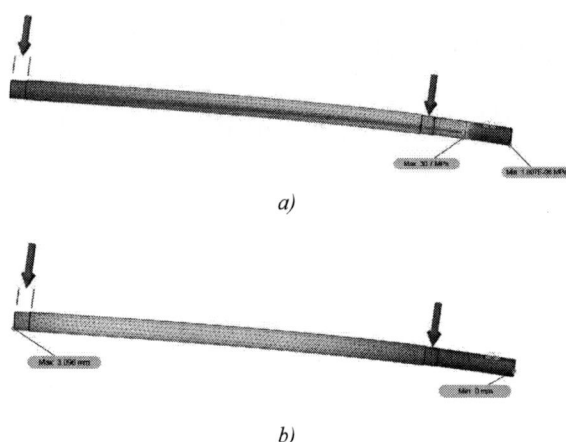

a)

b)

Fig. 2.   Shaft analysis a) maximum stresses; b) maximum displacement

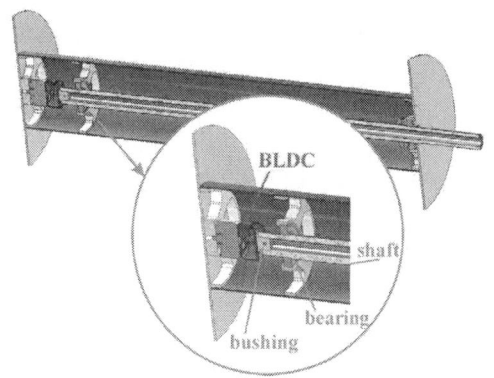

Fig. 3.   Bearing and brushless DC motor installation in the cylindric blade

B. Shaft analysis

The drag force obtained above affecting on the shaft of the cylinder at the bearing locations. All transformation simulations were implemented in the finite element analisys software [9]. The load acts of the place's location of the bearings. The shaft was fixed as showed the blue arrows in Fig.2. Fig.2a) shows simulations of the shaft with the maximum stress of 30.7 MPa has occurred. The yield strength of the aluminium material alloy is 320 MPa and provides a tenfold safety margin. The shaft weight equals 1.03 kg with the maximum safety margin. Fig. 2b) demonstrates simulations of the shaft with the maximum displacement of 3.096 mm. The frequency analysis was also done by using ANSYS software [10]. The shaft was analyzed in two modes: the shaft was fixed in place of its direct attachment to the wind turbine; the shaft was fixed on the bearings place's location. It allowed to find the natural frequencies of the shaft and represented as: 42.31 Hz; 271Hz; 760.3Hz for the first mode, 374 Hz, 1006 Hz at the second mode.

C. Bearings

The bearings solution must take into account not only shaft geometric parameters but also the maximum rotation speed of the cylinders which does not exceed 6000 rpm. According to this, a single-row radial bearing 60/28 2RS was selected. Based on [11] long shafts require cylinders constructive solutions without "back-to-back" installations due to shaft elongation in connection with increasing the

temperature. The shaft elongation, however, is not expected when applied to the Magnus wind turbine. The bearings were installed "back-to-back" as shown in Fig.3. The left bearing adjoins to the shaft with the inner ring, and to the flange with the outer ring. The outer ring of the right bearing is also inserted into the flange, and the inner one adjoins to the nut. The nut allows minimizing auto oscillations to be moved.

D. Bearing flanges

The bearing flanges design require the calculation of centrifugal force acting on the cylinder during rotation of the rotor. The centrifugal force can be expressed as

$$F_C = \frac{m \cdot V^2}{r},$$ (2)

here m - mass, kg; V - the speed, m/s; r - the rotation radius, m. The approximate weight of polyvinyl chloride cylinders is 0.8 kg. The flanges mass is 0.5 kg. The rotating part total mass is 1.3 kg. The average cylinder radius is 0.5 m, with a rotation speed of 20 rad/s, we obtain a linear velocity V=10 m/s. Based on mentioned above the centrifugal force equals 300 N.

The bearing flange was designed by using CAD software and was simulated in two modes - affecting maximum stresses and maximum displacement which are represented in Fig.4 a) and b) respectively. The parameters of the bearing flange are the outer diameter is 125 mm, the inner diameter is 52 mm, the width is 15 mm, and the mass is 69 g. The maximum flange stress is 6.6 MPa, while the yield strength of ABS-plastic is 35 MPa, which provides a five-fold safety. The maximum load occurs on the flange surface, which it makes possible to reduce the filling percentage of the part during 3D printing. The maximum flange edge displacement is 0.15 mm.

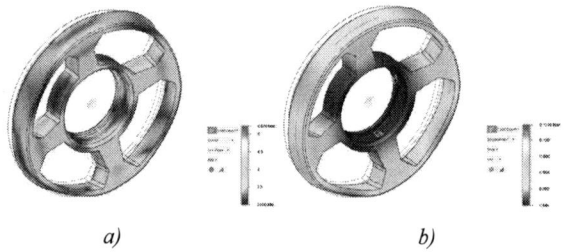

a)   *b)*

Fig. 4.   Bearing flange analysis a) maximum stresses; b) maximum displacement

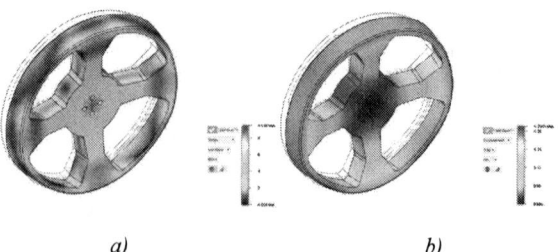

a)   *b)*

Fig. 5.   Bearing motor flange analysis a) maximum stresses; b) maximum displacement

A similar simulation for bearing motor flange was done and presented in Fig.5. This flange is attached to the BLDC motor implemented inside the cylinder. The forces are identical to the bearing flange, with one exception. It is needed taking into account the BLDC motor torque equal to 0.6 N/m. The maximum flange stress is 9.5 MPa, while the yield strength of ABS-plastic is 35 MPa, which provides a three-fold safety. The maximum flange edge displacement is 0.24 mm.

As a result, the total cylindric blade weight is approximately 2.5 kg. This parameter can be reduced by reducing the thickness of the aluminum shaft, as well as choosing a different material for the cylinder.

III. CYLINDRIC BLADE CONTROL SYSTEM

From the control system point of view, the cylindric blade consists of a BLDC motor with additional inertia depending on the cylinder implementation. The measured parameters of the BLDC motor and cylinder's moment of inertia are shown in Table 1.

A BLDC motor can be modelled in the same way as a three-phase synchronous motor [12]. Some dynamic characteristics differ due to the presence of a permanent magnet on the rotor. The flux linkage from the rotor is determined by the magnet material. As a result, a saturation of the magnetic flux linkage is typical for this type of motor. The BLDC motor, like any other three-phase motor, is powered by a three-phase voltage source. The source does not have to be sinusoidal. As long as the peak voltage does not exceed the maximum voltage limit of the motor, a square wave or another wave shape can be used. Similarly, the armature winding model for the BLDC motor is represented as follows:

$$
\begin{aligned}
V_a &= Ri_a + \frac{d}{dt}\lambda_a, \\
V_b &= Ri_b + \frac{d}{dt}\lambda_b, \\
V_c &= Ri_c + \frac{d}{dt}\lambda_c, \\
T_e &= B\omega + J\frac{d}{dt}\omega + T_L,
\end{aligned}
\tag{3}
$$

here - V_a, V_b, V_c are phase's voltages, V; i_a, i_b, i_c are phase's currents, A; $\lambda_a, \lambda_b, \lambda_c$ are flux linkages, Wb; R is stator resistance , Ω; L is stator inductance , H; T_e is electromagnetic torque , N.m; B is viscous friction coefficient, N.m.s/rad; ω is mechanical angular speed, rad/sec; J is moment of inertia, kg.m²; T_L is load torque, N.m.

By using a specific transformation known as Clarke forward transformation [13], that model can be simplified for simulation, analysis, and control design. This transformation turns the (A-B-C) phase system model representation to the (α-β) system. When proper physical parameters are taken into account, the stated transforming process is valid for both current, voltage, and flux components. It denotes a reduction of three phases or three acceptable phase axes into two (α-β) axes. The axes are stationary to the stator coordinate system.

TABLE I. T-MOTOR MN4014 KV400 MEASURED PARAMETERS

Parameter	Value	Unit
Inductance	$27.9 \cdot 10^{-6}$	H
Resistance	$85 \cdot 10^{-3}$	Ω
Motor's moment of inertia	$7.57 \cdot 10^{-6}$	kg·m²
Number of pole pairs	12	
Velocity constant	400	rpm/V
Viscous friction coefficient	$0.437 \cdot 10^{-6}$	N/m·s
Cylinder's moment of inertia	$6.476 \cdot 10^{-3}$	kg·m²

The transformed model can be expressed as follows:

$$
\begin{aligned}
V_a &= Ri_\alpha + L\frac{d}{dt}i_\alpha + \lambda_m\dot\omega\cos(\theta_e), \\
V_b &= Ri_\beta + L\frac{d}{dt}i_\beta + \lambda_m\dot\omega\sin(\theta_e), \\
J\dot\omega &= -B\omega + \lambda_m\left(\sin(\theta_e)i_\beta - \cos(\theta_e)i_\alpha\right) - T_L, \\
\dot\theta_e &= \omega,
\end{aligned}
\tag{4}
$$

here λ_m is the amplitude of the flux linkages established by the permanent magnet as viewed from the stator phase winding, Wb; θ_e is electrical angular position, rad.

To simulate the BLDC motor and control method, MATLAB/Simulink software was used. We used the mathematical model of BLDC motor in (α-β) coordinates as we explained earlier. Field oriented control (FOC) method was chosen to control the motor. The aim of vector control is to convert these nonlinear equations into linear equations and decouple the three-phase stator currents into flux and torque components [14].

The simulation has six main blocks. They are BLDC motor, Park's forward transformation, Park's inverse transformation, and three PI controllers' blocks shown in Fig. 6.

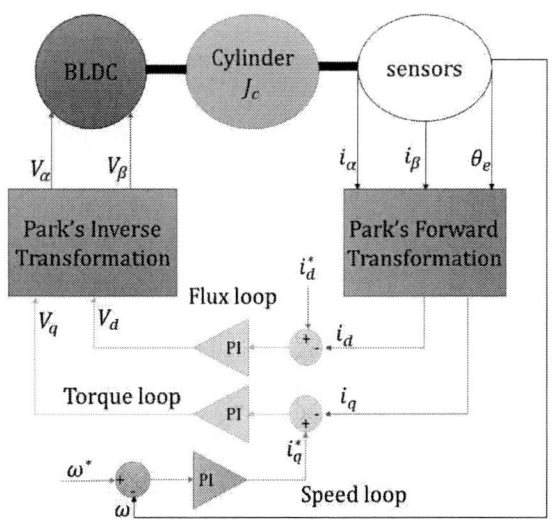

Fig. 6. Field Oriented Control of BLDC Motor

TABLE II.	CONTROLLERS' GAINS	
Loop	*Kp*	*Ki*
Torque	0.5	10
Flux	0.5	10
Speed	0.2	5

In the two-phase (α-β) representation, the rotor position is not considered. However, knowledge of the rotor position is necessary to ensure that the magnetic field generated by the stator is 90 electrical degrees ahead of the rotor magnetic field. Therefore, the (α-β) coordinate system is rotated from the stator reference to the rotor reference by θ_e the electrical position of the rotor. This is done using Park's forward transformation [15].

At this point, the two quantities in the (*d-q*) coordinate system represent the torque (aligned with the *q*-axis) and the resulting rotor flux (aligned with the *d*-axis) of the motor. Since these quantities are orthogonal to each other, they can be controlled independently, typically by separated PI controllers, to produce the desired (*d-q*) voltages. To apply the desired voltages, the (*d*) quantities must be converted back to the (α-β) reference system. This is done by rotating the (*d*) quantities by θ_e as given by Park's inverse transformation [14]. The reference value of the i_d^* was taken as 0, because it creates additional losses in the motor.

Finally, the speed of the motor was controlled by a PI controller with an angular speed feedback from the BLDC motor block.

The machine under test is an outrunner BLDC motor (T-motor MN4014 KV400) whose parameters are listed in Table I. The obtained PI controllers' parameters are shown in Table II.

The simulation step response of the BLDC motor is shown in Fig. 7. The load torque is equal to 0 [N.m]. The analysis of the transient response is shown overshoot is 6%, rising time 0.005 sec, settling time 0.24 sec.

The wind turbine could be controlled taking into account the shadow effect [16], but the typical control approach is to use maximum power tracking control (MPPT) [17]. The signals to the cylindrical blades come from the outer power control loop, produced by the MPPT controller with a fixed initial strep size. The controller uses an adaptive MPPT algorithm with a one second interval and a maximum speed increment of 30 rad/s. The response of the controller to a step change in wind speed is shown in Fig. 8. At the initial steps the MPPT controller uses the maximum increment, but after reaching the maximum generation point it starts to decrease the increment until reaching steady-state. The simulated cylinder's response to the MPPT control signal is shown on Fig.8.

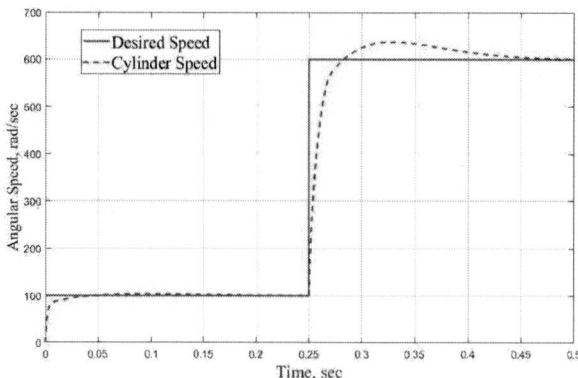

Fig. 7. Angular speed response of step signal

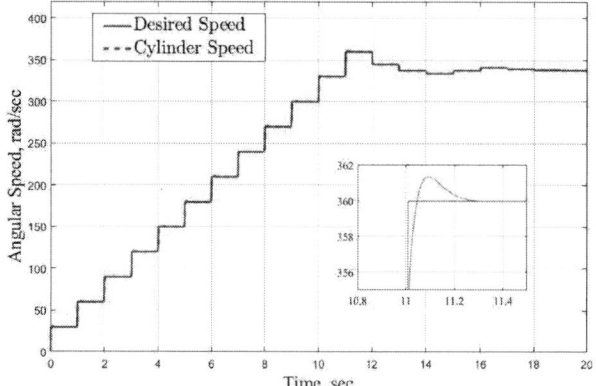

Fig. 8. Angular speed response of MPPT signal

IV. CONCLUSION

This research focused on modifying the Magnus wind turbine design satisfaction and synthesizing the control system of the cylindric blades for the further synthesizes of the power control system. The topology of such a technical system is not regulated by the industry and this augurs several iterations of such wind turbines. The recommendations for calculating shaft, bearings and bearing flanges for the maximum energy efficiency was done. The control system of the cylindric blades was synthesised by using field-oriented control. The implemented cylinder speed control system has shown the possibility of operating blades control system collectively to the inner power control loop.

REFERENCES

[1] N. Poliakov, G. Demidova, P. Zolov, K. Vorobev, D. Lukichev and R. Strzelecki, "Distributed Energy Laboratory Concept Focused on Power Electronics Units," 2021 International Conference on Electromechanical and Energy Systems (SIELMEN), 2021, pp. 387-392, doi: 10.1109/SIELMEN53755.2021.9600330.

[2] A. Lukin et al., "Experimental Prototype of High-Efficiency Wind Turbine Based on Magnus Effect," 2020 27th International Workshop on Electric Drives: MPEI Department of Electric Drives 90th Anniversary (IWED), 2020, pp. 1-6, doi: 10.1109/IWED48848.2020.9069565.

[3] A. Lukin et al., "Generative Design in Development of Mechanical Components for Magnus Effect-Based Wind Turbine," 2020 XI International Conference on Electrical Power Drive Systems (ICEPDS), 2020, pp. 1-5, doi: 10.1109/ICEPDS47235.2020.9249344.

[4] Nobuhiro Murakami, Jun Ito (2009). Magnus type wind power generator (U.S. Patent No.US7504740B2). U.S. Patent and Trademark Office.

[5] Xiaojing Sun, Yueqing Zhuang, Yang Cao, Diangui Huang, and Guoqing Wu, "A three-dimensional numerical study of the Magnus wind turbine withdifferent blade shapes," J. Renewable Sustainable Energy 4, 063139 (2012); doi: 10.1063/1.4771885

[6] D. S. Aneesh, R. Stanly, S. B. Sagaram and S. S. Suneesh, "Numerical analysis of Magnus wind turbine," 2016 7th International Conference on Mechanical and Aerospace Engineering (ICMAE), 2016, pp. 191-195, doi: 10.1109/ICMAE.2016.7549533.

[7] Ahmad Sedaghat,"Magnus type wind turbines: Prospectus and challenges in design and modelling," Renewable Energy, Vol. 62, 2014, pp.619-628, doi: 10.1016/j.renene.2013.08.029

[8] Aldo Vieira da Rosa and Juan Carlos Ordóñez,"Fundamentals of Renewable Energy Processes, 4th ed., Elsevier, 2021, pp.721-794

[9] X. Ye, Y. Tian, L. Zhu and L. Shen, "Deformation Analysis of the Shaft and Rotor of External Circulation Piston Pump Based on ANSYS," 2011 Second International Conference on Digital Manufacturing & Automation, 2011, pp. 347-351, doi: 10.1109/ICDMA.2011.92.

[10] Liming Yu and Xiliang Chen, "Technique to deal with the frequency dependent character of viscoelastic material basing on ANSYS modal analysis," 2011 Second International Conference on Mechanic Automation and Control Engineering, 2011, pp. 5336-5338, doi: 10.1109/MACE.2011.5988197.

[11] H. Li, H. Geng, S. Tang, H. Lin, H. Lv and L. Yu, "Numerical method for analysing the static characteristics of sliding bearing based on Ansys," 2017 IEEE International Conference on Mechatronics and Automation (ICMA), 2017, pp. 125-130, doi: 10.1109/ICMA.2017.8015800.

[12] Paul C. Krause; Oleg Wasynczuk; Scott D. Sudhoff, "Brushless dc Motor Drives," in Analysis of Electric Machinery and Drive Systems , IEEE, 2002, pp.557-601, doi: 10.1109/9780470544167.ch15.

[13] A. Anuchin, D. Savkin, Y. Khanova and D. Grishchuk, "Real-time model for motor control coursework," 2015 IEEE 5th International Conference on Power Engineering, Energy and Electrical Drives (POWERENG), 2015, pp. 427-430, doi: 10.1109/PowerEng.2015.7266355.

[14] B. P. Reddy and A. Murali, "SoC FPGA-based field oriented control of BLDC motor using low resolution Hall sensor," IECON 2016 - 42nd Annual Conference of the IEEE Industrial Electronics Society, 2016, pp. 2941-2945, doi: 10.1109/IECON.2016.7793092.

[15] A. Dianov A, A. Anuchin, "Adaptive Maximum Torque per Ampere Control of Sensorless Permanent Magnet Motor Drives," Energies. 2020; 13(19):5071. https://doi.org/10.3390/en13195071

[16] A. Anuchin and A. Chepiga, "Wind Turbine Control System with Compensation of Wind Flow Fluctuations and Tacking into Account Shadow Effect," 2020 27th International Workshop on Electric Drives: MPEI Department of Electric Drives 90th Anniversary (IWED), 2020, pp. 1-5, doi: 10.1109/IWED48848.2020.9069585.

[17] C. V. Govinda, S. V. Udhay, C. Rani, Y. Wang and K. Busawon, "A Review on Various MPPT Techniques for Wind Energy Conversion System," 2018 International Conference on Computation of Power, Energy, Information and Communication (ICCPEIC), 2018, pp. 310-326, doi: 10.1109/ICCPEIC.2018.8525219.

APPENDIX

TABLE III. CYLINDER PARAMETERS

Parameter	Symbol	Value	Unit
Surface area	S	0.0968	m^2
Wind speed	v	40	
Air density	ρ	1.2754	kg/m^3
Drag coefficient	c_x	0.5	

Adjusting the Phase Switching-on Positions of the Traction Switched Reluctance Motor in the Low Speed Range

Alexander Krasovsky
Department of Electrical Ingineering and Industrial Electronics
Bauman Moscow State Technical University
Moscow, Russia
krasovsky@bmstu.ru

Elena Vostorgina
Department of Electrical Ingineering and Industrial Electronics
Bauman Moscow State Technical University
Moscow, Russia
elenka.job@gmail.com

Abstract—**The paper evaluates the best-known approaches to adjusting the phase switching position of a traction switched reluctance motor (SRM) in order to reduce the current consumed from the power supply to provide a given value of average torque in the low-speed zone. Two fundamentally different variants of solving this problem were identified - off-line and on-line optimization. Off-line optimization requires a considerable amount of preliminary work, and the results obtained are applicable only to a specific motor type or similar. Known analytical dependencies for application in on-line optimization are either inaccurate or cumbersome and inconvenient for practical use. The algorithm proposed by the authors is based on the known idea of automatic on-line control over the first peak of phase current and moving it by control means to the angular position of the beginning of overlapping of stator and rotor poles. The algorithm recommended by the authors in the implementation requires minimal computing resources of the software part of the control system because it includes only the simplest calculations. Modeling in MATLAB-SIMULINK environment showed that under various perturbations, the transition of the SRM phase inclusion position control system to a new steady state is completed in no more than three cycles of motor switching.**

Keywords—variable speed drives, brushless machines, torque control, voltage control, power system modeling.

I. INTRODUCTION

The traction drive of electric vehicles (EV) is typically powered by the battery [1–3]. An important indicator of EV is the mileage without recharging the battery, which depends on its capacity. As the battery capacity increases, its dimensions and weight increase. Therefore, usually seek to achieve some compromise between the battery capacity and mileage of the EV.

Increasing the specific energy performance of the traction electric drive implies achieving the specified or necessary values of power or torque on the motor shaft at the minimum values of current consumption from the power supply. In modern electric drive, such control mode is called Maximum Torque Per Ampere (MTPA). The MTPA mode is widely used to control induction and synchronous permanent magnet motors. In them, it is realized in the regulation of currents along with the axes d and q according to special algorithms [4, 5].

The switched reluctance motor (SRM) is one of the competitive options for the automotive industry since, in the absence of expensive permanent magnets; it has a simple design, high reliability, and wide regulation capabilities [6, 7]. Research results show that the SRM, with appropriate design, is not inferior to a permanent magnet motor in terms of specific performance. Significant progress has been achieved in the development of methods to reduce the noise and torque pulsations characteristic of SRM [8, 9].

The peculiarity of SRM operation is sequential switching of phase windings in certain angular positions of the rotor. This makes it possible to use an additional control channel - by changing according to a certain law the positions of motor phase windings switching-on and switching-off, along with the traditional for most modern electric drives control channel associated with changes in the amplitude of phase currents.

Studies have shown that in the low-speed SRM zone, changing the phase switching position within certain limits has little effect on the average value of the developed torque T_{av}, but has an effect on the Root Mean Square (RMS) value of the current I_{RMS} consumed from the power supply. Changing the phase-off position, on the other hand, affects both the T_{av} torque value and the I_{RMS} current value. Thus, for transport applications, when T_{av} is controlled by proper selection of the phase switching position, the I_{RMS} current value for a given T_{av} value can be minimized. The I_{RMS} current value determines the copper losses of the SRM, so reducing it for a given value of T_{av} torque can improve the energy efficiency of the drive and minimize its size and weight.

A lot of literature is devoted to the selection of optimal SRM switching parameters, the analysis of which allows us to distinguish two fundamentally different variants of solving this problem - off-line and on-line optimization. In the first variant, the switching parameters are determined by some criteria in advance and are set most often in the form of tables. This can be obtained by open-loop control with various average torque levels and speed. The disadvantage of such a control algorithm is a rather tedious amount of preliminary work and a rigid binding of the obtained results to a specific type of motor. Therefore, the authors of this article do not consider this approach further.

978-1-6654-6787-2/22 $31.00 © 2022 IEEE

In the second control variant, the switching parameters in each cycle are determined continuously for each switching cycle taking into account the electromagnetic parameters of the SRM as well as the set values of torque and speed, which can be corrected by applying additional analytical transformations, or automatically with the use of various feedbacks.

In the next section of this paper, we consider the peculiarities of switching processes in the SRM when operating in the low-speed zone. We also discuss the best-known approaches to continuous control of the phase switching-on positions. It is shown that the nonlinearity of the SRM is the main cause of errors when using various analytical dependence to determine the phase switching positions. Further, a new algorithm of automatic regulation of phase switching position is proposed. In the final part of the article, we presented some results of testing the competitiveness of the proposed algorithm by simulation.

II. PECULIARITIES OF SRM SWITCHING PROCESSES IN THE LOW-SPEED ZONE

The low-speed operating range of the SRM is limited to the speed change interval within which the power supply voltage U_{DC} exceeds the induced in-phase motion EMF. To prevent excessive phase current I_{ph} rise, it is constrained using a relay current controller. Fig. 1 shows a typical type of curves: phase voltage U_{ph}, inductance L and current I_{ph} as a function of the angular position of the rotor Θ. When switching on at the rotor angular position Θ_{on}, a voltage pulse with an amplitude equal to the power supply voltage U_{DC} is applied to the phase.

When selecting the position Θ_{on} it must be taken into account that in the motor mode of SRM each phase generates a significant torque only in the zone of partial overlap of the stator and rotor poles (between the Θ_{unal} start position and the full overlap of the poles Θ_{al}). The phase inductance L limits the rate of phase current rise I_{ph} during the switch-on phase.

Therefore, in order to allow the current I_{ph} to build up to the required value I_{ref}, the phase is switching-on with an angular advance γ_{on} relative to the unal position Θ_{unal}. The current I_{ph} under the influence of the increased phase voltage U_{ph} with a small value of the phase inductance L increases rapidly to the value I_{ref} and then pulses relative to this value with the amplitude ΔI, set by the width of the dead zone of the phase current relay controller.

Ideally, to minimize the current drawn from the power supply, the current I_{ph} should reach the I_{ref} value at the Θ_{unal} position (blue solid line in Fig. 1). We then refer to this position Θ_{on} as the optimum position. From the energy point of view, the most disadvantageous mode is the mode in which the angular position Θ_{on} is additionally shifted to the lead relative to its optimal position by some angular interval $\Delta\gamma_{on}$, as shown in Fig. 1. In this case, the phase current I_{ph} rises to the set level I_{ref} before the start of pole overlapping (green solid line in Fig. 1).

This results in a motion interval $\Delta\gamma_m$ (hereafter referred to as the offset) within which the phase does not develop torque, although a full operating current at the I_{ref} level is flowing through it. The choice of the Θ_{com} phase switching-off position depends on additional conditions, for example,

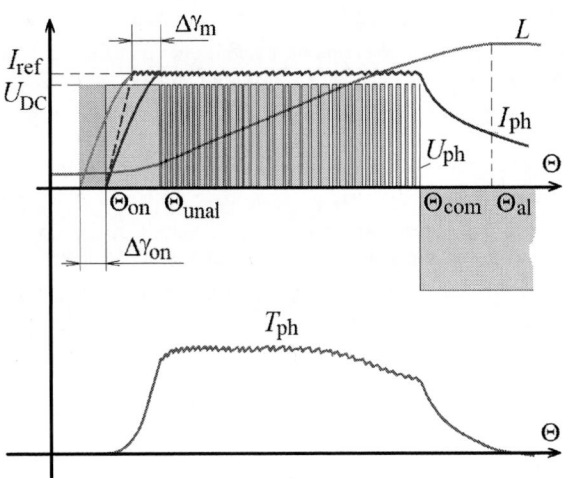

Fig 1. Changing the parameters of the view in the switching cycle in the low-speed zone.

this position is chosen taking into account minimizing torque ripple, or obtaining the maximum torque per switching cycle, etc. Usually, the Θ_{on} and Θ_{com} positions for all SRM phases have the same orientation relative to the corresponding phase inductance curves $L(\Theta)$. The mutual spatial offset between adjacent phase curves $L(\Theta)$ depends on the design features of the particular SRM instance.

In the simplest control algorithm, the position Θ_{on} is set optimal only for the nominal mode of SRM and then does not change when changing its mode of operation. However, when the speed of SRM decreases, the motion EMF decreases, which entails an increase in the rate of current rise (blue dotted line in Fig. 1) and the current I_{ph} reaches the value I_{ref} before the start of pole overlapping. This results in a certain movement interval $\Delta\gamma_m$, within which the phase does not develop torque at full operating current at I_{ref}. This leads to a decrease in the energy efficiency of the drive.

Obviously, if the speed changes within the same limits and other conditions are equal, the angular displacement interval $\Delta\gamma_m$ depends on the level of the given current limitation I_{ref}. In order to obtain quantitative estimates, the authors of this article conducted a study on the simulation model of the SRM. The simulation model was based on the magnetic characteristics of a real four-phase SRM obtained experimentally.

The experimental four-phase motor with a rated power of 5 kW has the following main constructive parameters: the number of stator poles 8, the number of rotor poles 6, the outer diameter of the active part of the stator is 206 mm, the outer diameter of the rotor is 116 mm, and the air gap is 0.4 mm. The description of the simulation model is given in the authors' previous works, for example, in [10].

To give generality to the simulation results, we used the transition to relative values of the variables. The basic values of the angular intervals of rotor motion, except for the phase-inclusion interval, are equal to the angular interval of partial overlap of the interacting teeth of the stator and rotor, i.e. $\gamma_{base} = \gamma_{overlap} = (\Theta_{al} - \Theta_{unal})$. The base value of the interval of switching on the phase $\gamma_{on\,base}$, is equal to the angular interval between the angle of the beginning of overlapping of the teeth Θ_{unal} and the angle Θ_0, where Θ_0 is the angle of intersection of the continuation of the growing section of the dependence $L(\Theta)$ with the Θ - axis, i. e. at $L(\Theta)=0$.

The base voltage value is equal to the rated voltage of the motor $U_{\text{base}} = U_{\text{rated}} = U_{\text{DC}}$. The base values of current, torque, and speed are taken as their values at the point of transition of the magnetic system of the SRM from linear mode to local saturation of interacting teeth $I_{\text{base}} = I_{\text{sat}}$, where I_{sat} is the current at which is begins local saturation of teeth.

Thus, the base value of the interval of switching on the phase $\gamma_{\text{on base}}$, ensures the constancy of the phase current with the amplitude I_{sat} in the overlapping zone of the stator and rotor teeth at the rated voltage U_{rated} and the corresponding speed value ω_{rated}. All relative values of parameters and variables are subsequently marked with an asterisk (*).

Fig. 2 presented the dependencies obtained from the simulation. They shows the change in bias $\Delta\gamma^*_m$ in fractions of optimal value as a function of speed ω^* at an unchanged switching-on position of phase SRM. The dependencies $\Delta\gamma^*_m(\omega^*)$ are plotted for several characteristic levels of current limiting I^*_{ref}. We have determined the optimum value γ^*_{on} in advance for a basic speed value ($\omega^*=1$) and the corresponding current value I^*_{ref}.

As you can see, when you change I^*_{ref} from 1.5 to 3.0 at speed $\omega^* = 0.1$ the value $\Delta\gamma^*_m$ increases from 0.25 to more than 0.45, that is, the deviation of the switching interval from its optimum value for a given level of current limitation can exceed 45%. It is important to emphasize again that at this angular interval the phase practically creates no torque, and the current consumed from the source only heats additionally the SRM windings.

The search for analytical dependence convenient for practical use for their implementation in algorithms for changing the angular interval γ_{on} as a function of the current values of speed ω and current I_{ref} is complicated by the nonlinearity of SRM. Therefore, many authors ignore this nonlinearity and recommend, for example, using a simplified dependence [11] for γ_{on} in the form of

$$\gamma_{\text{on}} = \frac{L_{\text{unal}} I_{\text{ref}} \omega}{U_{\text{DC}}} \qquad (1)$$

In (1) it is assumed that the phase inductance is constant in the interval from Θ_{on} to Θ_{unal} and is equal to its value at the Θ_{unal} position, i.e. $L(\Theta) = L_{\text{unal}} = const$. Expression (1) is convenient for practical use, but is an approximation and gives a significant error in determining $\Delta\gamma_{\text{on}}$. The authors of this article quantified this error on the simulation model of SRM in Matlab environment in their previous work [12]. In particular, we found that in $I^*_{\text{ref}} < 0,5$ for all values of speed ω^* estimated values γ_{on} derived from the correlation (1) and the simulation results taking into account the nonlinearity of SRM practically coincide. A further increase in the value of I^*_{ref} the expression (1) gives inflated values of the inclusion

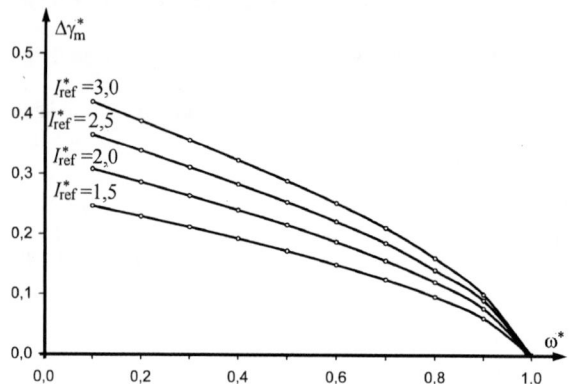

Fig 2. Changing of displacement $\Delta\gamma_m$ as a function of speed at an unchanged Θ_{on} position.

interval γ_{on} (in some modes up to 20% and more). In the same paper, we proposed a generalized analytical relationship to correct the results obtained on the basis of the ratio (1).

In real conditions, the phase inductance of the SRM non-linearly depends not only on the phase current and rotor position but also on environmental conditions, in particular on temperature. It is practically impossible to take all this into account analytically. All these factors can only be adequately taken into account by switching to automatic regulation of the SRM phase switching position with the use of various feedbacks.

The purpose of this research work is to briefly analyze the known solutions in this direction and to develop proposals for the implementation of a convenient for practice simple method of automatic control of the SRM phase switching position.

III. A Brief Analysis of Known Solutions

There are many publications on the on-line control of SRM phase switching positions. However, most of them discuss approaches to determining more or less complex analytical relationships for the on-line calculation of the phase switching-on delay interval, taking into account the speed values ω, the current limiting I^*_{ref} and the electromagnetic parameters of the motor, and the load.

The authors consider various criteria for selecting the optimum switching parameters. In particular, in quite a number of works, this problem is considered only from the position of minimizing torque ripples, while the problem of increasing energy efficiency of drive operation is either touched upon in passing or is not considered at all. Such works include publications [13, 14] and a number of others.

Mention should be made of paper [15], which also deals with the selection of SRM switching angles. A simple control law is proposed for controlling the speed of switched reluctance motor. In the region of the low speeds, the phase currents are controlled by chopping. PI controller generates the reference current which is used for current chopping. The switching-on angle is fixed, and the switching-off angle is made to decrease with an increase in speed. The authors' specific proposals relate more to the high-speed zone of SRM operation with single-pulse phase currents.

The problem of simultaneously reducing torque ripple and increasing the energy efficiency of SRM operation in the low-speed zone is discussed in the paper [16]. However, the authors consider the regulation of the phase disconnection position to be more significant. The authors recommend that the initial switching-on position be determined according to the expression (1), and then they recommend using its adaptive correction algorithm by shifting the phase inclusion position programmatically in both directions to achieve the optimum position. However, the authors do not disclose details of the implementation of the algorithm and have only provided a block diagram.

At a principled level, a similar approach to controlling the switching-on - and switching-off phase positions of the SRM is proposed in the paper of [17, 18]. For on-line estimation of switching-on and switching-off angles, the position sensors detect the rotor position and the rotor angular speed is then calculated by that. The current switching-on and switching-off positions of the phase are then determined using analytical relationships. The results obtained in this way correspond only to the specific operating mode of a particular motor type. In the paper [18], the authors propose to use as a basis the calculated data obtained by Field Reconstruction Method.

One of the first papers to substantiate the feasibility of automatic control of the switching positions of SRM phases in a closed control structure can be considered the article Yilmaz Sozer и David A. Torrey [19]. Using a simplified SRM model as an example, the authors have shown the cause of the phase switching-on position error and proposed a control system functional diagram to compensate for it. However, the regulating system is presented by the authors only in a general view, the principle of formation of control signals and commands is not described which complicates its practical implementation. A similar approach to SRM control in a generator mode of operation the same authors recommended in paper [20].

Most modern methods of controlling switching processes in switched reluctance machines, especially for the automotive industry, are quite complex and require the high computing power of the control system. Thus, in paper [21] the authors propose a closed loop control of the mean torque using Fourier series with continuous control of current setting, phase on and phase off angles. However, this involves using a powerful processor with frequency 150 – 300 MHz and dsPIC30f6010A.

In the method of controlling the average torque of the three-phase SRM in the paper [22], the authors suggest using a sliding mode control. At the same time, the authors propose to use rather cumbersome analytical dependencies to continuously calculate the motor commutation parameters. In fact, a similar approach is implemented in the paper [23]. The authors of this paper propose even more cumbersome

analytical relations for the continuous calculation of SRM switching angles. Moreover, their proposals relate only to the high-speed zone of the drives.

Such a review could be continued further, but the conclusion, in our opinion, is obvious. According to the authors of this paper, a simple and reliable online method for controlling the switching position of the SRM can be based on the "retro" approach proposed by Yilmaz Sozer и David A. Torrey [19].

IV. PROPOSED CONTROL ALGORITHM

For automatic control of the switching-on position of the phase Θ_{on} when the set average torque T_{av} and speed change, it is necessary to fix the angular position of the rotor Θ_m in which the current I_{ph} reaches the value I_{ref} in each cycle of SRD phase switching. After that, it is necessary to calculate the phase shift of this position $\Delta\gamma_m$ with respect to the angular position of the beginning of the overlapping poles Θ_{unal}.

Then, using the value of $\Delta\gamma_m$ using a *PI* controller or a predetermined nonlinear transformation, generate an error signal for the phase $\Delta\gamma_{on}$ position, which will be used later to correct the phase on position. The correction signal must be proportional to the error of switching-on position $\Delta\gamma_{on}$.

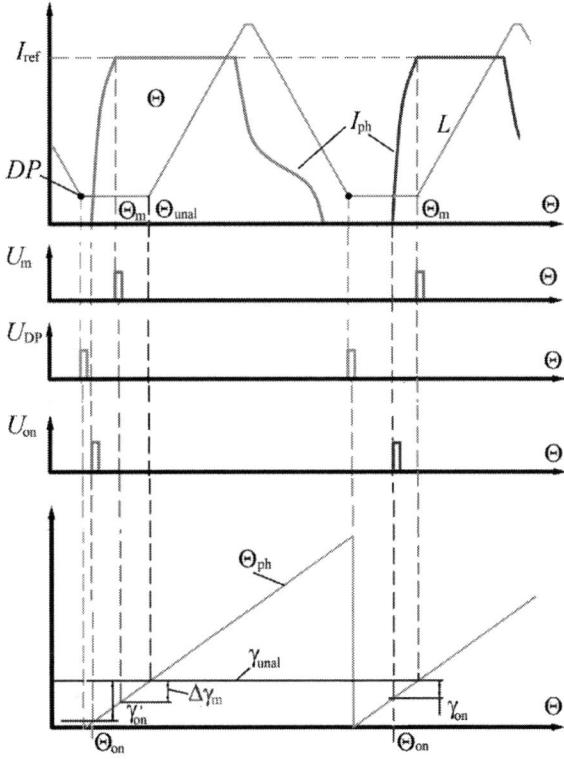

Fig 3. Explanation of the formation of control commands in the operation of the proposed algorithm.

Fig. 3 shows the signals we have to generate when implementing the algorithm and their spatial arrangement relative to the phase inductance curves $L(\Theta)$ and phase current $I_{ph}(\Theta)$ for two consecutive SRD phase switching cycles. Curves $L(\Theta)$ and $I_{ph}(\Theta)$ are shown in simplified form. In the proposed SRM control algorithm, the following sequence of signals is formed. Rotor position sensor (DP) generates a pulse signals U_{DP} at the beginning of the angular interval of the total misalignment of the poles of the stator and rotor in the position Θ_{DP}. The signal U_{DP} triggers a special block or subprogram to form a saw tooth signal $\Theta_{ph}(\Theta)$, which is essentially a sweep of the angular position of the phase within one rotor revolution.

Angle sweep signal $\Theta_{ph}(\Theta)$ is compared on the comparator with a constant that numerically determines the phase shift γ'_{on} in the initial Θ'_{on} position of the first switching cycle with respect to the U_{DP} signal. Phase shift value γ'_{on} can be determined, for example, analytically by ratio (1), or to be asked for any other considerations.

At the moment the comparator input signals are equal $\Theta_{ph}(\Theta)$ and γ_{on} the comparator generates a pulse signal on its output which determines the initial angular position of the phase inclusion Θ_{on}. In the Θ_{on} position, the phase is then connected to the power supply, i.e. the phase current I_{ph} starts to increase. When the current I_{ph} reaches the maximum value I_{ref}, phase current detector generates a pulse signal U_m that determines the angular position Θ_m. The pulse signal in position Θ_m determines the angular position at which the phase current reaches the value I_{ref}. Based on the angular position of the current maximum found Θ_m the offset value is determined $\Delta\gamma_m$:

$$\Delta\gamma_m = \Theta_{unal} - \Theta_m \qquad (2)$$

A non-zero value of the interval $\Delta\gamma_m$ indicates that the initial switch-on position of phase Θ_{on} is set with error on, which must be compensated for in subsequent switching cycles. The traditional PI controller solution can be used for this, with appropriate gear ratios. With a non-linear position Θ_{on} the control system, the transmission coefficients of the PI controller must also generally vary according to non-linear laws as a function of speed and a given average torque value. Clearly, more research is needed to determine the methodology for determining the regulator parameters. The authors of this paper used a simplified approach, first defining the non-linear relationship between $\Delta\gamma_m$ and $\Delta\gamma_{on}$ as a function of speed and setpoint I_{ref}.

The successive repetition of the control algorithm under consideration over several switching cycles makes it possible

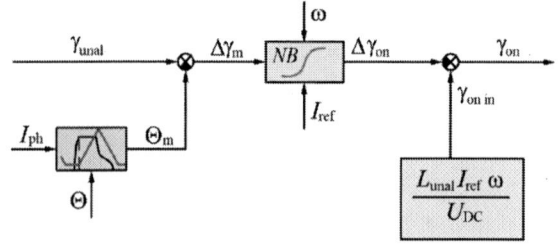

Fig 4. A functional diagram of part of the SRM control system.

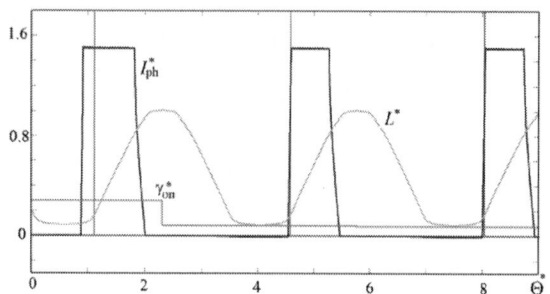

Fig 5. Operation of the SRM control system when compensating for phase switching-on position error.

to bring the offset value $\Delta\gamma_m$ practically to zero and then automatically maintain it to a minimum during SRM operation. A functional diagram of part of the SRM control system, further explaining the operation of the proposed control algorithm, is shown in Fig. 4. The designation of all signals used in the algorithm and their phase positions are explained earlier in the text and correspond to Fig. 3. *NB* – non-linear block implemented as a table to convert $\Delta\gamma_m$ to $\Delta\gamma_{on}$.

V. SIMULATION RESULTS

We have examined the effectiveness of the proposed SRM phase position control algorithm using two typical examples.The oscillogram shown in Fig. 5 illustrates the operation of the SRM control system when compensating for the phase switching position error. Under the condition of the model experiment, with the speed equal to $\omega^*=0.8$ and $I^*_{ref}=1.5$, the phase switching interval γ^*_{on} in the first switching cycle of SRM is deliberately set with a significant shift towards advance relative to the optimum position (approximately three times the optimum value).

As can be seen, in the first switching cycle the phase current reaches the setpoint much earlier than the beginning of the overlapping of the cooperating poles (the position of the beginning of the overlapping of the poles is marked in Fig. 5 with red vertical lines). In the second switching cycle, however, the angular positions in which the phase current reaches the set point and the start of pole overlapping are almost identical. In the third switching cycle, however, the phase switching-on position error is fully compensated.

Oscillogram in Fig. 6 shows the operation of the phase-on position control algorithm when processing a sudden change in current I^*_{ref}. According to the condition of the

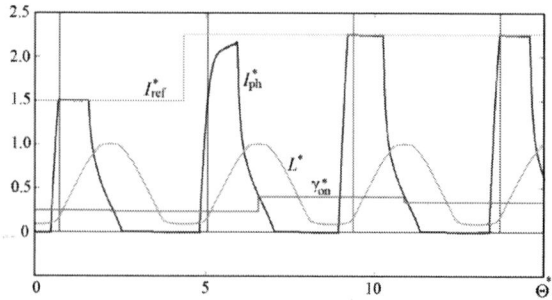

Fig 6. Oscillograms confirming the operation of the proposed algorithn with a change of current limiting I_{ref}.

978-1-6654-6787-2/22 $31.00 © 2022 IEEE

model experiment, at a speed equal to $\omega^*=0.8$ in the first switching cycle of the SRM phase, the current is $I^*_{ref}=1.5$. Further, according to the experimental condition, between the first and second cycles of phase switching, the current limiting level abruptly changes from $I^*_{ref}=1.5$ to $I^*_{ref}=2.25$. Obviously, ceteris paribus, this should cause a transient process to shift the optimal position of the inclusion of the vase to the side of greater advance in relation to the beginning of the pole overlap in the first cycle of switching phases of the SRM.

From the waveform in Fig. 6, it can be seen that the process of regulating the position of switching-on the phase begins already in the second cycle of switching the phase. Note that the process of working out an abrupt change has an oscillatory character. This is clearly seen from the oscillogram. In the third switching cycle, the position of switching on the phase is shifted towards the lead with some excess, but already in the subsequent switching cycle, the phase is switched on in the optimal position.

VI. CONCLUSION

When a traction drive is powered from an autonomous power supply, improving its energy efficiency is particularly important. Even a relatively small reduction in the current draw at a given torque value contributes to the drive's energy efficiency, as it extends the run time without recharging the battery. In SRM-based drives, we can achieve this by eliminating movement intervals where its phases do not generate torque, but draw full current from the power supply.

The algorithm proposed by the authors, which is simple to implement, automatically controls the phase switching positions in each switching cycle, which helps to solve this problem. Experiments on the model have shown that if initially the phase switching position is set with an error, the control system moves the phase switching position to almost the optimum position in maximum of three cycles of motor phase switching.

REFERENCES

[1] John G. Hayes, G. Abas Goodarzi, "Electric Powertrain. Energy Systems, Power Electronics and Drives for Hybrid, Electric and Fuel Cell Vehicles", John Wiley & Sons Ltd, 2018, pp. 529.

[2] Jiuchun Jiang, Caiping Zhang, "Fundamentals and applications of lithium-ion batteries in electric drive vehicles", John Wiley & Sons Singapore Pte. Ltd, 2015, pp. 277.

[3] Helmut Weiss, Thomas Winkler, "Large Lithium-Ion Battery-Powered Electric Vehicles - From Idea to Reality", 2018 Elektro, DOI: 10.1109 / Elektro.2018.8398241.

[4] C.B. Butt, M.A. Hoque, M.A. Rahman, "Simplified fuzzy logic based mtpa speed control of ipmsm .drive", IEEE Trans. Ind. Appl., vol. 40, no. 6, pp. 1529-1535, Nov., 2004.

[5] S.H. Kim, S.K. Sul, "Maximum torque control of an induction machine in the field weakening region", IEEE Trans. Ind. Appl., vol. 31, no. 4, pp. 787-794, Jul. /Aug., 1995.

[6] Zhi Yang, at al., "Comparative study of interior permanent magnet, induction, and switched reluctance motor drives for EV", IEEE Transactions on Transportation Electrification, vol. 1, no. 3, pp. 245-254, 2015.

[7] R. Pindoriya, at al., "Comparative analysis of permanent magnet motors and switched reluctance motors capabilities for electric and hybrid electric vehicles", IEEMA Engineer Infinite Conference (eTechNxT), pp. 1–5, 2018.

[8] Tetsuya Kojima and Rik W. De Doncker, "Optimal Torque Sharing in Direct Instantaneous Torque Control of Switched Reluctance Motors", 2015 IEEE Energy Conversion Congress and Exposition (ECCE) pp. 327–333.

[9] Dong, M., "Research on reduction the torque ripple in switched reluctance motor", (2016) Proceedings - 2016 IEEE International Symposium on Computer, Consumer and Control, IS3C 2016, art. no. 7545380, pp. 1071-1074. doi: 10.1109/IS3C.2016.270.

[10] Alexander Krasovsky, "Simulation and analysis of improved direct torque control of switched reluctance machine", Indonesian Journal of Electrical Engineering and Computer Science, Vol. 18, No. 1, May 2019, pp. 251~260. DOI 10.11591/ijeecs.v18.i1.pp. 251-260.

[11] Alecksey Anuchin, Maxim Lashkevich, Dmitry Aliamkin; Fernando Briz, "Achieving maximum torque for switched reluctance motor drive over its entire speed range", 2017 International Symposium on Power Electronics (Ee), DOI: 10.1109/PEE.2017.8171676.

[12] Alexander Krasovsky, Sergey Vasyukov, Elena Vostorgina, "Obtaining the MTPA mode in the Three-phase Traction SRM with a Flat Topped Shape of Phase Current", 2021 International Conference on Industrial Engineering, Applications and Manufacturing (ICIEAM), DOI: 10.1109/ICIEAM51226.2021.9446405.

[13] Masatsugu Takemoto, Akira Chiba, Tadashi Fukao, «A method of determining the advanced angle of Square-Wave Currents in a Bearingless Switched Reluctance Motor", IEEE Transation on industry applications, vol. 37, NO. 6, november/december 2001.

[14] Gaoliang Fang;Jennifer Bauman "Optimized switching angle-based torque control of switched reluctance machines for electric vehicles", 2020 IEEE Transportation Electrification Conference & Expo (ITEC), DOI: 10.1109/ITEC48692.2020.9161705.

[15] Gyanedra Kumar Sah at al., "A new simplified Control Strategy for Switched Reluctance Motor", Conference 2017 Resent Developments in Control Automation and Power Ingineering (RDCOPE), October 2017, DOI:10.1109/RDCAPE.2017.8358274.

[16] Mademlis C., Kioskeridis I. "Performance Optimization in Switched Reluctance Motor Drives With Online Commutation Angle Control // IEEE Transactions On Energy Conversion. – 2003. – Vol. 18, No. 3. – pp. 448-457.

[17] S. Fatemi, H. Cheshmehbeigi and E. Afjei, "Self-tuning approach to optimization of excitation angles for switched-reluctance motor drives", 2009 European Conference on Circuit Theory and Design, 2009, pp. 851-856, doi: 10.1109/ECCTD.2009.5275117.

[18] C. Lin and B. Fahimi, "Optimization of commutation angles in SRM drives using FRM", 2012 IEEE Transportation Electrification Conference and Expo (ITEC), Dearborn, MI, 2012, pp. 1-6.

[19] Y. Sozer, D.A. Torrey, E. Mese, "Automatic control of excitation parameters for switched-reluctance motor drives", IEEE Transactions on Power Electronics (Volume: 18, Issue: 2, March 2003), Page(s): 594 – 603, DOI: 10.1109/TPEL.2003.809352.

[20] [E. Mese; Y. Sozer; J.M. Kokernak; D.A. Torrey, "Optimal excitation of a high speed switched reluctance generator", APEC 2000. Fifteenth Annual IEEE Applied Power Electronics Conference and Exposition (Cat. No.00CH37058), DOI: 10.1109/APEC.2000.826128.

[21] H. Cheng, H. Chen, and Z. Yang, "Average torque control of switched reluctance machine drives for electric vehicles", IET Electric Power Applications 2015, 9, (7), pp.459-468.

[22] Gurmeet Singh, Bhim Singh, "An Analytical Approach for Optimizing Commutation Strategy of Switched Reluctance Motor Drive for Light Electric Vehicle", 2020 IEEE International Conference on Power Electronics, Smart Grid and Renewable Energy (PESGRE2020)

[23] Roberto Rocca, Fabio Giulii Capponi, "Optimal Advance Angle for Aided Maximum-Speed-Node Design of Switched Reluctance Machines", IEEE Transactions on Energy Conversion, Vol. 35, Issue: 2, June 2020, pp 775 – 785. DOI: 10.1109/TEC.2019.2959823.

The Accuracy Increasing of Voltage Stabilization on the Input of PWM Inverter in the Electric Drive with Energy Storage based on Supercapacitors

Polyakov Vladimir
Ural Federal University named after the first President of Russia B.N. Yeltsin
Ekaterinburg, Russia
v.n.polyakov@urfu.ru

Plotnikov Iurii
Ural Federal University named after the first President of Russia B.N. Yeltsin
Ekaterinburg, Russia
yu.v.plotnikov@urfu.ru

Abstract — **The paper deals with the problem of the voltage stabilization in the DC link of a frequency converter with an PWM inverter, which is part of AC drive with a frequency method of speed control. To stabilize the voltage in the DC link, energy storage technology is used, which is implemented with using an energy storage device based on supercapacitors, which is connected to the DC link by means of bi-directional DC-DC converter. The work uses the concept of stabilizing the DC bus voltage by compensating the PWM inverter load current with the external current of the energy storage device. The energy storage device is equipped with a single-loop tracking current control system. The current control system is based on the combined control method. The aim of the work is to improve the accuracy of voltage stabilization in the DC link of frequency converter. The sensitivity of the DC link voltage to various disturbing factors is investigated. The theoretical analysis of the error coefficient of the current control system is carried out. The simulation of the electric drive is carried out without and taking into account the discreteness of the DC-DC converter. The research results can be useful in the development of supercapacitor energy storage control systems for various applications.**

Keywords — *Capacitive energy storage, Variable speed drives, DC-DC power conversion, Supercapacitors, Power system modeling, Frequency control.*

I. INTRODUCTION

Due to the well-known advantages [1], energy storage devices on supercapacitors are used in regulated general application, industrial [4, 6], traction [3, 7], automobile [5, 10], marine [9] and other electric drives [2, 8, 12]. In this case, energy storage devices allow to save the electrical energy, received from motor during braking, with its subsequent using during start-up in order to increase the energy efficiency, provide an uninterrupted power supply in case of a short-term disappearance of the mains voltage, and also to solve the problem of smoothing the peak power [11, 13].

One of the main tasks [6], which is solved with using the energy storage devices on supercapacitors as part of electric drives, is to stabilize the voltage at the input of an PWM inverter with a given accuracy to eliminate overvoltage during braking. In this case, the charging / discharging current and the voltage at the terminals of the supercapacitors block must be maintained within the specified limits. An example is the using of storage devices to stabilize the voltage in the electric railway power supply system [13]. In [14], simultaneously with energy saving, increasing the energy efficiency, and peak power control, the problem of voltage stabilizing of the pantograph is solved, which can drop up to 20% during the vehicle accelerations.

To solve this task, in the energy storage device an automatic control system (ACS) is used. Among the control systems, the multi-loop control systems with a cascade connection of controllers are most widely used [7, 14, 19]. Usually, they have two or three circuits. In the double-loop version, the internal loop is the current control loop on the supercapacitors block, and the external loop is the voltage control loop on the input of autonomous inverter [7, 14]. In the case of a three-loop design of the ACS, the intermediate loop is a voltage control loop at the terminals of the supercapacitor block [7]. Less commonly, the single-loop automatic control system of the current on the supercapacitor block [14] or of the external current of the DC-DC converter [15, 16] is used. All these types of control systems for energy storage with varying degrees of accuracy solve the problem of voltage stabilization at the input of the PWM inverter.

The aim of this work is to improve the accuracy of voltage stabilization at the input of an autonomous inverter of a frequency converter based on the principle of combined control when building a single-loop ACS current for energy storage on supercapacitors. The material of the article is a further development of the concept of voltage stabilization at the input of a PWM inverter by compensating the load current of the inverter with an external current of the energy storage, which was formulated by the authors in [16].

Section II of the article provides a description of the control object and formulation of research problem. Section III is devoted to the analysis of the sensitivity of the voltage at the input of an PWM inverter in relation to disturbing factors. Section IV describes the single-loop combined ACS of the energy storage device and analyzes the accuracy of the system by means of using the error coefficients method. In section V, by means of mathematical simulation, an accuracy calculation of voltage stabilization at the input of

---

The study was supported by a grant from the Russian Science Foundation № 22-29-01305, https://rscf.ru/en/project/22-29-01305/

an PWM inverter is given. Finally, in the Section VI the conclusions from the research were formulated.

II. OBJECT OF STUDY

The simplified structures of the power section of the electric drive and energy storage circuit with a single-loop ACS are shown in Fig. 1 and Fig. 2.

The power part of the electric drive includes a squirrel-cage induction motor (AC Motor), a frequency converter with PWM inverter (DC/AC inverter) and an input uncontrolled rectifier (Bridge Rectifier). An LC filter with inductance L_d and capacitance C_d is installed in the DC link. The electric drive is equipped with its own automatic control system (Drive Control Unit).

The power part of the energy storage device consists of a block of supercapacitors with a total capacity C_{sc}, a smoothing reactor with an inductance L and a bi-directional DC-DC converter. The DC-DC converter is based on the simplest topology with two fully controlled switches [15].

In the drive circuit with energy storage, there are motor stator current sensors (SCS), voltage inverter current (VICS) and voltage (IVS) sensors of a PWM inverter, as well as an external DC-DC converter current sensor (ICS). The energy storage device has an automatic control system of DC-DC converter external current. The setpoint signal for automatic control system is the signal of PWM inverter current $I^*_{dc}=I_{in}$ from the VICS sensor.

III. THE ANALYSIS OF INPUT VOLTAGE SENSITIVITY OF THE PWM INVERTER TO THE EXTERNAL CONTROL AND DISTURBANCE ACTIONS

Fig. 3 shows the equivalent circuit of the DC link for motor (a) and generator (b) modes of the motor.

As can be seen from the diagrams, the accuracy of voltage U_{in} stabilization generally depends on both the time variation of the voltage E_d, the initial conditions $U_{in(0)}$ and $I_{d(0)}$, the structure and parameters R_d, C_d, T_d of the DC link filter, as well as the load current compensation error $\Delta I=I_{dc}-I_{in}$.

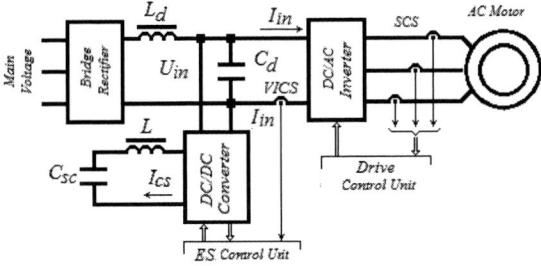

Fig. 1. Functional diagram of power part of electric drive with energy storage

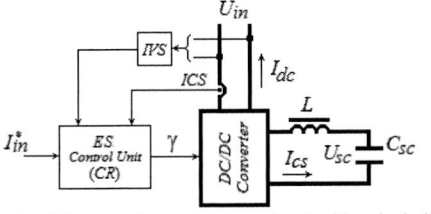

Fig. 2. Functional diagram of energy storage circuit with a single-loop ACS of DC-DC converter external current

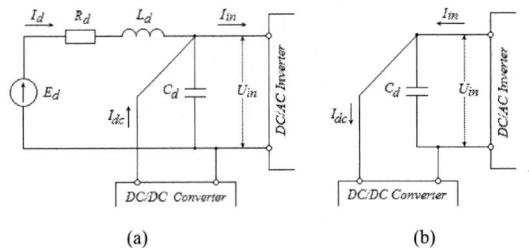

Fig. 3. Equivalent circuit of DC link:
a – motor mode; b - generator mode

These equivalent circuits correspond to the structural schemes, which are shown in Fig. 4.

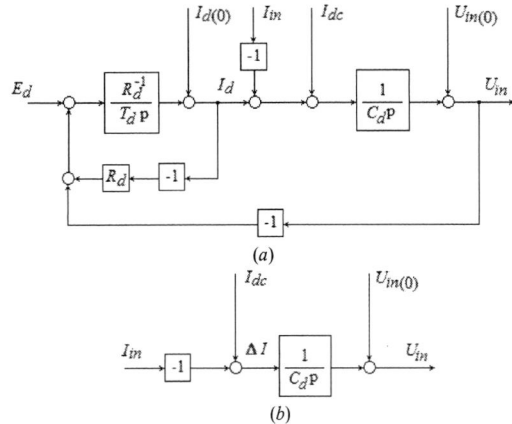

Fig. 4. Block diagrams of frequency converter DC link
a – motor mode; b - generator mode

The degree of influence E_d, $U_{in(0)}$, $I_{d(0)}$ and ΔI can be identified by analyzing the following transfer functions:

a) *For motor mode of electric drive*

$$W_{E_d}(\mathrm{p}) = \frac{U_{in}(\mathrm{p})}{E_d(\mathrm{p})} = \frac{1}{T_{cd}T_d\mathrm{p}^2 + T_{cd}\mathrm{p}+1} ; \qquad (1)$$

$$W_{U_{in(0)}}(\mathrm{p}) = \frac{U_{in}(\mathrm{p})}{U_{in(0)}(\mathrm{p})} = \frac{T_{cd}\mathrm{p}(T_d\mathrm{p}+1)}{T_{cd}T_d\mathrm{p}^2 + T_{cd}\mathrm{p}+1} ; \qquad (2)$$

$$W_{I_{d(0)}}(\mathrm{p}) = \frac{U_{in}(\mathrm{p})}{I_{d(0)}(\mathrm{p})} = \frac{R_d T_d\mathrm{p}}{T_{cd}T_d\mathrm{p}^2 + T_{cd}\mathrm{p}+1} ; \qquad (3)$$

$$W_{\Delta I}(\mathrm{p}) = \frac{U_{in}(\mathrm{p})}{\Delta I(\mathrm{p})} = R_d\frac{T_d\mathrm{p}+1}{T_{cd}T_d\mathrm{p}^2 + T_{cd}\mathrm{p}+1} , \qquad (4)$$

b) *For generator mode of electric drive*

$$W_{U_{in(0)}}(\mathrm{p}) = \frac{U_{in}(\mathrm{p})}{U_{in(0)}(\mathrm{p})} = 1 ; \qquad (5)$$

$$W_{\Delta I}(\mathrm{p}) = \frac{U_{in}(\mathrm{p})}{\Delta I(\mathrm{p})} = R_d\frac{1}{T_{cd}\mathrm{p}} , \qquad (6)$$

where T_d and T_{cd} – time constants, $T_d=L_d/R_d$ and $T_{cd}=R_dC_d$. $\mathrm{p}=d/dt$ - differentiation operator.

The voltage deviation U_{in} due to the actions of external disturbances can be represented as follows:

a) *For motor mode*

$$U_{in.ea}(t) = W_{E_d}(\mathrm{p})E_d(t) + W_{\Delta I}(\mathrm{p})\Delta I(t) ,$$

b) *For generator mode*

978-1-6654-6787-2/22 $31.00 © 2022 IEEE   103

$$U_{in.ea}(t) = W_{\Delta I}(\text{p})\Delta I(t).$$

These equations characterize the responses of the DC link to external actions at zero initial conditions.

With nonzero changes in the initial conditions relative to the basic voltage deviation, we have the second component, determined by the operator equation:

a) *For motor mode*

$$U_{in.sc}(t) = W_{I_{d(0)}}(\text{p})I_{d(0)}1(t) + W_{U_{in(0)}}(\text{p})U_{in(0)}1(t),$$

b) *For generator mode*

$$U_{in.sc}(t) = W_{U_{in(0)}}(\text{p})U_{in(0)}1(t),$$

where, $W_{Id(0)}(\text{p})$, $W_{Uin(0)}(\text{p})$- transfer functions for the initial conditions; $I_{d(0)}$ and $U_{in(0)}$ – constants, which are equal to the deviations of the initial conditions relative to the basic conditions; $1(t)$ - single step function.

These components characterize free motion with nonzero deviations of the initial conditions.

The particular transients in the DC link to deviations of external actions and initial conditions are shown in Fig. 5 and 6.

Here, partial transients $u_{in}(t)$ are presented in relative units from the base value, which, for reasons of obtaining a greater degree of generality in the analysis of external actions and initial conditions influences on DC link voltage U_{in}, is referred to the amplitude value of the rated phase voltage of the stator winding of an induction motor with a nominal power of 55 kW [16].

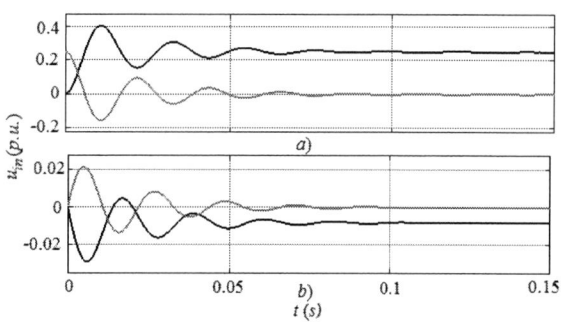

Fig. 5. Partial transients in the DC link in a motor mode to deviations of external actions and initial conditions:

a) On the control action E_d at zero deviations of the initial conditions (black line); on the initial voltage deviation $U_{in(0)}$ (red line);

b) On the disturbance action ΔI at zero deviations of the initial conditions (black line); on the initial current deviation $I_{d(0)}$ (red line);

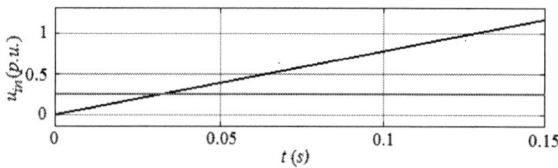

Fig. 6. Partial transients in the DC link for a generator mode to deviations of external actions and initial conditions: on the disturbance action ΔI at zero deviations of the initial conditions (black line); on the initial voltage deviation $U_{in(0)}$ (red line).

In this case, external actions and initial conditions were also set in accordance with their values in the nominal mode of the motor.

Analysis of transfer functions (1) - (6), as well as Fig. 5 and 6, shows that the DC link voltage U_{in} is most sensitive to a changes of the initial condition $U_{in(0)}$ and disturbance action ΔI in the generator mode. Since the changes in the initial condition $U_{in(0)}$ is less susceptible to changes in real conditions, the quality of the current ACS of the energy storage becomes one of the determining factors affecting on the accuracy of voltage stabilization during the operation of the electric drive system. Thus, the accuracy improving of the current control system for an energy storage device is an actual investigation task.

IV. MATHEMATICAL MODEL OF ENERGY STORAGE CONTROL SYSTEM

The continuous averaged equivalent model of a single-loop ACS of the external current of an energy storage device based on supercapacitors is shown in Fig. 7. In this model, all variables are averaged over the sampling period of the DC-DC converter. In the mathematical model, all variables and parameters, except the time constants and current time, are presented in relative units. Combined control is used to increase the accuracy of tracking the setpoint signal i^*_{dc}. In this system, along with control by deviation, control by reference and disturbing actions is introduced [17].

A. Model of the Energy Storage Power Part

During simulation it is assumes that the energy storage is based on a bidirectional DC-DC converter with two fully controllable switches [11, 13].

Transfer function of the linear part of the model:

$$W_i(\text{p}) = \frac{T_{sc}\text{p}}{r_l T_{sc}\text{p}(T_l\text{p}+1)+1},$$

where T_l and T_{sc} are time constants, r_l – is the relative active resistance of the smoothing reactor.

The link with the characteristic $F_1(u_{cr}, i_{in})$ takes into account the difference in the transfer ratios of the DC-DC converter in the boost and buck modes.

Here u_{cr} is the control action of the DC-DC converter. The output of link $F_1(u_{cr}, i_{in})$ is determined by the following equation:

$$\gamma = \begin{cases} 1 - u_{cr} & \text{under } i_{in} \geq 0, \\ u_{cr} & \text{under } i_{in} < 0. \end{cases}$$

B. Model of the Control Part

At the output of the control part, a link with a characteristic $F_2(\gamma^*, i_{in})$ is installed, which performs the function according to the formula:

$$u_{cr} = \begin{cases} 1 - \gamma^* & \text{under } i_{in} \geq 0, \\ \gamma^* & \text{under } i_{in} < 0, \end{cases}$$

where γ^* is the input signal. Setting of this link provides a constant closed-loop current gain during mode changing of bidirectional DC-DC converter.

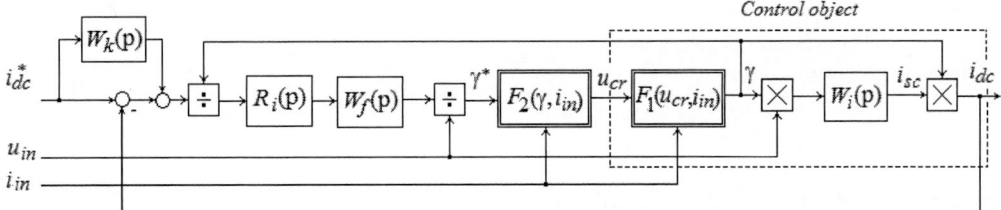

Fig. 7. Averaged continuous model of a single-circuit ACS of the external current for an energy storage device on supercapacitors.

The bandwidth of the current ACS is set by a low-pass filter. The filter transfer function is written as follows:

$$W_f(p) = \frac{1}{T_\mu p + 1},$$

where T_μ is the uncompensated time constant. A low-pass filter also protects the DC-DC converter from the influence of high-frequency noises.

To implement the deviation control principle, a current signal i_{dc} is used as feedback in the current controller. The setpoint for the current ACS is a signal from PWM inverter $i^*_{dc} = i_{in}$.

The transfer function of the current controller is described by following equation:

$$R_i(p) = W_i^{-1}(p)\frac{1}{T_i p} = r_l\left(\frac{T_l}{T_i} + \frac{1}{T_i p} + \frac{r_l^{-1}}{T_{sc} T_i p^2}\right),$$

where T_i – integration constant is chosen based on the tuning of the closed loop to the modular optimum, that means $T_i = 2T_\mu$.

To implement the control on the reference action, an additional signal is generated. This signal is taken from the correcting link with transfer function [17]:

$$W_k(p) = T_k p,$$

where T_k is the differentiation constant.

At the input of correction link the signal of i^*_{dc} is introduced.

To implement control by disturbance action, the compensating signals u_{in} and γ are introduced into the closed loop through the dividing elements.

C. Accuracy Investigation of the Combined Current Automatic Control System

To estimate the efficiency of combined control, let us first consider the case when, in addition to control by deviation, the control by disturbing actions u_{in} and γ is also introduced. To analyze the accuracy of such a system, the equivalent model, which is shown in Fig. 8, can be used.

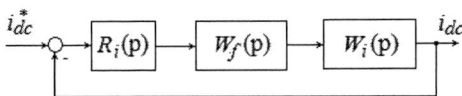

Fig. 8. Equivalent current ACS without the introduction of control on the reference action

The transfer function of the closed system under tuning the loop to the modular optimum is written as follows:

$$\Phi_i(p) = \frac{1}{T_i T_\mu p^2 + T_i p + 1} = \frac{1}{2T_\mu^2 p^2 + 2T_\mu p + 1}. \quad (7)$$

To estimate the accuracy of the setpoint i^*_{dc} tracking, the method of error coefficient is used [17].

For a system with transfer function (7), the first three of error coefficients are:

$$c_0 = 0;\ c_1 = T_i = 2T_\mu;\ \frac{c_2}{2} = -T_\mu T_i = -2T_\mu^2.$$

Consequently, the static error on the control action in the combined current ACS without using control by the reference action is equal to zero. The speed c_1 and acceleration c_2 error depend only from the uncompensated time constant T_μ, that is, from the performance of ACS.

With the introduction of control by the reference action, the equivalent transfer function of the ACS (Fig. 9) takes the following form:

$$\Phi_i(p) = \frac{T_k p + 1}{2T_\mu^2 p^2 + 2T_\mu p + 1}. \quad (8)$$

From expression (8) we can be seen that the introduction of control by the reference action does not change the characteristic equation of the system, operating by deviation, since the transfer function of the denominator for closed-loop system is the same as in (7). This means that the stability conditions will not be beaked.

The system error coefficients with the introduction of reference control take the following form:

$$c_0 = 0;\ c_1 = T_i - T_k = 2T_\mu - T_k;$$

$$\frac{c_2}{2} = T_i(T_\mu - T_i + T_k) = 2T_\mu(T_\mu - T_k).$$

The static control error coefficient c_0 is still zero. The speed c_1 and acceleration c_2 errors now depend on both from the T_μ and from the time constant of the correcting link T_k. By changing the value T_k, we can influence to the speed and acceleration errors. So, at $T_k = 2T_\mu$ we will get $c_1 = 0$, and at $T_k = T_\mu$ we will have $c_2 = 0$. Consequently, the using of control on the reference action increases the accuracy of the tracking ACS of the current for the energy storage device.

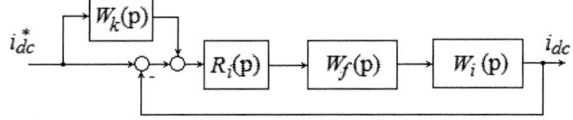

Fig. 9. Equivalent current ACS with the introduction of control by the reference action

V. The Accuracy Analysis of Voltage Stabilization at the Input of an Autonomous Inverter with Using the Mathematical Simulation

The analysis of the voltage stabilization accuracy at the input of the PWM inverter of the frequency converter was carried out both on the averaged model [16] and on the model, which takes into account the discreteness of the DC-DC converter [18] with the same structures of the current control system and the parameters of the electric drive power part.

A 55-kW asynchronous motor was used in the simulation. The parameters of the DC link model of the frequency converter and the energy storage device, as well as the basic values, can be found in the author's previous works [16]. The uncompensated time constant is assumed to be $T_\mu = 0.001$ s.

The current ACS accuracy is estimated for the test mode of the electric drive (See Fig. 10), where the following designations are introduced: t_{st}, t_{sta} and t_{br} - starting time, operation at constant speed and braking time; T_c - cycle time; ω - motor speed; m - electromagnetic torque; m_c – the mechanism load torque, reduced to the motor shaft; i_{in} - load current of the PWM inverter.

A. Mathematical Simulation of the Current ACS without Taking into Account the Discreteness of the DC-DC Converter

The transients in the current ACS of the energy storage without using the regulation by the reference action are shown in Fig. 11.

The maximum dynamic voltage error during motor acceleration is 3.5%, and during deceleration of the drive - 6%. It is important to note that the current ACS provides a zero static error when the current controller operates in the linear zone (interval from 3.5 to 5.5 sec., $u_{sc} < u_{sc.max}$).

At the time interval (1.5–2.5 sec.) the motor is accelerated, supercapacitors are discharged to the minimum voltage (Fig. 11, c), corresponding to the setting of the minimum supercapacitor voltage ($u_{sc.min}$). When the minimum voltage $u_{sc.min}$ is reached, a static current error is occurred. However, its influence on the voltage u_{in} is small, because in the motor mode the voltage sensitivity is reduced. At the same time, as can be seen from the figures, in the generator mode, despite on relatively small dynamic current error (Fig. 11, b), a significant overvoltage u_{in} is observed (Fig. 11, d), which is increased with a decreasing the performance of the tracking current ACS.

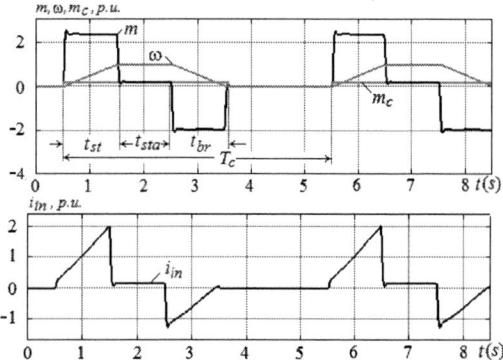

Fig. 10. Transients in electric drive system under test load.

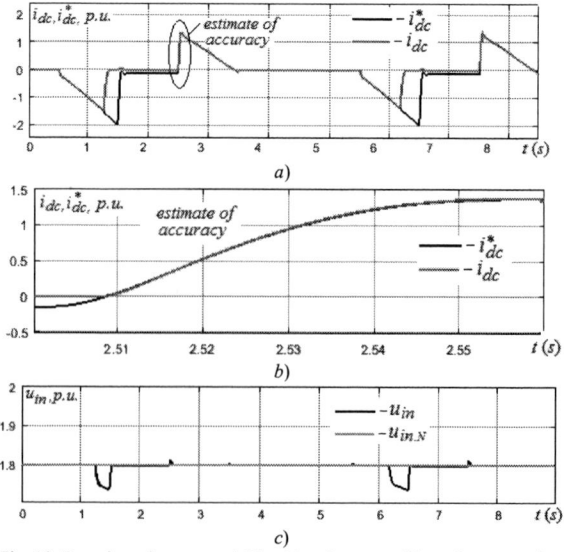

Fig. 11. The transients in the current ACS without using the control by reference action

Dynamic voltage error also appears in case of insufficient energy reserve on supercapacitors during acceleration of the electric drive. In this case, a voltage drop is observed (interval 1.25–1.5 sec.).

Fig. 12 shows the transients of the current ACS for energy storage with using the control by the reference action. Fig. 12 shows that the introduction of control by the reference action with $T_k = 2T_\mu$ decreases the dynamic error of the current ACS (Fig. 12, b), increasing the accuracy of voltage stabilization u_{in}. The overvoltage u_{in} is reduced to almost zero (Fig. 12, c). The presence of a voltage drop and a static error is explained by the reasons, identified in the study of the current ACS without using the control by the reference action.

Fig. 12. Transients in current ACS with using control by reference action

B. Mathematical Simulation of the Current ACS with Accounting the Discreteness of the DC-DC Converter

A feature of transients with taking into account the discreteness of the DC-DC converter is the presence of a static error in a single-loop current ACS. The static error is caused by the modulation component of the external current, which, at the given parameters of the energy storage device, depends from the value of the PWM frequency, at which the DC-DC converter operates, and from the performance of the current ACS.

In addition, at low PWM frequencies, the influence of the discontinuous current zone is appeared [18]. The gain of current loop in this case is changed, causing significant overshoot in the current. As the result a significant overvoltage on the DC bus of frequency converter is appeared. The increasing of the reactor inductance leads to a decreasing in the area of discontinuous currents at a fixed PWM frequency. However, in this case, due to the increasing of the inertia the current link in the supercapacitor circuit the voltage u_{in} reserve is requires. In this regard, in order to save the conditions, adopted in the simulation on the averaged model, in the numerical experiment with the discrete model of the DC-DC converter, a PWM frequency of 25 kHz was used.

Fig. 13 shows the processes in the combined current ACS with using the control by reference and disturbance actions.

As can be seen from the figure, the voltage stabilization accuracy is practically unchanged. The dynamic error is less than 5%.

VI. Conclusions

The DC link voltage u_{in} has a high sensitivity to external actions in the generator mode of the electric drive, which causes increased requirements to the accuracy of the single-loop ACS of the external current of the DC-DC converter for energy storage device.

The introduction of control by reference and disturbance actions increases the accuracy of a single-loop automatic control system of an external current of a DC-DC converter.

The discreteness of the DC-DC converter in the energy storage at a low PWM frequency significantly affects on the static and dynamic accuracy of the current ACS, which leads to decreasing the accuracy of voltage u_{in} stabilization.

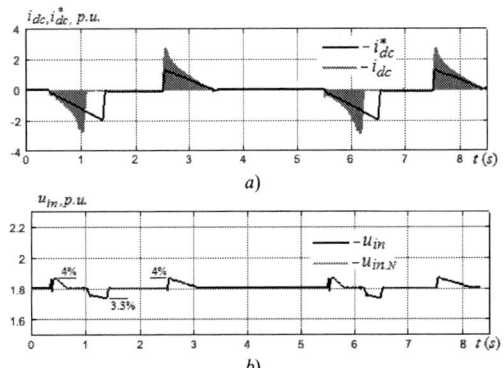

Fig. 13. Transients in the current ACS with taking into account the discreteness of the DC-DC converter with the introduction of control by reference and disturbance actions

References

[1] Pagano E.; Piegari L.; Tricoli P. Ultracap-based new electrical drives for off-road vehicles and rovers, Antriebssysteme für Off-Road-Einsätze, Garching, Germany (2003), VDI-Berichte 1793, pp. 97 – 109.

[2] Rufer, A.; Barrade, P.; Hotellier, D.; Hauser, S.: Sequential supply for electrical transportation vehicles: properties of the fast energy transfer between supercapacitive tanks, 38th Industry Applications Conference, vol. 3 (2003), pp. 1530 – 1537.

[3] Barrero, R., Tackoen, X., & Van Mierlo, J. Improving energy efficiency in public transport: Stationary supercapacitor based Energy Storage Systems for a metro network. IEEE Vehicle Power and Propulsion Conference, (2008) pp. 1-8.

[4] C. Attaianese; V. Nardi; G. Tomasso. High performances supercapacitor recovery system for industrial drive applications. Applied Power Electronics Conference and Exposition, 2004. APEC '04. Nineteenth Annual IEEE, vol. 3, pp. 1635 – 1641.

[5] Burke, A.F.; Hardin, J.E.; Dowgiallo, E.J. Application of ultracapacitors in electric vehicle propulsion systems, 34th International Power Sources Symposium (1990), pp. 328 – 333.

[6] P.J. Grbovic, P. Delarue, P. Le Moigne, and P. Bartholomeus. The ultracapacitor based controlled electric drives with braking and ride-through capability: Overview and analysis, IEEE Trans. Ind. Electron., vol. 58, no. 3, pp. 925–936, Mar. 2011.

[7] Petar J. Grbovi´c, Philippe Delarue, Philippe Le Moigne and Patrick Bartholomeus. Modeling and Control of the Ultracapacitor-Based Regenerative Controlled Electric Drives. IEEE Trans. Industrial Electronics, vol. 58, no. 8, August 2011.

[8] Burke, A.F. Prospects for ultracapacitors in electric and hybrid vehicles. 12th Battery Conference on Applications and Advances (1996), pp. 183 – 188.

[9] Chen, W., Ådnanses, A. K., Hansen, J. F., Lindtjørn, J. O., & Tang, T. (2010, September). Super-capacitors based hybrid converter in marine electric propulsion system. The XIX International Conference on Electrical Machines (ICEM 2010), pp. 1-6.

[10] Dowgiallo, E.J.; Hardin, J.E.: Perspective on ultracapacitors for electric vehicles, 10th Battery Conference on Applications and Advances (1995), pp. 153 – 157.

[11] A. Rufer, D. Hotellier, and P. Barrade. A supercapacitor-based energy storage substation for voltage compensation in weak transportation networks, IEEE Trans. Power Del., vol. 19, no. 2, pp. 629–636, Apr. 2004.

[12] D. Iannuzzi. Use of supercapacitors, fuel cells, and electrochemical batteries for electrical road vehicles: A control strategy. 33rd Annual Conference of the IEEE Industrial Electronics Society (IECON), 2007, pp. 539–544.

[13] S. Hase, T. Konishi, A. Okui, Y. Nakamichi, H. Nara, and T. Uemura. Fundamental study on energy storage systems for dc electric railway systems. Proceedings of the Power Conversion Conference-Osaka, 2002, vol. 3, pp. 1456–1459.

[14] D. Iannuzzi, D. Lauria. A new supercapacitor design methodology for light transportation systems saving. Energy Management Systems (2011): pp. 183-198.

[15] Plotnikov I., Polyakov V., Postnikov N. Single-Loop Control System for Energy Storage Device in the Frequency-Controlled Electric Drive. 2018 X International Conference on Electrical Power Drive Systems (ICEPDS). pp 1-5. 2018.

[16] V. Polyakov, I. Plotnikov, N. Postnikov. Dynamics Features of Electric Drive with a Supercapacitor Energy Storage Device Based on Single-Loop Automatic Current Control System. 2020 27th International Workshop on Electric Drives: MPEI Department of Electric Drives 90th Anniversary (IWED), 2020, pp. 1-6.

[17] Ya.Z. Tsypkin. Foundations of the theory of automatic systems. The main editorial office of physical and mathematical literature of the publishing house "Nauka", Moscow, 1977, 560 pages.

[18] Plotnikov, I., Braslavsky, I., Polyakov, V. The Accounting a Incomplete Controllability in the Mathematical Model of Bidirectional DC-DC Converter for Frequency Controlled Electric Drive with Supercapacitors. 2018 International Symposium on Power Electronics, Electrical Drives, Automation and Motion (SPEEDAM). 2018, pp 1 - 5.

[19] A. Anuchin, D. Shpak, M. R. Ahmed, E. Stolyarov, D. Surnin and J. P. Acedo, "Nested Loop Control of a Buck Converter under Variable Input Voltage and Load Conditions," 2020 55th International Universities Power Engineering Conference (UPEC), 2020, pp. 1-5.

978-1-6654-6787-2/22 $31.00 © 2022 IEEE

Additive Manufacturing of Cooling Systems Used in Power Electronics. A Brief Survey

Loránd Szabó
Department of Electrical Machines and Drives
Technical University of Cluj-Napoca
Cluj-Napoca, Romania
e-mail: Lorand.Szabo@emd.utcluj.ro, ORCID ID: 0000-0002-1096-4802

Abstract—**The power density of electronic devices continues to increase. While they are getting smaller, the amount of heat they generate is dramatically increasing. If using conventional cooling systems (heatsinks, heat pipes or various heat exchangers), their size and mass must be increased beyond limits, which is unacceptable especially in mobile applications (such as electric traction, aircraft industry, etc.). Therefore, new materials and fabrication technologies are urgently needed. Additive manufacturing, a revolutionary fabrication method, could be the key to these highly anticipated developments. This novel technology is perfectly fitted to the small-scale fabrication of specifically customized cooling systems. In this paper, the possibility of applying these new technologies to several cooling systems using enhanced thermally conductive materials is surveyed. At the same time, the main advantages brought about by this technological advance are highlighted.**

Keywords—*conducting materials, cooling, electronics cooling, environmentally friendly manufacturing techniques, three-dimensional printing.*

I. INTRODUCTION

Power electronic devices generate excess heat due to internal losses. Therefore, they require adequate thermal management to improve reliability and prevent failures from overheating. The power range of such devices continues to expand, while their size keeps getting smaller. As a result of the power boost, the required cooling systems are of increasing size and mass, which is a hindrance in any industrial sector, especially in mobile applications for land, air, and water transportation [1], [2]

Heatsinks are the simplest and most widely used cooling solutions in power electronics. Their thermal performance can be substantially enhanced by increasing their fin-like active convection surface area.

Thermal performance can also be enhanced by adding extrusion features to the fins that manipulate airflow by forcing unheated coolant towards hot surfaces or by causing turbulence to increase the mixing of hot and cold coolant.

These modifications can grow the area in contact with the fluid without increasing the external dimensions of the heat sink. Heat transfer can also be improved by reducing the temperature gradient of the heatsink fins, mainly by applying materials having greater heat conductance [2].

Unfortunately, the complex structures of the cooling systems that designers envisioned to fundamentally improve the cooling system used for power electronics are expensive and in almost all cases they cannot even be processed using traditional manufacturing methods [3]. The use of additive manufacturing technology can bring breakthroughs also in this engineering field.

Additive manufacturing (3D printing) is an advanced digital technology that combines high-precision mechanical equipment, digital control, and the use of advanced materials. Equipment using this method works upon digital data obtained through CAD, depositing various materials layer by layer to precisely form parts of any complex shape without material loss. Additive manufacturing techniques have been known for several years, but they were only used for very specific high-tech applications, such as space exploration, rapid prototyping. Relentless R&D and investment efforts over the past few years have made this technology essential for the fabrication of tailored-made components.

This advanced manufacturing technology has accomplished a revolutionary transformation from analog processing (mainly performed by subtractive manufacturing) to digital processing. The technology is in perfect harmony with the "smart factory" concept and other achievements of the Fourth Industrial Revolution (or Industry 4.0) [4], [5].

This survey paper discusses the possibility of improving the cooling systems of power electronic equipment. In the first part of this paper, the main current cooling methods are surveyed, and the basic requirements of the newly developed cooling system are detailed. In addition, emerging coolants that may be applicable are also reviewed. In the second part of the paper, some improvements of additive manufacturing to power electronics cooling systems are presented. This part of the paper covers information on promising 3D printable materials and representative additively manufactured heat sinks, pipes, and heat exchangers.

II. CURRENT COOLING APPROACHES

Thermal management is critical to the operation of modern power electronics. As the power of electronic devices has increased, the heat generated has boosted while the size of these devices has decreased significantly. As a consequence, they require more and more efficient cooling systems, which can deal with up to 500 W/cm^2 or even greater heat fluxes at the warmth source [2], [6]. Accordingly, cooling systems have also evolved from simple passive heatsinks to more complex air- or liquid-cooled pipes and heat exchangers. The ever-expanding variety of power

---

This work was supported by the grant "*Low loss amorphous materials for electrification enabled by additive manufacturing (LAMEAM)*" funded by the Romanian National Authority for Scientific Research and Innovation (CCCDI – UEFISCDI) in the frame of M-ERA.NET network.

978-1-6654-6787-2/22 $31.00 © 2022 IEEE

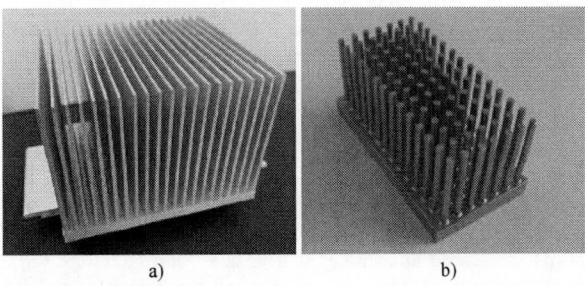

Fig. 1 .Passive heatsinks. a) Al extruded with straight fins [8]; b) Cu forged with needle fins [9].

electronic devices often requires customized cooling systems, which strongly motivates research in this field.

Basically, the used coolers are heat exchangers of great diversity. The basic purpose of these thermal management devices is to eliminate heat from designated warm places and convey it to a certain fluid. In the heat exchanger, the heat collected in the fluid is transferred to another fluid. The two fluids are never getting into direct interaction. The cooled original fluid gets backs to the place to be cooled and the heat transfer process is resumed. Upon the two involved fluids, gases (typically air) and liquids (water or dielectric fluids), the heat exchangers can be typical of air-to-air, liquid-to-liquid, or liquid-to-air. They can be passive, or active, where pumps or fans are applied to move the agent to be cooled [7].

The simplest air-to-air heat exchanger used to cool power electronic devices are the heatsinks, as those given in Fig. 1. They can be manufactured simply by extrusion, casting, stamping, or forging Al or Cu.

The main heat transfer possibilities can be the easiest to be exemplified on a simple passive cooling system with a heatsink (see Fig. 2).

In all these cases, the heat is transported from the hot spot to a colder medium, upon the basic property of heat to move in this direction. During conduction cooling, the heat is transferred by direct contact, as between the baseplate of the power electronic device and the heatsink. Usually, the thermal contact must be improved by using a good thermal conductive paste, gel, or pad, called gap filler [11]. By convection cooling, the heat is transferred through the action of the natural airflow in the upside direction. Radiated cooling means a heat transfer via electromagnetic radiation (i.e., energy waves) [10].

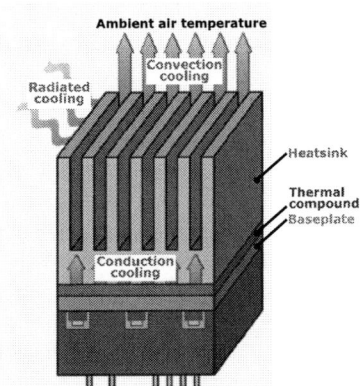

Fig. 2. Cooling a power module [10].

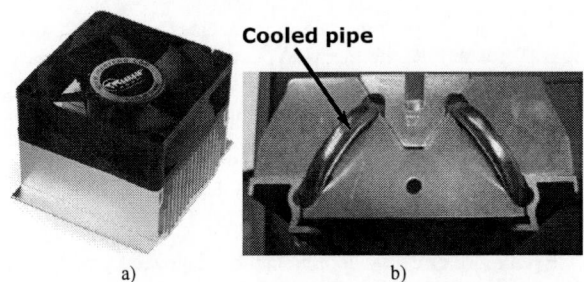

Fig. 3. Active heatsinks. a) with fan [12]; b) with cooled pipe [13].

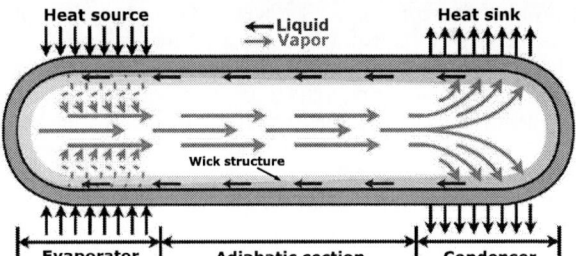

Fig. 4. Heat pipe structure [14].

The simple passive heatsinks described above can be easily made active by placing a fan on their top or integrating into them pumping water pipes to improve the heat transfer, as can be seen in Fig. 3.

The cooling systems using pumped water have some sustainable alternatives, for example, the diverse variants of the heat pipes, which are more reliable since they do not have internal moving parts. [15]. These are two-phase cooling devices that work upon the latent heat of vaporization. Basically, they are a vacuum-tight assembly comprising an external tubular metal envelope, the cooling fluid, and a specific wick structure. Upon their working principle, they have three basic compartments: evaporator, adiabatic transport section, and condenser, as can be seen in Fig. 4. Its thermal conductivity is more than 10 times greater than that of a simple heatsink.

In the evaporator part, the input heat turns into vapor the cooling liquid, which is traveling into the heat pipe to the condenser part, where it transforms back into a liquid while releasing the heat. The cooling agent then returns to the hot interface by means of capillary action, centrifugal force, or gravity, and the cooling cycle repeats [6], [16], [17], [18]. Together with the cooling agents, these liquid moving approaches are the key issues of heat pipe efficiency. If a wick structure is used (as in the case shown in Fig. 4), it must have an adequate configuration (axially grooved, of fine fiber, screen mesh, or sintered) made of porous metal or an organic polymer, as it can be seen in Fig. 5 [17]. The wick generates pumping capillary pressure assuring the movement of the cooling fluid [15].

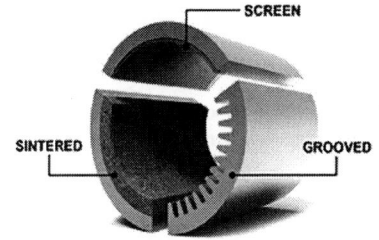

Fig. 5. Heat pipe wicks variants [18].

There are some other cooling devices similar to heat pipes. For example, the thermosyphons, which are two-phase cooling loops in which gravity is used to return the cooling fluid. In the so-called pumped two-phase heat exchangers, the coolant liquid is pumped over evaporator plates to take benefit of the superior heat transfer due to boiling. By using this approach the heat generated in power electronic devices can be transported at large distances [19], [20].

Even more effective cooling can be achieved by means of spray cooling [21]. Within this technology, tiny droplets of a dielectric coolant blown at high pressure through nozzles are sprayed directly on the hot surfaces of the power electronic devices. Upon a similar approach, the so-called jet impingement cooling method, the cooling liquid is directly washing the hot surfaces without droplet atomization. In case of extreme applications, the devices can be totally immersed in the colling liquid [6], [17], [18], [20]. In all these cases the selection of the used dielectric cooling agent is critical (see section "IV Emerging Cooling Agents" [16].

Due to its high heat transfer capacity and very compact size, cooling through microchannel heatsinks are also attractive alternatives since their heat dissipation can be up to 1000 W/cm² [6]. Upon this approach, in the heatsink, a grid of microchannels with forced convective liquid cooling are integrated. The used microchannels have hydraulic diameters not exceeding 200 μm [6], [16], [17], [20].

It should be mentioned that there are also several other cooling methods, both already widespread and emerging ones, not detailed here, as the thermoelectric cooling (by using the Peltier effect) [16], [22], cooling in vapor chamber [6], [23], etc.

III. Basic Requirements for New Cooling Systems

Choosing the right cooling agent is very important because it must be non-flammable, non-toxic, and must not corrode the cooling system. In addition, it has to be of very good thermophysical properties, such as high thermal conductivity, heat transfer coefficient and specific heat, and low viscosity. In addition, the coolant is expected to have a high boiling point and a low freezing temperature. Finally, it must be cheap [16].

As concerning the materials used for the manufacturing of the heatsinks, pipes, and heat exchangers, they must have very good mechanical strength and thermal conductivity. Therefore, traditionally, Al, Cu, and their different alloys were employed at a large scale. But these materials are of high density, hence they are not well-fitted for portable applications, as automotive, avionics, etc. Consequently, new materials to be used emerged, as ceramic and a great variety of other composites.

In the fabrication of the cooling systems, a paradigm shift took place, as the interest turned from traditional subtractive processing to additional manufacturing [24].

IV. Emerging Cooling Agents

Besides conventional dielectric (aliphatic, aromatic, silicone-, or fluorocarbon-based) and non-dielectric (usually aqueous solutions) cooling agents, several emerging cooling fluids are proposed in the literature [25]. Almost all such candidates are nanofluids, being suspensions of nanoparticles (as Al_2O_3, TiO_2, CuO, ZnO, etc.) in common cooling liquids

[21], [26]. As they exhibit very good thermal stability and thermophysical properties, they seem to be in the future real candidates to be widely used in cooling systems [16], [27].

Special attention must be paid to graphene-based nanofluids. Graphene is the "wonder material" of our century and can be an excellent solution also in this case [28], [29], [30]. This innovative material is basically a C allotrope (like graphite and diamond), being a flat, single 2D layer of C atoms structured as a hexagonal lattice. Graphene has some outstanding characteristics, as extreme thermal conductivity (more than ten times better than that of Cu), very high specific surface, and outstandingly mechanical properties, as strength and elasticity. All of these make it a very good choice for thermal management applications [31].

The permeation of graphene nanoparticles in a fluid can enhance the thermophysical properties (great thermal conductivity and specific heat capacity) of the base liquid [32], [33]. The use of graphene-based nanofluids results in diminished erosion and corrosion of the cooling surfaces and a very good collaging in the even very small-sized microchannels of the heatsinks and heat exchangers. Moreover, such liquids have a small friction coefficient, leading to better lubrication and lower required pumping power and pressure drop [34].

V. Improvements Brought by Additive Manufacturing to Cooling Systems

Over the past few years, additive manufacturing has gained a place in the industrial scene due to its many advantages, as handling very complex 3D geometries without any scratches, even joining multiple materials, etc. [35]. The advancements brought about by this revolutionary technology are also bearing fruit in the field of power electronics, especially in the development of very efficient, small, and lightweight cooling systems.

The American Society for Testing and Standards (ASTM) has collected and categorized the essential additive manufacturing technology typesd [36], [37]. Only those used also in 3D printing of power electronics cooling systems will be next mentioned.

Powder bed fusion (PBF) is one of the most widely used AM methods. Upon this technology, the powder is fused in a bed by melting the selected region. It has several variants, as selective laser melting (SLM), also called laser beam melting (LBM) [38], direct metal laser sintering (DMLS) [39], or electron beam fusion (EBM) [40]. Another 3D printing method, the fused deposition modeling (FDM) technology consists of layer-by-layer deposition of the molten material, and it is frequently used in the fabrication of heat exchangers, too [41].

Vat polymerization 3D printing method uses a tank of liquid photopolymer resin, out of which the part of the cooling system is built up layer by layer [42]. The liquid material is hardened by ultraviolet (UV) light only where it is required. For the fabrication of ceramic components concretely stereolithography (SLA) is used [43], [44].

Other such technologies to mention are binder jetting (BJ) [45], also called 3D inkjet printing, and cold spraying (CS). During binder jetting, selective dispense of a binder is applied for joining powder in a bed [46]. Cold spraying is a material jetting additive manufacturing technology applying material deposition and subsequent curing [47].

A. Promising Potential Materials

Traditionally, both Al and Cu are preferred materials for fabricating heatsinks and pipes for cooling systems due to their good thermophysical properties. Al, as well as Cu, can be relatively easily processed by a great variety of additive manufacturing technologies. Nevertheless, Al is more preferred since Cu rapidly (already during the printing process) interacts with the oxygen from the air and has low laser absorptivity. Practically not pure Al is applied, but some of its alloys with Mg and Si, as AlSi10Mg or Al 6061 [48].

Ceramics are among the most emerging materials of our days, and they are also adequate for additive manufacturing of cooling systems since they are good electrical insulators and have high thermal conductivity. Additionally, these materials have low density and they do not corrode [49], [50]. Theoretically, aluminum nitride ceramics should be the best choice for such applications due to their high, 320 W/(m·K), thermal conductivity coefficient. Unfortunately, still there are some problems related to their 3D printing caused by impurities causing defects during processing [51], [52]. Other choices are to use silicon carbide (SiC) or beryllium oxide (BeO) ceramics because they also have a high thermal conducting coefficient, 270 and 370 W/(m·K), respectively [53].

In this context, graphene is also a promising material. The thermal conducting coefficient of the additively processed graphene can exceed 1,000 W/(m·K), depending on the 3D printing technology applied, which is 7 and 4 times higher than that of Al and Cu, respectively. Its density of 1.1 g/cm³ is very small compared to those of Al and Cu [54]. In [55] the advantages of graphene for film envelope heat pipe were presented. This shows a specific thermal transfer coefficient about 3.5 times greater than that of conventional copper-based heat pipes.

By using additive manufacturing technologies very fine structures can be fabricated at high precision, also a low coefficient of thermal expansion (CTE) can be an essential requirement for the materials to be used for thermal management devices [56]. One of the choices is to use metal-matrix composites, as Invar (64Fe-36Ni) or Kovar (54Fe-29Ni-17Co) having very low CTE, but more modest thermal conduction [57], [58], [59].

Also, carbon-matrix composites can be considered for such applications. Besides the relatively high thermal conduction and low CTE, C is resistant to corrosion and has low density [60], [61]. It is totally compatible with diverse carbon fibers, so by applying graphitized carbon fibers its thermal conductivity can be improved [56].

Metal foams are a cellular structure composed of solid metal (aluminum or copper for a given application) with closed (sealed) or open (interconnected) pores. They are characterized by a very high porosity (as can be seen in Fig. 6), and they generally have the elementary physical properties of the base material from which they are made [62]. Metal foams with open cells (so-called metal sponges) are ideal for the manufacture of fluid-cooled heat exchangers. The porous structure creates a high-pressure drop in the coolant. This pressure drop can be set by changing the pore

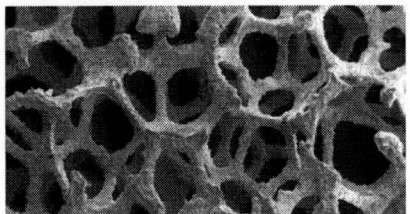

Fig. 6. Internal structure of a metal foam [64]

size. The heat exchanger made of foam metal has high strength and bending rigidity, good thermal conductivity, and low mass. They can be easily manufactured by using air plasma spray technology [63].

In power electronics, cooling is so critical issue that besides the above-mentioned high-priced materials even diamond as composite material having very high thermal conductivity is used. The reason is its very high thermal conductivity, 2000 W/(m·K), while Al and Cu have only 247 and 398 W/(m·K), respectively [56], [65].

Besides the above-mentioned heat-conducting materials which can be additively manufactured, there are materials specially developed for such processing. A typical example is the TCPoly Ice9 thermal conductive filament. It is composed of a non-insulating flexible plastic material mixed with carbon-based and ceramic fillers [66], [67].

When complex pipeline structures carrying liquid cooling agents are 3D printed, also easy to process stereolithographic resins can be applied. These are mixes of C, ceramic, or metal fillers with resin. Unfortunately, their thermal conductance does not reach the values mentioned above [68], [69].

B. Heatsinks Manufactured by 3D Printing

When the heatsink is manufactured by 3D printing, the main advantages of additive manufacturing (as the possibility to process very complex structures at high precision) can be fully applied. Even the disadvantages of these technologies concerning the obtained surface roughness and finish are used as an advantage since the active surface processed is greater [48].

Because additive manufacturing technology can handle 3D contours that are impossible or uneconomical to manufacture by traditional methods, researchers are interested in the development of innovative fin shapes for heatsinks. These fins have two basic roles: to act as active surfaces for the heat transfer by convection and to cover the pressure differential between the surrounding streams. Extra (usually extruded) features added to the fins can assist the air to flow in the right direction and induce adequate turbulence to increase the air mixing. Moreover, micro and macro channels can be performed in the body of the heatsinks by additive manufacturing technologies, thus enabling more complex 3D flow paths for enhancing heat convection in small places [6], [48]. The thermal performances of the heatsinks can be also enhanced by using more conductive materials (see section Promising Potential Materials), which are hard-to-process with traditional technologies [2].

By using the newest achievements in the field of computational fluid dynamics and optimizations, novel heatsink structures assuring directed fluid paths for superior heat convection were obtained by numerous research groups

978-1-6654-6787-2/22 $31.00 © 2022 IEEE

Fig. 7. 3D printed heatsink channel types [71].

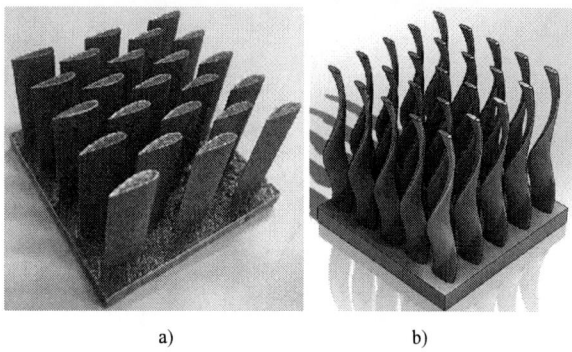

a)   b)

Fig. 8. Additively manufactures 3D printed heatsink structures with: a) airfoil fins [72]; b) twisted and tapered elliptical fins [48].

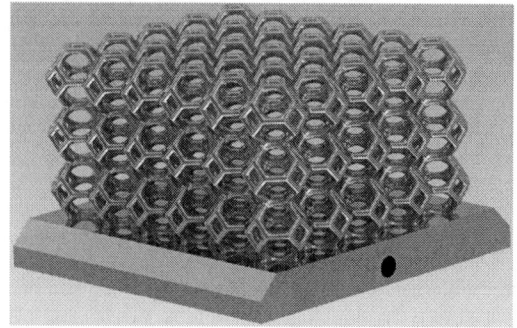

Fig. 10. Innovative 3D printed heatsink structure with truncated octahedron-shaped cooling elements [74].

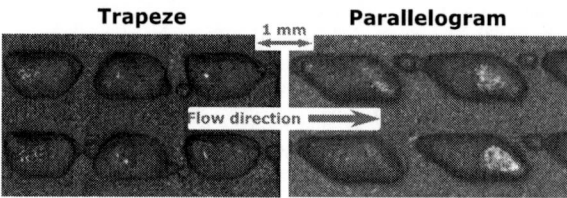

Fig. 11. Low-size heatsink fin profiles obtained by means of additive manufacturing [70].

involved in this field. Next, some related specific achievements are presented.

Additive manufacturing enabled not only the fabrication of heatsinks with very low fin wall sizes but also the processing of more complex channel patterns than the traditional straight-type ones, as can be seen in Fig. 7 [70], [71].

3D printing technologies also enabled the manufacturing of more complex heatsink fins with very good thermal transfer capabilities. In [72] a heatsink with airfoil-like fins processed by selective laser melting of AlSi10Mg metallic powder is presented (see Fig. 8a). A patented heatsink comprising of a matrix of complex twisted and tapered elliptical fin (shown in Fig. 8b) was presented in [48]. Due to the twisted fin shape, a flow pattern excellently mixing the cooling agent was achieved.

In Fig. 9 an additively manufactured innovative flexible and lightweight heat sink made of TCPoly thermal conductive filament is given [73].

A student competition prize winner heatsink structure is provided in Fig. 10b. The totally revolutionary lattice-like structure is composed of repeating truncated octahedron-shaped elements. Upon the performed analysis, it was proved to have very high heat transfer rate at low mass due to the good balance of the solid heat conduction material and the empty internal space between them [74].

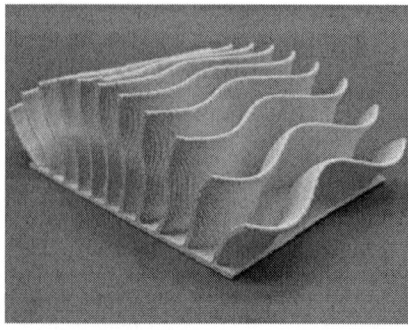

Fig. 9. Heatsink with flexible fins made via 3D printing of TCPoly thermal conductive filament [73].

The additive manufacturing technology (especially due to its very high processing accuracy) enabled the fabrication of small heatsinks with high cooling performances. The key was the integration into the cooling systems of very small-scale microchannels. A low stream of the cooling agent within these produces laminar flow. The obtained heat transfer coefficient is inversely proportional to the hydraulic diameter of the microchannel. Significant heat transfer coefficients can be achieved if small diameter microchannels can be processed [17].

Heatsinks having cooling fins as those given in Fig. 7 can be manufactured also at milli- and micrometer scale by using direct metal laser sintering technology, as can be seen in Fig. 11 [70].

Other achievements in this field were obtained by selective laser melting or cold spraying Al 6061 or AlSi10Mg [63] to manufacture pyramid or cube-shaped pin fins for heatsinks [75]. Upon other approaches, pure Al was used to cover an Al 6061 substrate via powder bed fusion [48], [76], or by means of fused depositing modeling [77],

C. Pipes and Heat Exchangers Manufactured by 3D Printing

The main requirements for an effective heat exchanger are the active cooling surface area, the pressure drop within, and the mass/volume of the entire cooling system. Additive manufacturing technologies can revolutionize the heat exchangers by using temperature, abrasion, and extreme chemical environment resistant materials (as for example ceramics) for the fabrication of functional- or structural-oriented complex structures.

In [51] a cooling pipe type heat exchanger with very great external cooling active surface (as it can be seen in Fig. 12) made of ceramics by using the lithography-based 3D printing technology was reported. The complex optimized thin branch-like structure cannot be manufactured with any traditional approach.

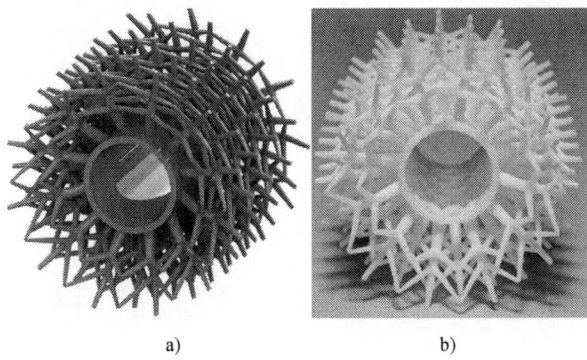

Fig. 12. Pipe-type heat exchanger fabricated through photopolymerization. a) 3D model; b) 3D printed prototype [51].

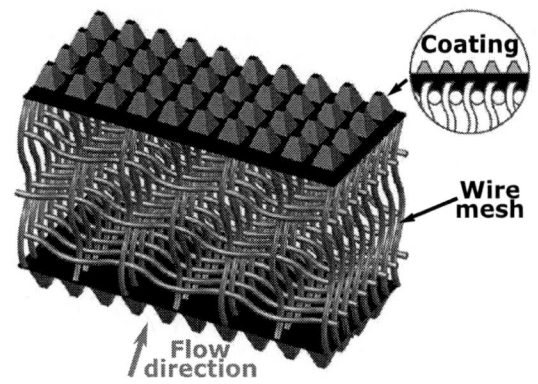

Fig. 13. Wire meshed heat exchanger 3D printed by using cold spray technology [63].

In plate- and tube-fin heat exchangers, cooling surfaces of various millimeter-sized fin shapes (planar, wavy, knitted, etc.) can be applied. Efficient cooling systems made from 3D arrangements of such finely processed surfaces can be constructed. Due to their full of twists and turns structures, they cannot be manufactured using traditional machining techniques [78], [79].

The 3D wire mesh heat exchangers have a complex, but the regular structure, resulting in low mass and density. An effective variant is detailed in [63], which has a knitted-like stacked wire-meshed innovative structure. Its bottom and top parts are closed by two surfaces comprising pyramidal fins, as can be seen in Fig. 13 [80].

Cold spray additive manufacturing enables the fabrication of forced cooled heat exchangers with a great variety of fin configurations. The Al-brazed plate-fin heat exchanger presented in [63] uses a separating wall dividing two fin array layers, which is splitting the two heat exchanging fluid streams, as it can be seen in Fig. 15, which results in a perfected heat transfer.

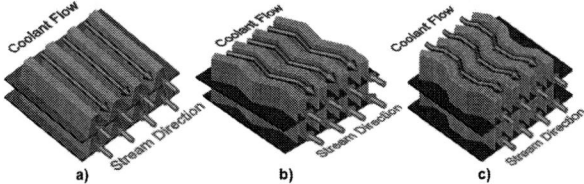

Fig. 15. Force cooled double-layer plate heat exchanger fin structure. a) plane; b) herringbone; c) offset strip-type [63].

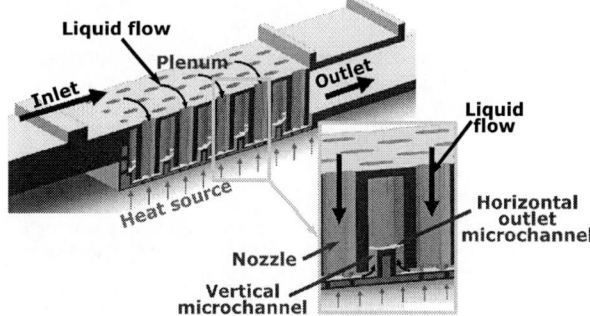

Fig. 14. The FINJET heatsink with microchannels. In the zoomed area the microchannel system can be easily seen [81].

One of the most exciting usages of microchannels in heatsinks developed for cooling power electronics devices is the patented FINJET cooling system, given in Fig. 14. This exploits the liquid jet impingement principle and well-illustrates the flexibility and effectiveness of the multi-metal stereolithography fabrication technology.

The liquid enters inside the heatsink in an inlet chamber (a plenum). Here the inlet stream practically stagnates due to the sufficiently large volume of the plenum. Therefore, in this area, the liquid has uniform pressure, which forces the liquid into the vertical channels, called jet nozzles.

At its bottom, the fluid is ejected in a way that impacts the hot base of the structure. The concentric walls of the vertical microchannel force the liquid to rise to a point where it is collected into a central horizontal outlet microchannel [81], [82].

The surfaces which are in contact with the cooling liquid are made of VALLOY-120, an erosion and corrosion resistant Ni-Co alloy, having 80% Ni and 20% Co. The base of the system is covered by Cu, acting as a good heat spreader. The microchannel-microjet FINJET cooling system even if it is only 1 mm high can handle heat fluxes over 1000 W/cm^2 [83].

VI. CONCLUSIONS

The ever-increasing demands for performance improvements in power electronics are driving the development of thermal management of these devices. The main research activities noted in this literature survey focus on improving key elements of thermal management systems, such as coolants, heat sinks, pipes, and heat exchangers.

Through its various effective techniques, additive manufacturing has proven to overcome the major limitations of conventional cooling systems in terms of heat transfer, geometry, mass, and overall efficiency. Furthermore, their use has expanded the imagination of engineers in developing unprecedented cooling system structures that are well-suited to both the consumer needs and to particular power electronic devices.

It can also be concluded that there are still some unresolved issues in the application of these new technologies. These still limit the mass production of 3D printed parts and promote future research in this emerging field.

REFERENCES

[1] Y. Shabany, *Heat Transfer: Thermal Management of Electronics.* Boca Raton (USA): CRC Press, 2009.

[2] A.S. White, D. Saltzman, S. Lynch, "Performance analysis of heat sinks designed for additive manufacturing," in *Proceedings of the International Electronic Packaging Technical Conference and Exhibition (InterPACK2020),* paper #V001T07A003, Anaheim (USA), 2020.

[3] S. Lee, "Optimum design and selection of heat sinks," *IEEE Transactions on Components, Packaging, and Manufacturing Technology: Part A,* vol. 18, no. 4, pp. 812-817, 1995.

[4] U.M. Dilberoglu, B. Gharehpapagh, U. Yaman, M. Dolen, "The role of additive manufacturing in the era of industry 4.0," *Procedia Manufacturing,* vol. 11, pp. 545-554, 2017.

[5] R. Jemghili, A.A. Taleb, K. Mansouri, "Additive manufacturing progress as a new industrial revolution," in *Proceedings of the 2nd International Conference on Electronics, Control, Optimization and Computer Science (ICECOCS '2020),* Kenitra (Morocco), 2020.

[6] Z. Zhang, X. Wang, Y. Yan, "A review of the state-of-the-art in electronic cooling," *e-Prime-Advances in Electrical Engineering, Electronics and Energy,* vol. 1, paper #100009, 2021.

[7] J.E. Hesselgreaves, R. Law, D. Reay, *Compact Heat Exchangers: Selection, Design and Operation.* Kidlington (UK): Butterworth-Heinemann, 2016.

[8] (2021, December 30). *Heat sink type-straight.* Available: https://commons.wikimedia.org/wiki/Category:Heat_sinks#/media/Fil e:Heat_sink_type-Straight.jpg

[9] (2021, December 30). *Needle fin heatsink.* Available: https://commons.wikimedia.org/wiki/Category:Heat_sinks#/media/Fil e:Moderate-Pin-Fin-Heatsink-Aluminium-.jpg

[10] M. Berman, "Techniques for cooling power and other electronic devices," *Embedded Technology Magazine,* no. 5, pp. 2-4, 2008.

[11] J.S. Lewis, "Reduction of device operating temperatures with graphene-filled thermal interface materials," *C - Journal of Carbon Research,* vol. 7, no. 3, paper #53, 2021.

[12] (2021, December 30). *Heatsink and Fan.* Available: https://cdn.sparkfun.com/assets/parts/5/3/9/6/10686-01.jpg

[13] (2021, December 30). *Heat pipe.* Available: https://upload.wikimedia.org/wikipedia/commons/f/f0/Heat_pipe.jpg

[14] (2022, January 2). *Heat pipes. The basics.* Available: http://www.heatpipe.nl/index.php?page=heatpipe&lang=EN

[15] High-Power Electronics Passive Cooling. *Power Electronics News (2021, March 7).* Available: https://www.powerelectronicsnews.com/high-power-electronics-passive-cooling/

[16] S.S. Murshed, C.N. De Castro, "A critical review of traditional and emerging techniques and fluids for electronics cooling," *Renewable and Sustainable Energy Reviews,* vol. 78, pp. 821-833, 2017.

[17] B.W. Williams, *Power Electronics: Devices, Drivers, Applications and Passive Components.* London (UK): Macmillan International Higher Education, 2006.

[18] H. Jouhara, N. Khordehgah, S. Almahmoud, B. Delpech, A. Chauhan, S.A. Tassou, "Waste heat recovery technologies and applications," *Thermal Science and Engineering Progress,* vol. 6, pp. 268-289, 2018.

[19] "Thermal solutions for power electronics," Advanced Cooling Technologies, Inc., Lancaster (USA), 2021. Available: https://www.psirep.com/sites/default/files/psi_u_document_files/powe relectronics-ebook-lowres.pdf

[20] C. Qian, A.M. Gheitaghy, J. Fan, H. Tang, B. Sun, H. Ye, *et al.,* "Thermal management on IGBT power electronic devices and modules," *IEEE Access,* vol. 6, pp. 12868-12884, 2018.

[21] M. Sanches, G. Marseglia, A.P. Ribeiro, A.L. Moreira, A.S. Moita, "Nanofluids characterization for spray cooling applications," *Symmetry,* vol. 13, no. 5, paper #788, 2021.

[22] K. Posobkiewicz, K. Górecki, "Influence of selected factors on thermal parameters of the components of forced cooling systems of electronic devices," *Electronics,* vol. 10, no. 3, paper #340, 2021.

[23] Y. Liu, X. Han, C. Shen, F. Yao, M. Zhang, "Experimental study on the evaporation and condensation heat transfer characteristics of a vapor chamber," *Energies,* vol. 12, no. 1, paper #11, 2019.

[24] J. Watson, K. Taminger, "A decision-support model for selecting additive manufacturing versus subtractive manufacturing based on energy consumption," *Journal of Cleaner Production,* vol. 176, pp. 1316-1322, 2018.

[25] S.M. Sohel Murshed, "Introductory Chapter: Electronics Cooling – An Overview," in *Electronics Cooling,* S. M. Sohel Murshed, Ed., London (UK): IntechOpen, 2016.

[26] M. Bahiraei, S. Heshmatian, "Electronics cooling with nanofluids: a critical review," *Energy Conversion and Management,* vol. 172, pp. 438-456, 2018.

[27] E.C. Okonkwo, I. Wole-Osho, I.W. Almanassra, Y.M. Abdullatif, T. Al-Ansari, "An updated review of nanofluids in various heat transfer devices," *Journal of Thermal Analysis and Calorimetry,* vol. 145, no. 6, pp. 2817-2872, 2021.

[28] L. Szabó, G.S. Szabó, R. Szabó, "Usage of Graphene in Power Systems. A Survey," in *Proceedings of 11th International Conference on Electrical Power Drive Systems (ICEPDS '2020),* Saint-Petersburg (Russia), 2020.

[29] C.K.A. Silva, A.K. Alves, "Graphene Application," in *Technological Applications of Nanomaterials,* A. Kopp Alves, Ed., Cham (Switzerland): Springer, 2022, pp. 195-206.

[30] R.A. Farade, N.I. Abdul Wahab, D.-E.A. Mansour, N.B. Azis, M.E.M. Soudagar, V. Siddappa, "Development of graphene oxide-based nonedible cottonseed nanofluids for power transformers," *Materials,* vol. 13, no. 11, paper #2569, 2020.

[31] Y. Fu, J. Hansson, Y. Liu, S. Chen, A. Zehri, M.K. Samani, *et al.,* "Graphene related materials for thermal management," *2D Materials,* vol. 7, no. 1, paper #012001, 2019.

[32] J. Gao, H. Wu, A. Li, Y. Yue, D. Xie, X. Zhang, "Graphene nanofluids as thermal management materials: molecular dynamics study on orientation and temperature effects," *ACS Applied Nano Materials,* vol. 2, no. 11, pp. 6828-6835, 2019.

[33] A. Esmaeilzadeh, M. Silakhori, N.N. Nik Ghazali, H.S.C. Metselaar, A. Bin Mamat, M.S. Naghavi Sanjani, *et al.,* "Thermal performance and numerical simulation of the 1-pyrene carboxylic-acid functionalized graphene nanofluids in a sintered wick heat pipe," *Energies,* vol. 13, no. 24, paper #6542, 2020.

[34] A. Arshad, M. Jabbal, Y. Yan, D. Reay, "A review on graphene based nanofluids: Preparation, characterization and applications," *Journal of Molecular Liquids,* vol. 279, pp. 444-484, 2019.

[35] F. Wu, A.M. El-Refaie, "Towards fully additively-manufactured permanent magnet synchronous machines: Opportunities and challenges," in *Proceedings of the International Electric Machines & Drives Conference (IEMDC '2019),* San Diego (USA), 2019, pp. 2225-2232.

[36] "ISO/ASTM 52900-15: Additive manufacturing - General principles - Terminology," ed. West Conshohockeas (USA): ASTM, 2015.

[37] E. MacDonald, R. Wicker, "Multiprocess 3D printing for increasing component functionality," *Science,* vol. 353, no. 6307, paper #aaf2093, 2016.

[38] H. Tiismus, A. Kallaste, T. Vaimann, A. Rassõlkin, A. Belahcen, "Electrical resistivity of additively manufactured silicon steel for electrical machine fabrication," in *Proceedings of the Electric Power Quality and Supply Reliability Conference (PQ '2019) & Symposium on Electrical Engineering and Mechatronics (SEEM '2019),* Kärdla (Estonia), 2019.

[39] B. Kocsis, I. Fekete, L. Varga, "Metallographic and magnetic analysis of direct laser sintered soft magnetic composites," *Journal of Magnetism and Magnetic Materials,* vol. 501, paper #166425, 2020.

[40] F. Calignano, "Additive manufacturing (AM) of metallic alloys," *Crystals,* vol. 10, no. 8, paper #704, 2020.

[41] K.V. Wong, A. Hernandez, "A review of additive manufacturing," *ISRN Mechanical Engineering,* vol. 2012, paper #208760, 2012.

[42] M. Pagac, J. Hajnys, Q.-P. Ma, L. Jancar, J. Jansa, P. Stefek, *et al.,* "A review of vat photopolymerization technology: Materials, applications, challenges, and future trends of 3D printing," *Polymers,* vol. 13, no. 4, paper #598, 2021.

[43] S. Zakeri, M. Vippola, E. Levänen, "A comprehensive review of the photopolymerization of ceramic resins used in stereolithography," *Additive Manufacturing,* vol. 35, paper #101177, 2020.

[44] J. Huang, Q. Qin, J. Wang, "A review of stereolithography: Processes and systems," *Processes,* vol. 8, no. 9, paper #1138, 2020.

978-1-6654-6787-2/22 $31.00 © 2022 IEEE

[45] P.K. Gokuldoss, S. Kolla, J. Eckert, "Additive manufacturing processes: Selective laser melting, electron beam melting and binder jetting—Selection guidelines," *Materials,* vol. 10, no. 6, paper #672, 2017.

[46] N. Urban, A. Kühl, M. Glauche, J. Franke, "Additive manufacturing of Neodymium-Iron-Boron permanent magnets," in *Proceedings of the 8th International Electric Drives Production Conference (EDPC '2018),* Schweinfurt (Germany), 2018.

[47] J.C. Najmon, S. Raeisi, A. Tovar, "Review of Additive Manufacturing Technologies and Applications in the Aerospace Industry," in *Additive Manufacturing for the Aerospace Industry,* F. Froes and R. Boyer, Eds., Amsterdam (Netherlands): Elsevier, 2019, pp. 7-31.

[48] B.M. Nafis, R. Whitt, A.-C. Iradukunda, D. Huitink, "Additive manufacturing for enhancing thermal dissipation in heat sink implementation: A review," *Heat Transfer Engineering,* vol. 42, no. 12, pp. 967-984, 2021.

[49] (2021, December 29). *3D printing is a hot topic: FACT Industries produces materials for heat sink.* Available: https://eitrawmaterials.eu/3d-printing-is-a-hot-topic-fact-industries-produces-materials-for-heat-sink-to-big-companies/

[50] W. Nawrot, K. Malecha, "Additive manufacturing revolution in ceramic microsystems," *Microelectronics International,* vol. 37, no. 2, pp. 79-85, 2020.

[51] U. Scheithauer, E. Schwarzer, T. Moritz, A. Michaelis, "Additive manufacturing of ceramic heat exchanger: opportunities and limits of the lithography-based ceramic manufacturing (LCM)," *Journal of Materials Engineering and Performance,* vol. 27, no. 1, pp. 14-20, 2018.

[52] N. Botter, Y. Avenas, J.M. Missiaen, D. Bouvard, R. Khazaka, "Thermal analysis of power module with double sided direct cooling using ceramic heat sinks," in *Proceedings of the 11th International Conference on Integrated Power Electronics Systems (CIPS '2020),* Berlin (Germany), 2020.

[53] C. Trento. (2022, January 4). *What Are the Ceramic Materials With High Thermal Conductivity?* Available: https://www.samaterials.com/content/what-are-the-ceramic-materials-with-high-thermal-conductivity.html

[54] N. Wang, Y. Liu, L. Ye, J. Li, "Highly thermal conductive and light-weight graphene-based heatsink," in *Proceedings of the 22nd European Microelectronics and Packaging Conference & Exhibition (EMPC '2019),* Pisa (Italy), 2019.

[55] L. James, J. Erbacher, "Graphene-based pipes for cooling power electronics," Mesago Messe Frankfurt GmbH, Frankfurt (Germany), 2021. Available: https://www.power-and-beyond.com/graphene-based-pipes-for-cooling-power-electronics-a-994351

[56] D.D.L. Chung, "Materials for thermal conduction," *Applied Thermal Engineering,* vol. 21, no. 16, pp. 1593-1605, 2001.

[57] D. Serebrennikov, A. Bykov, A. Trigub, N. Kolyshkin, A. Freydman, A. Aborkin, *et al.,* "Near zero thermal expansion in metal matrix composites based on intermediate valence systems: Al/SmB$_6$," *Results in Physics,* vol. 21, paper #103843, 2021.

[58] S.S. Sidhu, S. Kumar, A. Batish, "Metal matrix composites for thermal management: a review," *Critical Reviews in Solid State and Materials Sciences,* vol. 41, no. 2, pp. 132-157, 2016.

[59] M.M.A. Baig, S.F. Hassan, N. Saheb, F. Patel, "Metal matrix composite in heat sink application: reinforcement, processing, and properties," *Materials,* vol. 14, no. 21, paper 6257, 2021.

[60] K.K. Chawla, "Carbon Fiber/Carbon Matrix Composites," in *Composite Materials,* K. K. Chawla, Ed., Cham (Switzerland): Springer, 2019, pp. 297-311.

[61] T. Windhorst, G. Blount, "Carbon-carbon composites: a summary of recent developments and applications," *Materials & Design,* vol. 18, no. 1, pp. 11-15, 1997.

[62] F. García-Moreno, "Commercial applications of metal foams: Their properties and production," *Materials,* vol. 9, no. 2, paper #85, 2016.

[63] A. Farjam, Y. Cormier, P. Dupuis, B. Jodoin, A. Corbeil, "Influence of alumina addition to Aluminum fins for compact heat exchangers produced by cold spray additive manufacturing," *Journal of Thermal Spray Technology,* vol. 24, no. 7, pp. 1256-1268, 2015.

[64] (2022, January 4). *Metal Foam in Scanning Electron Microscope.* Available: https://commons.wikimedia.org/wiki/File:Image-Metal_Foam_in_Scanning_Electron_Microscope,_magnification_10x_b.GIF

[65] Y. Li, H. Zhou, C. Wu, Z. Yin, C. Liu, Y. Huang, *et al.,* "The interface and fabrication process of diamond/Cu composites with nanocoated diamond for heat sink applications," *Metals,* vol. 11, no. 2, paper #196, 2021.

[66] (2022, January 4). *TCPoly thermal conductive filaments for 3D printing.* Available: https://www.compositesworld.com/products/tcpoly-thermal-conductive-filaments-for-3d-printing-

[67] K. Timbs, M. Khatamifar, E. Antunes, W. Lin, "Experimental study on the heat dissipation performance of straight and oblique fin heat sinks made of thermal conductive composite polymers," *Thermal Science and Engineering Progress,* vol. 22, paper #100848, 2021.

[68] L.E. Paniagua-Guerra, B. Ramos-Alvarado, "Efficient hybrid microjet liquid cooled heat sinks made of photopolymer resin: thermo-fluid characteristics and entropy generation analysis," *International Journal of Heat and Mass Transfer,* vol. 146, paper #118844, 2020.

[69] C. Heinle, D. Drummer, "Potential of thermally conductive polymers for the cooling of mechatronic parts," *Physics Procedia,* vol. 5, pp. 735-744, 2010.

[70] S.S. Panse, S.V. Ekkad, "Forced convection cooling of additively manufactured single and double layer enhanced microchannels," *International Journal of Heat and Mass Transfer,* vol. 168, paper #120881, 2021.

[71] A. Datta, D. Sanyal, A. Agrawal, A.K. Das, "A review of liquid flow and heat transfer in microchannels with emphasis to electronic cooling," *Sādhanā,* vol. 44, no. 12, paper #234, 2019.

[72] J.Y. Ho, K.K. Wong, K.C. Leong, T.N. Wong, "Convective heat transfer performance of airfoil heat sinks fabricated by selective laser melting," *International Journal of Thermal Sciences,* vol. 114, pp. 213-228, 2017.

[73] (2022, January 5). *Thermal Management Applications.* Available: https://tcpoly.com/products/thermalmanagement/

[74] M. Johnson, S. Thilakaratne, S. Arora. (2022, January 4). *How to Design a Heat Sink for Additive Manufacturing.* Available: https://blogs.sw.siemens.com/simcenter/how-to-design-a-heat-sink-for-additive-manufacturing/

[75] T. Kang, Y. Ye, Y. Jia, Y. Kong, B. Jiao, "Enhanced thermal management of gan power amplifier electronics with micro-pin fin heat sinks," *Electronics,* vol. 9, no. 11, paper #1778, 2020.

[76] K.L. Kirsch, K.A. Thole, "Pressure loss and heat transfer performance for additively and conventionally manufactured pin fin arrays," *International Journal of Heat and Mass Transfer,* vol. 108, pp. 2502-2513, 2017.

[77] R.A. Felber, G. Nellis, N. Rudolph, "Design and modeling of 3D-printed air-cooled heat exchangers," in *Proceedings of the International Refrigeration and Air Conditioning Conference (IRACC '2016),* West Lafayette (USA), 2016.

[78] H. Fugmann, E. Laurenz, L. Schnabel, "Wire structure heat exchangers – New designs for efficient heat transfer," *Energies,* vol. 10, no. 9, paper #1341, 2017.

[79] C. Walter, S. Martens, C. Zander, C. Mehring, U. Nieken, "Heat transfer through wire cloth micro heat exchanger," *Energies,* vol. 13, no. 14, paper #3567, 2020.

[80] Y. Cormier, P. Dupuis, A. Farjam, A. Corbeil, B. Jodoin, "Additive manufacturing of pyramidal pin fins: height and fin density effects under forced convection," *International Journal of Heat and Mass Transfer,* vol. 75, pp. 235-244, 2014.

[81] R. Kempers, J. Colenbrander, W. Tan, R. Chen, A. Robinson, "Experimental characterization of a hybrid impinging microjet-microchannel heat sink fabricated using high-volume metal additive manufacturing," *International Journal of Thermofluids,* vol. 5, paper #100029, 2020.

[82] A. Robinson, W. Tan, R. Kempers, J. Colenbrander, N. Bushnell, R. Chen, "A new hybrid heat sink with impinging micro-jet arrays and microchannels fabricated using high volume additive manufacturing," in *Proceedings of the 33rd Thermal Measurement, Modeling & Management Symposium (SEMI-THERM '2017),* San Jose (USA), 2017, pp. 179-186.

[83] A. Robinson, R. Kempers, J. Colenbrander, N. Bushnell, R. Chen, "A single phase hybrid micro heat sink using impinging micro-jet arrays and microchannels," *Applied Thermal Engineering,* vol. 136, pp. 408-418, 2018.

978-1-6654-6787-2/22 $31.00 © 2022 IEEE

Equalization of Losses in a Multilevel Frequency Converter in Fault Condition

Yulia Kazemirova
Department of Electric Drives
Moscow Power Engineering Institute
Moscow, Russia
KazemirovaYK@yandex.ru

Alecksey Anuchin
Department of Electric Drives
Moscow Power Engineering Institute
Moscow, Russia
anuchin.alecksey@gmail.com

Dmitry Aliamkin
Department of Electric Drives
Moscow Power Engineering Institute
Moscow, Russia
aliam2002c@yandex.ru

Maxim Lashkevich
Department of Electric Drives
Moscow Power Engineering Institute
Moscow, Russia
MaxSoftPage@yandex.ru

Alexandr Zharkov
Department of Electric Drives
Moscow Power Engineering Institute
Moscow, Russia
zarckov@mail.ru

Alexey Kovyazin
Tchaikovsky Branch
Perm National Research Polytechnic University
Perm, Russia
kav3@etz-vektor.ru

Abstract—The cascaded multilevel frequency converter is one of the most used topologies of medium voltage drives. One of the main advantages of the topology is the ability to decommission any failed cell without interrupting the operation of the drive. The PWM technique with walking cell adapts to a different number of cells in the phase. However, in case of cell failure, the switching losses of each cell in the phase grow with respect to other healthy phases. This paper considers the strategy for equalization of switching losses of all cells of the frequency converter in faulty conduction. The proposed strategy was verified using a model in Matlab Simulink and losses were estimated in PSIM.

Keywords—medium voltage frequency converter, multi-level inverter, walking cell pulse-width modulation, faulty power cell, equalization of losses.

I. INTRODUCTION

Multilevel inverters (MI) is the industry standard topology for medium voltage (MV) high power drivers. They have several advantages, such as voltage and current wave-forms of superior quality with good harmonic composition (THD), the low losses in the switching devices due to low switching frequency. Cascaded multilevel frequency converter has series-connected low-voltage cells [1]. Output voltage is formed by summing the output voltages of each cell. The considered MV frequency converter contains 24 low-voltage cells combined into 3 chains of 8 low-voltage cells Fig. 1. Each single cell contains the active front end (AFE), DC link, full-bridge inverter Fig. 2 and bypass. The circuit includes the ability to decommission any failed cell by bypass circuit and continue to work with reduced power. The bypass allows shunt the failed cell from operation.

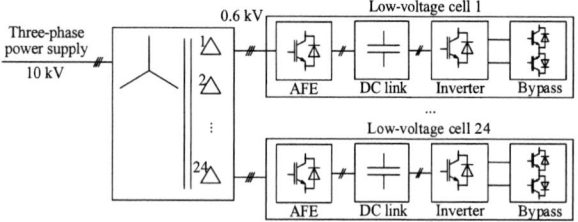

Fig. 1. Functional diagram of the cascaded multilevel frequency converter.

This research is supported by the Russian Science Foundation grant (Project №21-19-00696).

Fig. 2. Configuration of the inverter of a single cell.

The one of widely used PWM strategy in industrial applications for multilevel inverters is the phase-shift PWM (PSWM) that implements phase-shifting of the carrier signals of the cell [2]. The main disadvantage of this technique is normally slow response time. Also, the level-shift PWM (LSPWM) is applied for invrters control. It does not imply the equalization of power consumed from the cells [3]. This drawback can be solved by algorithms with cell rotation [4, 5] that bases on LPWM. This algorithms require the updating of carrier signals. Dynamically updating the carrier signals results in unpredictable distortions during the transition process.

The walking cell PWM strategy was developed for equal losses distribution in the steady-state [6]. This algorithm uses rotation of operation condition of the cells. Hoewer, it does not require the updating of carrier signals. Thus, operation condition of the cells is changing from zero, positive or negative DC link voltage each PWM cycle. In case of single cell failure there is no need to shift PWM carriers of the left cells and shift the voltage reference, because their voltages will be automatically redistributed by adjusting a little average commutation frequency of each cell. The walking cell PWM adapts for faulty cells in a phase and it was verified in Simulink Model [7]. A switching frequency of one cell depends on PWM frequency and the number of cells in a phase. So, cell failure leads to an increase switching losses of remaining cells in the phase.

The paper considers the modification of the walking cell PWM for switching losses minimization in the fault conduction as the continuation of papers [6, 7].

II. WALKING CELL PULSE-WIDTH MODULATION PRINCIPLES

The walking cell PWM technique uses rotation of cell's operation modes. Thus, the operating modes from PWM, +Vdc, and zero voltage of the cell are shifted by one each PWM cycle Fig. 3 the voltage reference varies slowly during several PWM cycles. Three cells in the phase are turned on applying a full positive voltage to the load and one operates with PWM reference.

The operation frequency of each cell equals to f_{PWM}/N, where N is the number of cells in the leg. Thus, the effective PWM frequency is low, while the response time is small, which can speedup the current loop.

The each PWM cycle, the number of cells operating with full voltage n is calculated, taking into account their current DC link voltages. And then, the duty cycle of the cell, which operates with PWM reference, is calculated based on the following equations:

$$\gamma_{pwm} = \left(v_{ref} - \sum_{i}^{N} v_{DCi} \right) / v_{DCpwm}, \qquad (1)$$

where γ_{pwm} is the duty cycle of the cell, which operates in PWM mode, i is the cell number, v_{ref} is the reference voltage of converter, V_{DC} is the DC link voltage and v_{DCpwm} is the DC link voltage of the cell which operates in PWM mode which is used to maintain the commanded output voltage [12].

III. THE STRATEGY FOR SWITCHING LOSSES EQUALIZATION

A. Walking Cell PWM Strategy in a Faulty Condition

The various PWM strategies allow decreasing switching losses in the inverter [8, 9]. A referenced voltage vector can be applied with clamping to the negative or positive bus bar of the DC link. The walking cell PWM forms zero voltage at the output of the cell in two ways. Zero output voltage is formed when bottom switches (Q_2, Q_4) are closed and top switches (Q_1, Q_3) are opened and vice versa. Thus, this method equally distributes switching losses and has a low response time.

The space-vector PWM improves the DC link voltage utilization [10]. In case of continuous SPWM the zero sequence is added to reference sinusoidal voltage (v_{aRef}, v_{bRef}, v_{cRef}). So, if a zero sequence is added or subtracted to all the values of phase potentials, then the values of phase-to-phase voltages do not change. The recalculated reference voltage (v'_{aRef}, v'_{bRef}, v'_{cRef}) are obtained by the following equations:

$$\left. \begin{array}{l} v_{min} = min\left(v_{aRef}, v_{bRef}, v_{cRef} \right); \\[4pt] v_{max} = max\left(v_{aRef}, v_{bRef}, v_{cRef} \right); \\[4pt] v_0 = -\dfrac{v_{max} - v_{min}}{2} - v_{min}; \\[4pt] v'_{aRef} = v_{aRef} + v_0; \\[4pt] v'_{bRef} = v_{bRef} + v_0; \\[4pt] v'_{cRef} = v_{cRef} + v_0. \end{array} \right\} \qquad (2)$$

The form of the obtained reference voltage is the same as in conventional SPWM.

If the cell in the phase is bypassed, balance output phase-to-phase voltages are required. The most common technique for balance phase-to-phase voltage is adding a fundamental zero-sequence component [11]. This method allows to maximize phase-to-phase voltage available in faulty conditions. However, it does not take into account losses distribution and it is based on phase-shift PWM. In [7] the operation of the walking cell in the fault conduction is considered. The principle of the algorithm is not changing, when one or several cells are bypassed. It removes one cell from operation and rotation mode of the cell remains between healthy cells (Fig. 4). The average PWM frequency grows in this case, which causes growth of the switching losses in the cells. This method requires a limited maximum of reference phase voltage by:

$$V_{max\_lim} = \frac{2}{\sqrt{3}} \sum_{i}^{N-n} v_{DCi}, \qquad (3)$$

where $i = 0,\ 1\ldots N$ is the number of healthy cells in the phase, n is the number of faulty cells in the phase, v_{DCi} is the voltage of the DC link of ith cell.

The technique of the walking cell PWM adapts to any different combination of faulty cells in any phases. But losses in healthy cells increase.

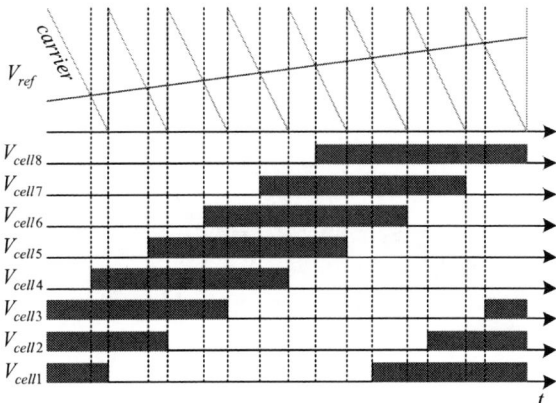

Fig. 3. Walking-cell PWM algorithm with trailing edge modulation.

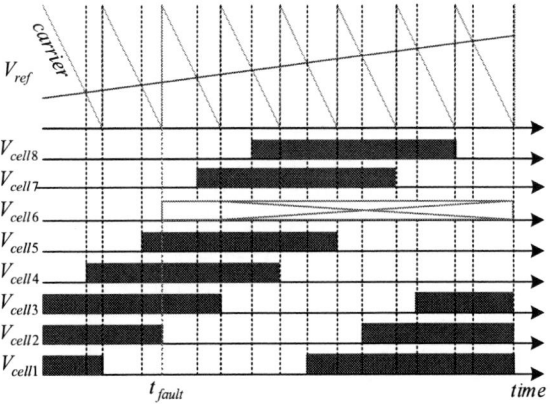

Fig. 4. Control algorithm with walking cells PWM and trailing edge modulation, when the fault of the cell occurs.

B. Strategy for Minimization of Switching Losses in Fault Conduction

The following modification of the PWM module is proposed to maintain constant switching frequency of all cells in the fault conduction. If a bypassed cell has no zero mode during the PWM period, then during this period all cells in the phase doesn't change their state (yellow squared in Fig. 5). Thus, the output phase voltage of the converter differs from the reference voltage. It leads to distortion phase-to-phase voltages. It is necessary to add difference between the referenced voltage and output voltage during the PWM cycle to the reference voltages of other phases. If a bypassed cell is in full positive or negative mode, then the reference voltage of other phases should be recalculated by the value of DC link voltage. If a bypassed cell is in PWM mode, then the reference voltage of other phases should be recalculated by the value of v_{cor}:

$$ v_{cor} = v'_{ref} - \sum_{i}^{N-n} v_{DCi} \tag{4} $$

Fig. 6 shows normal operation mode of the inverter (left side of the figure) and faulty mode (right side of the figure), when one cell in phase A is failed. The figure shows phase-to-phase voltages remain sinusoidal form.

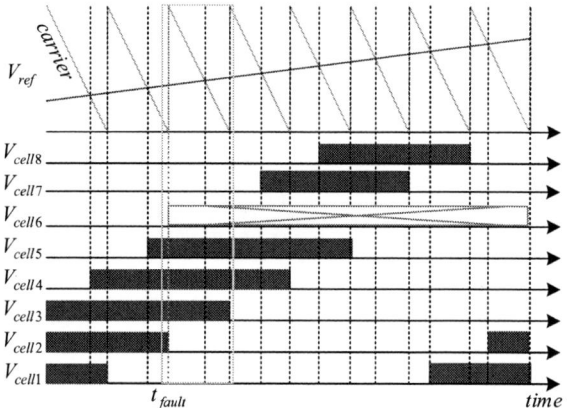

Fig. 5. Proposed control algorithm with walking cells PWM with strategy for minimization of switching losses and trailing edge modulation, when the fault of the cell occurs.

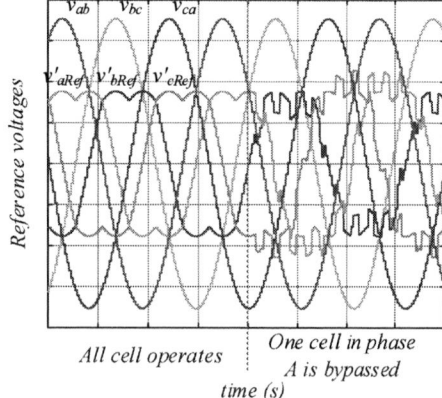

Fig. 6. Reference voltage of continuous SPWM when all cell operates and when one cell in phase A is failed.

This method has limitations. The correction value of reference voltage should not exceed the limited maximum V_{max_lim}.

IV. MODELING RESULTS

The simulation model was implemented in the MATLAB Simulink environment using SimPowerSystem. The model includes a three-phase synchronous motor with excitation from the permanent magnets, rated power of 8 MW, and rated speed equals to 3000 rpm. The motor is powered by the converter consisting of 8 low-voltage cells in each phase. The operation of the AFEs is not considered in this paper. The AFE maintains the voltage of the DC link equals to 1020 V. The control algorithm was written in C using the MATLAB Simulink S-Function blocks, which simplifies transferring the program code to the microcontroller for future embedding into the real control system.

The rated steady-state operating conditions of the converter were simulated for power of 4 MW. The research was conducted at less than the full power of the converter due to keeping reserve power. The PWM carrier frequency equals to 4 kHz. So, in normal steady-state operation of each cell switches equal to 500 Hz. At the instance of time 0 s the output voltages of each cell are shown when the strategy for losses distribution is not used. At the instance of time 0.3 s the cell number 2 is bypassed and the converter continues working. And in the time 0.35 s cells 4 and 7 are bypassed. The result of the simulation shows that switching frequency of the celles increase; therefore, switches losses increase too.

The inverter model was implemented in PSIM for losses estimation. PSIM is a simulation software specifically designed for power electronics and motor drives. The thermal module of PSIM allows the calculation of semiconductor devices losses. The control system remained unchanged in Matlab Simulink, but all power electronics part was modeled in PSIM. The inverter part is made of power modules 2MBI1400VXB-170E-50 Fuji Electric. Results of PSIM modulation are shown in Fig. 8.

Fig. 9 and Fig. 10 show operation of the converter with the strategy for minimization of switching losses. The losses are not changed in faulty condition. The phase voltages of the inverter are shown in Fig. 11. The phase-to-phase voltages are shown in Fig. 12, the balance of them is ensured in the faulty conduction.

V. CONCLUSIONS

In this paper the PWM strategy is investigated, which was developed to equalize loss distribution in the faulty conductions. The proposed algorithm was verified using MATLAB Simulink model and PSIM model. The results in PSIM show, that losses are not changed in fault conduction.

The algorithm adapts for different faulty cell numbers in a phase. This PWM technique is suitable for cascaded frequency converters with any number of low-voltage cells excluding overheating due to the increase in frequency.

978-1-6654-6787-2/22 $31.00 © 2022 IEEE

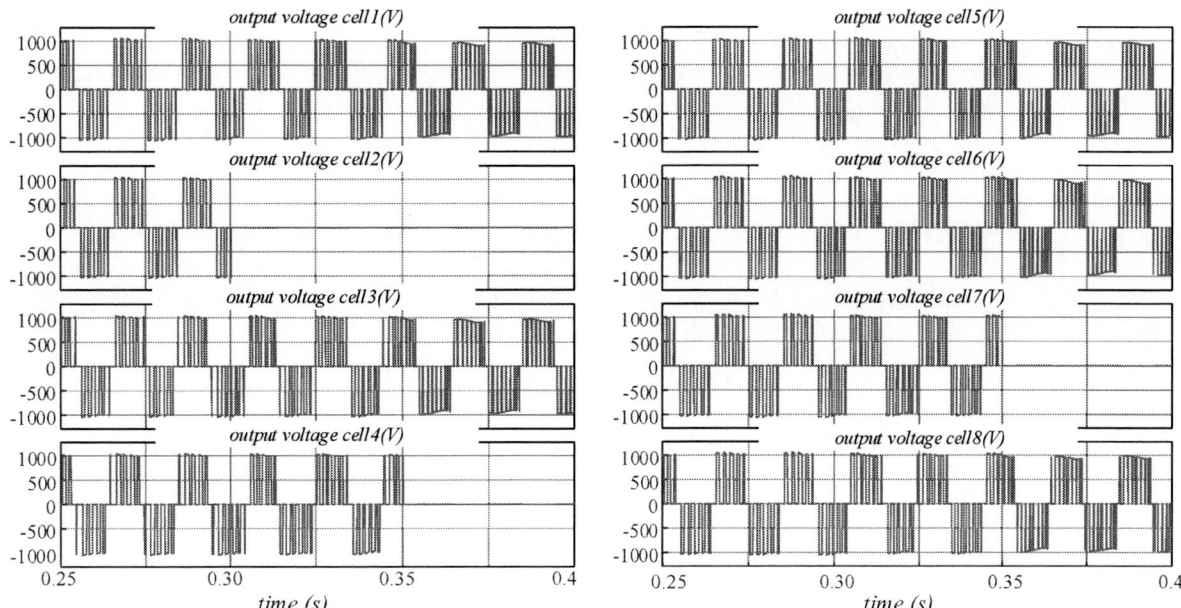

Fig. 7. Output voltages of each of eight cells in phase A without strategy for equalization of switching losses.

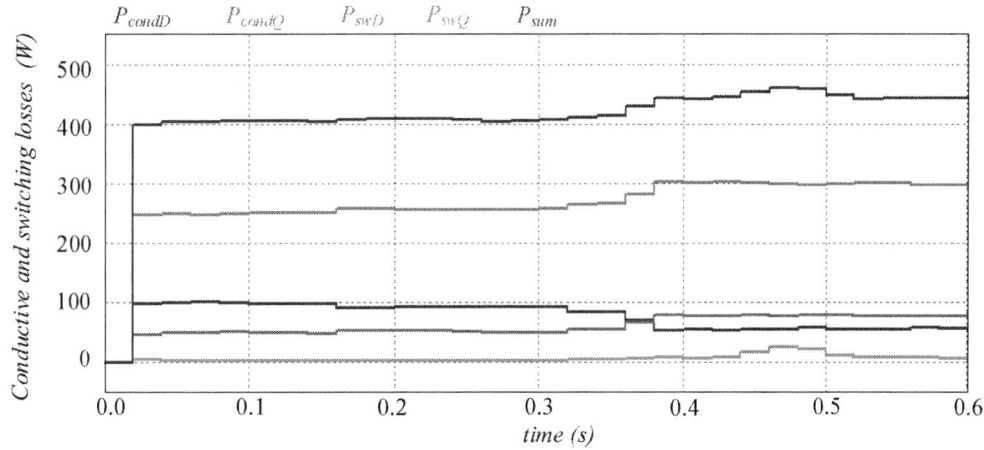

Fig. 8. Losses of switches and diodes of one IGBT module, when strategy for equalization of switching losses is not applied (P_{condD}—conduction losses of diodes, P_{condQ}—conduction losses of switches, P_{swD}—switching losses of diodes, P_{swQ}—switching losses of switches, P_{sum}—total module losses).

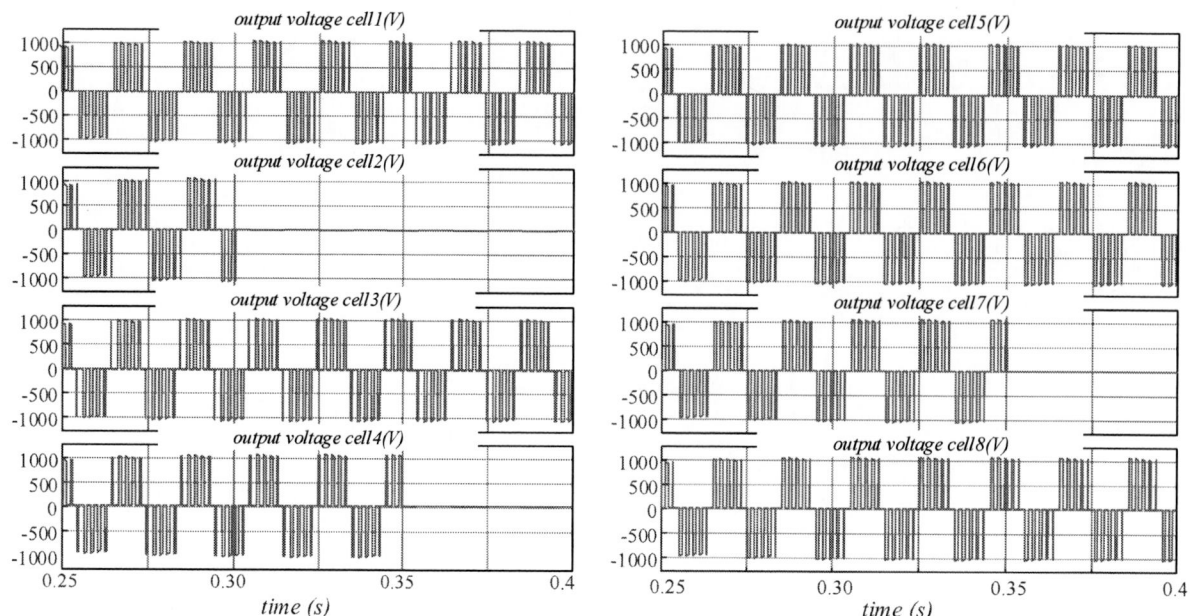

Fig. 9. Output voltages of each of eight cells in phase A with the strategy for equalzation of switching losses.

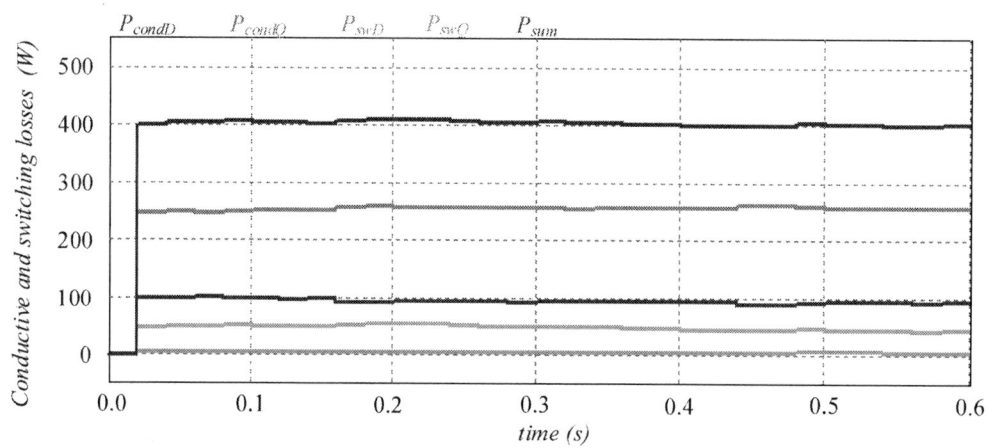

Fig. 10. Losses of switches and diods of one IGBT module, when strategy for equalization of switching losses is implemented (P_{condD}—conduction losses of diodes, P_{condQ}—conduction losses of switches, P_{swD}—switching losses of diodes, P_{swQ}—switching losses of switches, P_{sum}—total module losses).

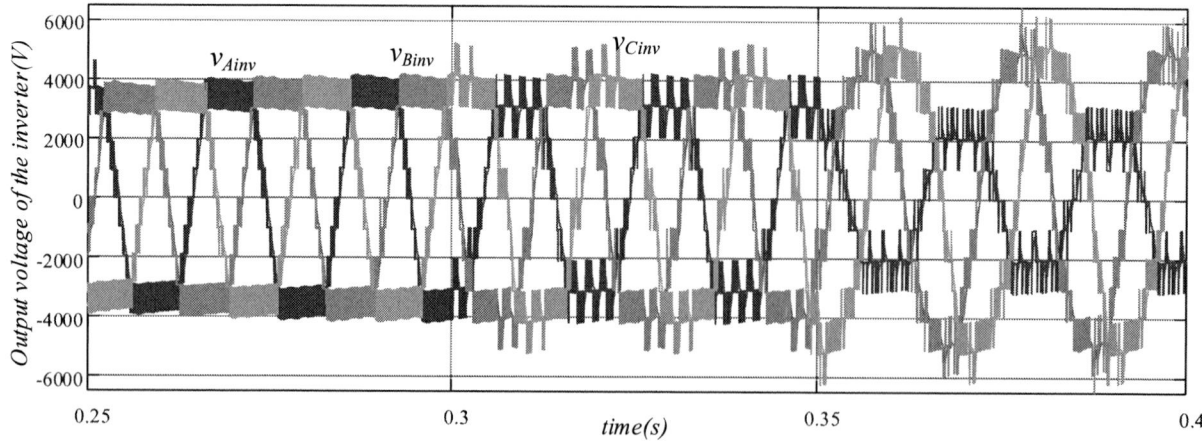

Fig. 11. Output phase voltage of the inverter.

978-1-6654-6787-2/22 $31.00 © 2022 IEEE

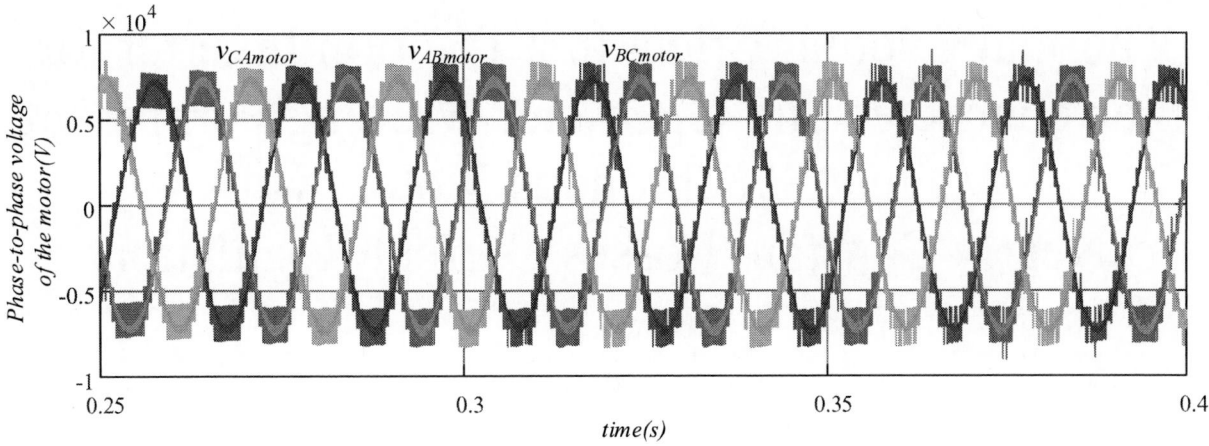

Fig. 12. Phase-to-phase voltage of the motor when cell #2 is bypassed at 0.3 s and cells #4 and 7 are bypassed at 0.35 s.

REFERENCES

[1] I. D. Pharne and Y. N. Bhosale, "A review on multilevel inverter topology," 2013 International Conference on Power, Energy and Control (ICPEC), Sri Rangalatchum Dindigul, 2013, pp. 700-703, doi: 10.1109/ICPEC.2013.6527746.

[2] S. Singh, A. Agnihotri, S. Bind and S. Kumar, "Performance Analysis of A New Carrier Rotation Method for Cascaded H-bridge Multilevel Inverter," *2020 IEEE First International Conference on Smart Technologies for Power, Energy and Control (STPEC)*, 2020, pp. 1-6, doi: 10.1109/STPEC49749.2020.9297729.

[3] T. Nondahl, J. Liu, Z. Cheng, N. Zargari, "Cascaded H-bridge (CHB) Inverter Level Shift PWM with Rotation" US patent 8,982,593 B2, Rockwell Automation Technologies, Inc., 2015.

[4] Y. Kazemirova, A. Anuchin, A. Kovyazin, M. Lashkevich, D. Aliamkin and S. Grishin, "PWM Strategy for Equal Distribution of Losses Between Low-Voltage Cells in an MV Frequency Converter," *2020 55th International Universities Power Engineering Conference (UPEC)*, 2020, pp. 1-6, doi: 10.1109/UPEC49904.2020.9209791.

[5] B. P. McGrath and D. G. Holmes, "A comparison of multicarrier PWM strategies for cascaded and neutral point clamped multilevel inverters," in 2000 IEEE 31st Annual Power Electronics Specialists Conference. Conference Proceedings (Cat. No.00CH37018), Galway, Ireland, 2000, vol. 2, pp. 674–679, doi: 10.1109/PESC.2000.879898.

[6] Y. Kazemirova, A. Anuchin, A. Zharkov, M. Lashkevich, A. Kovyazin and D. Savkin, "Analysis of Walking Cell PWM Strategy in Multilevel Frequency Converter in Fault Condition," *2021 9th International Conference on Modern Power Systems (MPS)*, 2021, pp. 1-5, doi: 10.1109/MPS52805.2021.9492601.

[7] B. P. McGrath and D. G. Holmes, "Multicarrier PWM strategies for multilevel inverters," in IEEE Transactions on Industrial Electronics, vol. 49, no. 4, pp. 858-867, Aug. 2002, doi: 10.1109/TIE.2002.801073.

[8] M. C. Di Piazza and M. Pucci, "Efficiency analysis in induction motor drives with discontinuous PWM and electrical loss minimization," 2014 International Conference on Electrical Machines (ICEM), 2014, pp. 736-743, doi: 10.1109/ICELMACH.2014.6960263.

[9] A. M. Bazzi and P. T. Krein, "Review of Methods for Real-Time Loss Minimization in Induction Machines," in IEEE Transactions on Industry Applications, vol. 46, no. 6, pp. 2319-2328, Nov.-Dec. 2010, doi: 10.1109/TIA.2010.2070475.

[10] H. W. van der Broeck, H. . -C. Skudelny and G. V. Stanke, "Analysis and realization of a pulsewidth modulator based on voltage space vectors," in *IEEE Transactions on Industry Applications*, vol. 24, no. 1, pp. 142-150, Jan.-Feb. 1988, doi: 10.1109/28.87265.

[11] L. Maharjan, T. Yamagishi, H. Akagi and J. Asakura, "Fault-Tolerant Operation of a Battery-Energy-Storage System Based on a Multilevel Cascade PWM Converter With Star Configuration," in IEEE Transactions on Power Electronics, vol. 25, no. 9, pp. 2386-2396, Sept. 2010, doi: 10.1109/TPEL.2010.2047407.

[12] S. Milovanović and D. Dujić, "Comprehensive Spectral Analysis of PWM Waveforms With Compensated DC-Link Oscillations," in IEEE Transactions on Power Electronics, vol. 35, no. 12, pp. 12898-12908, Dec. 2020, doi: 10.1109/TPEL.2020.2996418.

Thermal Cycling Effect in a Traction Inverter for Star-connected and Open-end Winding Permanent Magnet Synchronous Motors with Nearly Constant Losses Current Regulation

Yousef Ali
Department of Electric Drives
Moscow Power Engineering Institute
Moscow, Russia
joseph86.ali@gmail.com

Egor Kulik
Department of Electric Drives
Moscow Power Engineering Institute
Moscow, Russia
kulikegorka@gmail.com

Alecksey Anuchin
Department of Electric Drives
Moscow Power Engineering Institute
Moscow, Russia
anuchin.alecksey@gmail.com

Duy Hiep Do
Department of Automation and Control
Engineering
Thuyloi University
Hanoi, Vietnam
hiep.d.do@gmail.com

Abstract—One of the causes of faults in the electric drive is the failure of the power semiconductor module. Semiconductors degrade under the effect of thermal cycles. Thermal aging of materials occurs due to temperature difference between the elements of the power module as well as difference of their thermal expansion coefficients, which eventually leads to the destruction of the power module construction. The influence of thermal cycles is most expressed in traction drives, which are characterized by modes with low speeds and variation of commanded torque in time. This article considers a BMW i3 electric drive model and an open-end winding drive with Nearly Constant Losses Current Regulation for evaluating thermal cycles in a power module. The new European driving cycle was used as a test cycle for both drives. It is shown how the proposed algorithm makes it possible to reduce the thermocycling effects in semiconductor devices.

Keywords—open winding machine, thermocycling, electric vehicles, power electronics aging, active thermal control.

I. INTRODUCTION

Due to the great interest in electric vehicles over last decade, the propulsion system of the EV are the subject of research and optimization [1]. Electric motors are the hearts of EV. There are of different types such as interior permanent magnet synchronous machine [2], induction machine [3], synchronous DC-excited machine [4], synchronous homopolar machine [5], [6], and more. Having relatively high price the permanent magnet synchronous motors (PMSMs) have the best performance in EVs [7] because of small size and weight, wide speed range, high power density, and high efficiency[8]. However, conventional three-phase star-connected motor drives have poor fault-tolerant capability. Therefore, the open-end winding motor drives (see Fig. 1) are starting to be used more widely [9] and [10]. The open-end winding (OEW) motor drive topology is constructed by opening the neutral point of the conventional three-phase star-connected motor. Now the OEW motor has three independent phases which can be driven from dual inverter[11]. that give the system

extra degree of freedom, which can be utilized by the control system in the various ways.

Most researches related to open-end winding motor were focused on the torque quality improvement by elimination of the zero-sequence current (ZSC) with common dc bus, and fault tolerance control[12],[13]. Few researchers reviewed the lifetime of the power semiconducting and the problem of thermal cycling in traction applications. Power semiconductor modules consist of several layers of copper and substrate. Due to their different thermal expansion coefficients the temperature fluctuation leads to the aging and failure described in [14], [15]. To increase the lifetime of power semiconductors the amplitude and duration of thermal cycles have to be reduced.

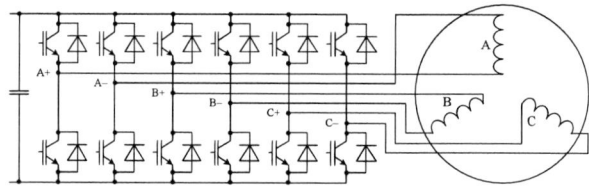

Fig. 1. Open-end winding motor fed from dual inverter.

This paper presents comparison of thermal cycling effect in a traction inverter for star-connected and open-end winding PMSMs with nearly constant losses (NCL) current regulation strategy which has been proposed in [16]. The model of the traction motor from BMWi3 electric vehicle was taken to make simulations. The method of NCL current regulation for OEW PMSM was proposed to minimizes variation of ohmic losses; hence, the temperature fluctuation of power modules. The simulation results show that the proposed method allows eliminating temperature drops in the power module and thereby increasing its lifetime.

This research is supported by RFBR (Project №20-38-90270)

978-1-6654-6787-2/22 $31.00 © 2022 IEEE

II. MODEL OF ELECTRIC VEHICLE

The mathematical model of electric vehicle is based on the mathematical description of mechanical, electrical, and energy parameters of its electric propulsion system.

The mathematical model of electric vehicle includes: model of PMSM (star-connected or open-end winding), three phase power inverter (three-phase inverter for star-connected motor and three-phase dual power inverter for open-end winding motor (Fig. 1), field-oriented control system of the PMSM, driving cycle profile (New European Driving Cycle (NEDC)) and vehicle mechanical model that provide load torque which is a function of speed and characteristics of the vehicle and road. The NEDC is a standard profile which is usually criticized for its unreal behaviour. More realistic dependances of the speed as a function of time can be obtained by experiments [17].

A. Dynamic Model of the Electric Vehicle

The various forces and the moving vehicle are presented on the following in Fig. 2.

In order to move the vehicle forward, the motor generates the driving force F_d acting on the wheels and while the vehicle is resisted by the rolling resistance force F_r, aerodynamic drag F_{ad} and gravity force F_g which are summarized as road load F_{rl}.

By using Newton's second law the equation of mechanics can be expressed as follows:

$$F_d - F_{rl} = m\frac{dv}{dt}. \quad (1)$$

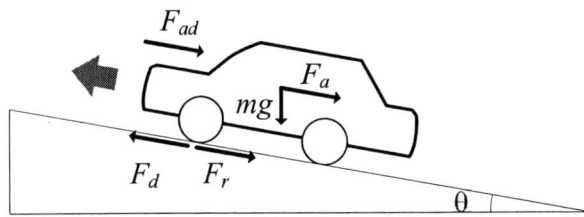

Fig. 2. Forces acting on a moving vehicle.

1) Rolling resistance force:

The rolling resistance force acts on the level of tires and is opposed to the free movement of the vehicle. It is caused by the deformation of the tires on the roadway which generates a rolling friction. It depends on the vehicle mass m, the acceleration of gravity g and the rolling resistance coefficient C_R:

$$F_r = mgC_R. \quad (2)$$

2) Gravitational force:

The gravitational force affects immediately the vehicle on the slope. It depends on the inclination of the slope θ and the vehicle mass m.

$$F_g = mg\sin(\theta) \quad (3)$$

3) Aerodynamic Force:

This is the air resistance force. It varies depending on the vehicle speed and depends on non-linear phenomena within the fluid mechanics. It is proportional to the voluminal density of the air ρ, to the front cross section S_f of the vehicle, the drag coefficient C_d of the air and vehicle speed V:

$$F_{ad} = \frac{\rho C_d S_f v^2}{2}. \quad (4)$$

B. Electric Drive Model and Control System

The control system structure is designed to handle with two types of PMSM (open-end winding or star-connected winding) for the motor with parameter of the one used in BMW i3 electric vehicle.

1) Star-connected motor drive

The control system is FOC, the inverter is two level three-phase one based on IGBT, and the electric motor is star-connected interior PMSM as shown in Fig. 3.

The input of the control system is the speed command ω_{ref} (NEDC), which defines the speed reference. The speed loop is controlled by speed controller (SC) and provides torque reference to the maximum torque per ampere (MTPA) block, which computes d- and q-axis current references at low speeds.

The current loop control is implemented by two current controllers for the d-, q- (CCd and CCq). The d- and q- axes voltage references are converted into the αβ stationary reference frames and then into three-phase voltage references. The current feedback is read from two phase current sensors due to the fact that sum of three currents in the star-connected motor is equal to zero. These currents are transformed into the αβ reference frames and then to the dq reference frames in order to provide the feedback to the current controllers.

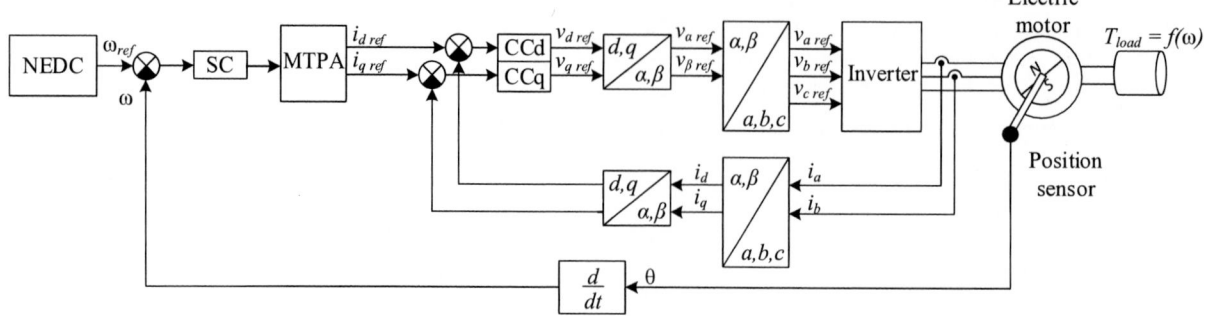

Fig. 3. The functional diagram of the control system structure for star-connected motor.

2) Open-end winding motor drive

In this control system structure, the control system is FOC, the inverter is two level three phase dual inverters based on SIC MOSFET. This inverter can be also considered to be three-level inverter due to number of voltage levels can be applied for a single phase. The parameters of the electric motor are the same, but the neutral point of the motor is opened (open-end winding PMSM) as shown in Fig. 4.

The open-end winding motor has independent phases and current in each phase can be regulated independently. This means that unlike star-connected machine the sum of currents is not equal to zero. The sum of the current is called zero sequence current. In open-end winding motor conventional sinewave current regulation can be used as well as nearly constant losses current regulation which were well described in[16].

The Nearly Constant Losses Current Regulation method is applicable only at low speed where the back-EMF is low and transition of the currents from positive to negative values is fast enough comparing to the entire revolution. When speed increases the conventional sinewave current regulation should be used. Therefore, the control system structure is designed to handle with the two current regulation methods and capability to switch between them.

The input of the control system is the speed command ω_{ref} taken form NEDC, which defines the speed reference.

The speed loop is controlled by speed controller (SC) which provides torque reference to the NCL algorithm block. Using the provided torque reference this block converts it into $i_{d\,ref}$, $i_{q\,ref}$ and $I_{\gamma\,ref}$ by using NCL current regulation method at low speeds. At high speeds the zero sequence current is set to zero.

The current loop control is implemented by three current controllers for the d-, q-, and zero-sequence components (CCd, CCq, and CCγ) of the motor currents. The d- and q-axes voltage references are converted into the αβ stationary reference frames and then into three-phase voltage references. Each of these references is applied to its open-end winding. The current feedback is read from each of three phase current sensors due to the fact that their sum is not equal to zero having additional zero sequence component. These three currents are transformed into the αβ reference frames and then to the dq reference frames in order to provide the feedback to the current controllers.

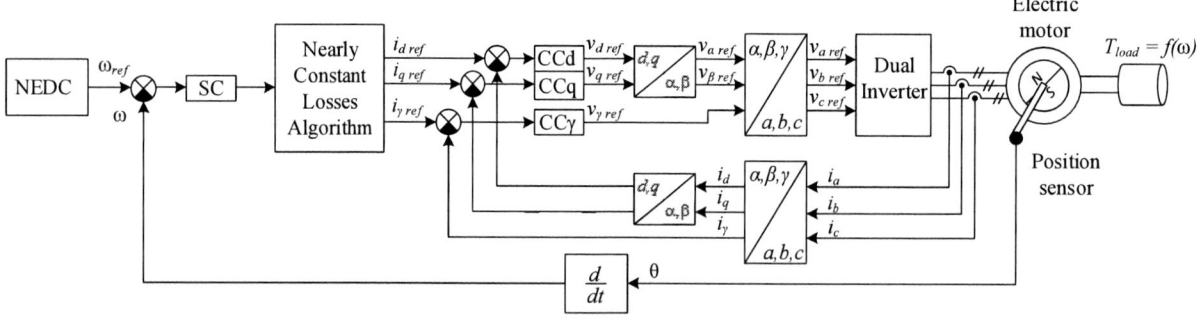

Fig. 4. The functional diagram of the control system structure for open-end winding motor.

C. Estimation of Semiconductor Thermal Cycle

A rainflow counting algorithm is used to analyze thermal cycles [18]. The method sequentially extracts thermal cycles in temperature from the module temperature graph. This calculation is carried out in the MATLAB environment.

The model for estimating the junction temperature is shown in Fig. 5. To estimate the losses, data from the

datasheet is included in the model of switching device. Data on the thermal impedance of the module is also provided by the manufacturer. Since water cooling is used, it is assumed that it does not have thermal inertia and the temperature of the module case is constant.

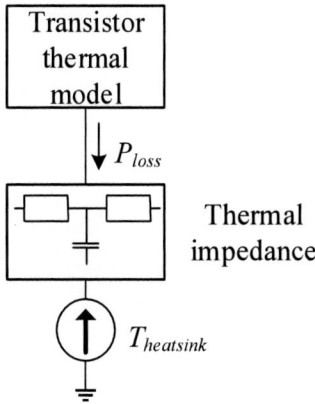

Fig. 5. Thermal model of the switching device.

III. SIMULATION RESULTS

The parameters of the vehicle and road load are listed in Table I and motor parameters are shown in Table II [19]. For purposes of the simulation and for reproducing a road section with different driving conditions New European Driving cycle (NEDC) is used, The NEDC consists of four repeated ECE-15 urban driving cycles (UDC) and one Extra-Urban driving cycle (EUDC). The simulation was implemented for a ECE-15 (first 195 s).

The simulation of the system was performed for a star-connected motor and an open-end winding motor.

The open-end winding motor is fed from a dual inverter made of six full SiC MOSFET power modules CAB760M12HM3 [20]. Unlike IGBT these switches have reverse conductivity, so, the current flows though the transistor most of the time and it heats the diode only during dead-time.

For star-connected motor the motor is fed from an inverter made of IGBT power module FS800R07A2E3 [21] where the thermal process will be different from SiC MOSFET because utilization of the freewheeling diode when current flows in the reverse direction.

The simulation model was made in PLECS simulation software. The system with star-connected motor and the system with open-end winding motor examined as well. Heatsink temperature was set constant equal to 50°C. The thermal model of the switch was implemented using partial-fraction model [22] comprising equivalent resistors and capacitors representing thermal resistance and thermal capacitances of the power module elements. The PWM frequency is 10 kHz for the both inverters.

TABLE I. Vehicle Parameters

Parameter	Value	Units
Vehicle mass	1600	kg
Rolling resistance coefficient CR	0.01	
Aerodynamic drag factor Cd	0.29	
Vehicle front section Sf	2.38	m^2
Slope of road	0.2	Rad
Radius of the wheels	0.35	m
Air density	1.225	Kg/m^3
Gear ratio	9.665:1	

TABLE II. Motor Parameters

Parameter	Value	Units
Stator resistor	5	mOhm
Ld stator inductance	0.0886	mH
Lq stator inductance	0.285	mH
Poles pairs	6	
Rated output power	75	kW
Maximum output power	125	kW
Rated speed	4500	rpm
Maximum speed	11400	rpm
Rated torque	88	Nm
Maximum torque	250	Nm
Maximum current	400	A
Motor mass	69	kg

The model evaluates the current waveforms of each phase, the power of the instant losses in the IGBT and SiC MOSFET, and the junction temperature of top switch of the phase A.

The results of simulation for star-connected motor with inverter based on IGBT are shown in Fig. 6. The detailed curves of the junction temperature, motor torque, vehicle speed and phase currents during start are shown in Fig. 7.

The results of simulation for open-winding motor with dual inverter based on SiC MOSFET are shown in Fig. 8 where the control system uses the nearly constant losses current regulation method at low current frequency and switches to conventional sinewave current regulation at higher speeds. The detailed curves of the junction temperature, motor torque, vehicle speed and phase currents are shown in Fig. 9.

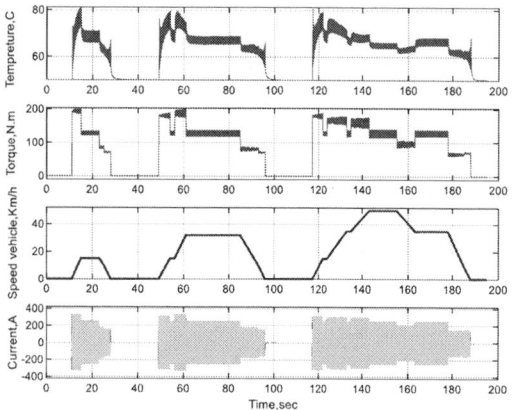

Fig. 6. The conventional sinewave current regulation in star-connected motor.

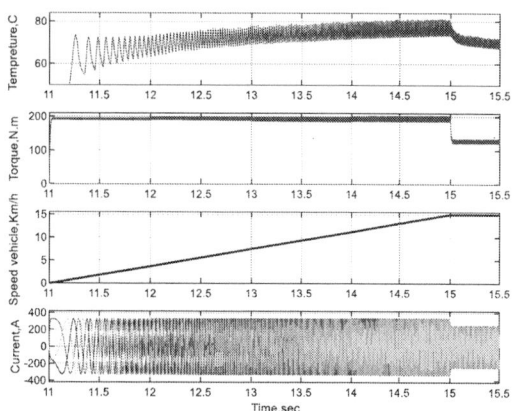

Fig. 7. The detail results of conventional sinewave current regulation in star-connected motor.

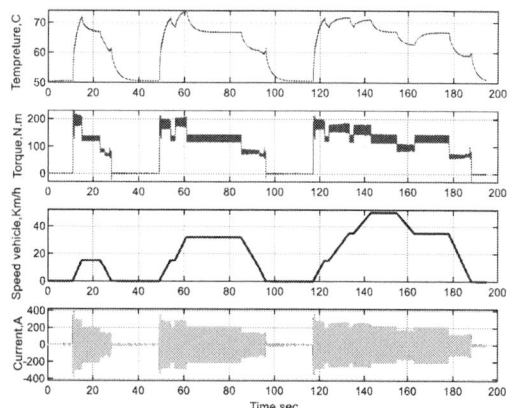

Fig. 8. NCL current regulation at low speed and conventional sinewave current regulation at high speed in open-end winding motor.

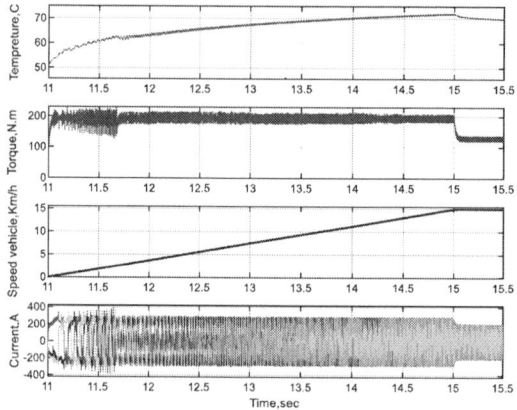

Fig. 9. The detail results of NCL current regulation at low speed and conventional sinewave current regulation in open-end winding motor.

By comparing the simulation results, the sinusoidal current regulation with star-connected winding has a big temperature deviation of the junction in power modules (IGBT). By applying the NCL current regulation at low speed and sinusoidal current regulation at high speed with open-end winding motor the deviation of the temperature is minimized.

The thermal cycles for star-connection drive are shown in Fig. 10. According to the manufacturer [18] temperature fluctuations of more than 10 degrees should be considered. There are many cycles from 10 to 40 degrees.

The thermal cycles for OEW drive are shown in Fig. 11. The crystal has no temperature drops of more than 10 degrees, apart from fluctuations caused by traffic.

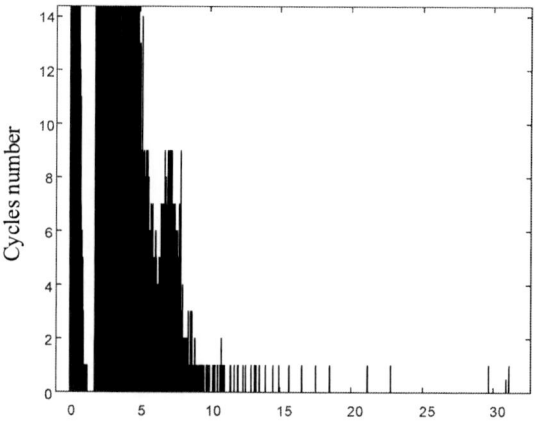

Fig. 10. The thermal cycles for star-connection drive.

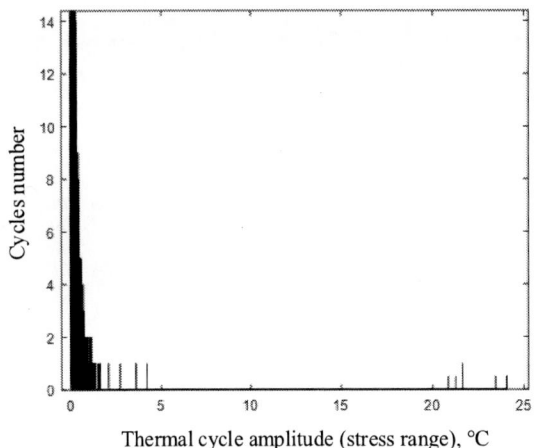

Fig. 11. The thermal cycles for OEW drive with NCL control.

IV. CONCLUSIONS

Simulation, on the example of a traction inverter, shows that the proposed method of current regulation with a open-end winding makes it possible to reduce unwanted thermal cycles in the power module. The examples of two types of traction inverters and motors were considered. The utilization of the nearly constant losses current regulation strategy helps to increase the service life of the power module and reduce the financial cost of maintenance. Further research will be devoted to the implementation of the algorithm on a real drive to test the influence of torque ripples.

REFERENCES

[1] B. O. Varga, A. Sagoian, and F. Mariasiu, "Prediction of electric vehicle range: A comprehensive review of current issues and challenges," Energies, vol. 12, no. 5, 2019, doi: 10.3390/en12050946.

[2] L. Jarzebowicz and S. Mirchevski, "Modeling the impact of rotor movement on non-linearity of motor currents waveforms in high-speed PMSM drives," 2017 19th European Conference on Power Electronics and Applications (EPE'17 ECCE Europe), 2017, pp. P.1-P.8, doi: 10.23919/EPE17ECCEEurope.2017.8098920.

[3] I. Nuca, V. Cazac, A. Turcanu and M. Burduniuc, "Development of Traction System with Six Phase Induction Motor for Urban Passenger Vehicle," 2020 International Conference and Exposition on Electrical And Power Engineering (EPE), 2020, pp. 749-754, doi: 10.1109/EPE50722.2020.9305555.

[4] I. Boldea, G. D. Andreescu, C. Rossi, A. Pilati and D. Casadei, "Active flux based motion-sensorless vector control of DC-excited synchronous machines," 2009 IEEE Energy Conversion Congress and Exposition, 2009, pp. 2496-2503, doi: 10.1109/ECCE.2009.5316509.

[5] M. Lashkevich, A. Anuchin, D. Aliamkin and F. Briz, "Control strategy for synchronous homopolar motor in traction applications," IECON 2017 - 43rd Annual Conference of the IEEE Industrial Electronics Society, 2017, pp. 6607-6611, doi: 10.1109/IECON.2017.8217153.

[6] V. Dmitrievskii, V. Prakht, A. Anuchin and V. Kazakbaev, "Traction Synchronous Homopolar Motor: Simplified Computation Technique and Experimental Validation," in IEEE Access, vol. 8, pp. 185112-185120, 2020, doi: 10.1109/ACCESS.2020.3029740.

[7] G. Liao, W. Zhang, and C. Cai, "Research on a PMSM control strategy for electric vehicles," Adv. Mech. Eng., vol. 13, no. 12, p. 16878140211051462, 2021.

[8] Y. K. Chin and J. Soulard, "A permanent magnet synchronous motor for traction applications of electric vehicles," in IEEE International Electric Machines and Drives Conference, 2003. IEMDC'03., 2003, vol. 2, pp. 1035–1041.

[9] Q. An, J. Liu, Z. Peng, L. Sun, and L. Sun, "Dual-space vector control of open-end winding permanent magnet synchronous motor drive fed by dual inverter," IEEE Trans. Power Electron., vol. 31, no. 12, pp. 8329–8342, 2016.

[10] B. A. Welchko, T. A. Lipo, T. M. Jahns, and S. E. Schulz, "Fault tolerant three-phase AC motor drive topologies: a comparison of features, cost, and limitations," IEEE Trans. power Electron., vol. 19, no. 4, pp. 1108–1116, 2004.

[11] X. Lin, W. Huang, W. Jiang, Y. Zhao, D. Dong, and X. Wu, "Direct torque control for three-phase open-end winding PMSM with common DC bus based on duty ratio modulation," IEEE Trans. Power Electron., vol. 35, no. 4, pp. 4216–4232, 2019.

[12] W. Hu, C. Ruan, H. Nian, and D. Sun, "Simplified modulation scheme for open-end winding PMSM system with common DC bus under open-phase fault based on circulating current suppression," IEEE Trans. Power Electron., vol. 35, no. 1, pp. 10–14, 2019.

[13] N. Hunter, T. Cox, P. Zanchetta, S. A. Odhano, and L. Rovere, "Torque quality improvement of an open-end winding PMSM," in 2018 IEEE Energy Conversion Congress and Exposition (ECCE), 2018, pp. 4254–4261.

[14] M. Andresen, K. Ma, G. Buticchi, J. Falck, F. Blaabjerg, and M. Liserre, "Junction Temperature Control for More Reliable Power Electronics," IEEE Trans. Power Electron., vol. 33, no. 1, pp. 765–776, 2018, doi: 10.1109/TPEL.2017.2665697.

[15] A. Anuchin, V. Ostrirov, Y. Prudnikova, M. Yakovenko, F. Briz, and M. Podlesny, "Thermal stabilization of power devices for compressor drive with start/stop operation mode," in 2016 57th International Scientific Conference on Power and Electrical Engineering of Riga Technical University (RTUCON), 2016, pp. 1–5.

[16] E. Kulik, Y. Ali, A. Chepiga, D. H. Do, F. Getmanenko, and A. Anuchin, "Current Regulation with Nearly Constant Losses for an Open-end Winding Traction IPM Motor Operating at Low Speeds," in 2021 28th International Workshop on Electric Drives: Improving Reliability of Electric Drives (IWED), 2021, pp. 1–6.

[17] E. Kulik, X. T. Tran and A. Anuchin, "Estimation of the requirements for hybrid electric powertrain based on analysis of vehicle trajectory using GPS and accelerometer data," 2018 25th International Workshop on Electric Drives: Optimization in Control of Electric Drives (IWED), 2018, pp. 1-5, doi: 10.1109/IWED.2018.8321394.

[18] AN2019-05, "Use of Power Cycling Curves for IGBT4", Infineon

[19] J. Hendershot and T. Burress, "Motoring Generating Simulation & Test Results for the Current BWM i3 Electric Vehicle Traction Machine," 2017.

[20] "Wolfspeed:", CAB760M12HM3 *datasheet*, 2020, [online] Available: www.wolfspeed.com.

[21] "Infineon:", FS800R07A2E3 *datasheet*, 2014, [online] Available: www.infineon.de.Infineon

[22] Infineon Technologies AG, "Transient Thermal Measurements and thermal equivalent circuit models (AN 2015-10)," Infineon, pp. 1–13, 2015.

978-1-6654-6787-2/22 $31.00 © 2022 IEEE

2022 29th International Workshop on Electric Drives: Advances in Power Electronics for Electric Drives (IWED), Moscow, Russia. Jan 26 – 29, 2022

Comparative Analysis of Disk Type Electrical Machines Designs

Flur Ismagilov
Department of Electromechanics
Ufa State Aviation Technical University
Ufa, Russia
ifr@ugatu.ac.ru

Vyacheslav Vavilov
Department of Electromechanics
Ufa State Aviation Technical University
Ufa, Russia
vavilov@niietkis.ru

Ildus Sayakhov
Department of Electromechanics
Ufa State Aviation Technical University
Ufa, Russia
isayakhov92@mail.ru

Alexey Zherebtsov
Department of Electromechanics
Ufa State Aviation Technical University
Ufa, Russia
zherebtsov@niietkis.ru

Razmik Dadoyan
Department of Electromechanics
Ufa State Aviation Technical University
Ufa, Russia
razmik.ad@mail.ru

Albina Nurieva
Department of Electromechanics
Ufa State Aviation Technical University
Ufa, Russia
a-nur_rabota@mail.ru

Abstract — **Modern electrical machines in the aviation industry are used in relatively new and non-standard applications. These new applications place significantly higher demands on electrical machines when it comes to power density and energy efficiency. Thus, electrical machines technologies for the aviation industry are aimed at increasing the power density and energy efficiency of each of the elements of the aircraft propulsion system. Disk type electrical machines with axial flux by permanent magnets are of interest for use in aircraft propulsion system with limited axial dimensions. In this work, numerical simulation of various designs of disk type electrical machines was carried out and the most effective design has been determined to ensure the maximum possible energy characteristics. It has been established that the weight of active parts of the ironless disk type electrical machine with the Halbach array has the smallest weight of active parts in comparison with other disk type electrical machine designs.**

Keywords — *aircraft, axial flux permanent magnet, electrical machines, comparative analysis*

I. INTRODUCTION

In the aerospace industry, reducing harmful emissions into the environment requires the introduction of innovative technologies in the field of all-electric and more electric aircraft. Currently, this is one of the main directions of development of the aviation industry throughout the world. This means that it is necessary to develop alternative propulsion systems, the efficiency of which will be much higher, while the weight will be lower than that of existing samples. As an alternative, it is possible to use a hybrid propulsion system (HPS) or an electric propulsion system (EPS) [1, 2].

In the HPS, the drive engines (internal combustion engines or gas turbine engines), effectively operating at the optimal mode, set in motion the shafts of the generators, which generate the necessary electricity. The generated energy is then fed to electric motors that drive the aircraft's main propulsion unit (propellers). This eliminates the mechanical connection between the power source and propellers, and also increases the efficiency of the entire installation [3,4].

In the EPS, rechargeable batteries (RB) feed the electric motors through their control system, which drive the aircraft propulsion units. The efficiency of such a propulsion system is much higher than the efficiency of the HPS. Currently, this type of EPS is used on ultralight aircraft with low takeoff weight. Modern aircraft have very limited overall dimensions in the places where electrical machines are installed, which imposes restrictions on the overall dimensions of the EMs themselves, especially with regard to the axial length. Therefore, when creating modern highly efficient electrical machines, it is necessary to rely on reducing the overall dimensions and increasing the power density[5, 6].

Disk type electric machines (EM) with axial flux by permanent magnets (AFPM) are of interest for use in aircraft propulsion system with limited axial dimensions. High electrical and magnetic loads make it possible to achieve high torque values in combination with small dimensions.

II. EM FOR AVIATION

Modern EMs in the aviation industry are being used in relatively new and non-standard applications: generators for HPS, motors that drive aircraft propellers, wheel motors for taxiing along the runway, and motors that provide vertical take-off and landing. These new applications place significantly higher demands on EMs when it comes to power density and energy efficiency, as both have a significant impact on HPS specific fuel consumption and EPS battery charge consumption. Thus, EM technologies for the aviation industry are aimed at increasing the power density and energy efficiency of each of the elements of the aircraft propulsion system [7, 8].

The general technology trend for EMs is that electrical machines over 100 kW are used in medium takeoff weight aircraft and large commercial aircraft. Electrical machines from 10 to 100 kW are mainly used for small and ultralight aircraft. Electrical machines with a power of less than 10 kW are mainly related to unmanned aerial vehicles and remote-controlled aircraft. EMs with power over 100 kW are mostly liquid-cooled, and EMs under 100 kW are air-cooled. In low-speed EMs, concentrated windings are most often used, and in high-speed machines, distributed windings are used to reduce higher spatial harmonics. All highly efficient EMs: radial EMs with internal and external rotors, as well as disk typeEMs, use permanent magnets (PM) [9, 10].

Disk type EMs in aviation are used as electric motors and generators of various prototypes of hybrid and electric propulsion systems of aircraft. The Launchpoint Technologies developed disk type electric machines for

This work was supported by Russian Science Foundation, project № 21-19-00454.

978-1-6654-6787-2/22 $31.00 © 2022 IEEE

hybrid propulsion systems of unmanned aerial vehicles, for driving landing gear wheels, and also for electric tail rotor drive for a Bell 206A/B helicopter (Fig. 1). These disc electric machines operate in the speed range from 2500 rpm to 12000 rpm and have a power density up to 8.2 kW/kg due to the design based on double rotors with Halbach array and an ironless stator. The disc type motors are air-cooled and create their own air flow, so they do not require an additional cooling system or fan [11].

The Emrax disk type motors (Fig. 2) have a design with two rotors and a yokeless and segmented armature, which allows developing high torque and high power over the entire speed range. These electric motors are used in propeller drives for electric gliders: Apis EA2 electric glider, Electric Taurus two-seat glider, Alexander Schleicher GmbH glider, DG Flugzeugbau GmbH glider, Binder Flugmotoren & Flugzeugbau GmbH glider. In addition, Emrax disc motors are installed on Eurosportaircraft aircraft and on the propulsion system of Efesto aircraft. These motors have a power density up to 9 kW/kg at power up to 380 kW with a speed range from 1840 rpm to 4000 rpm liquid-cooled, but also air-cooled [12]. Also, two Emrax motors of the same size can be combined with each other to increase power and torque.

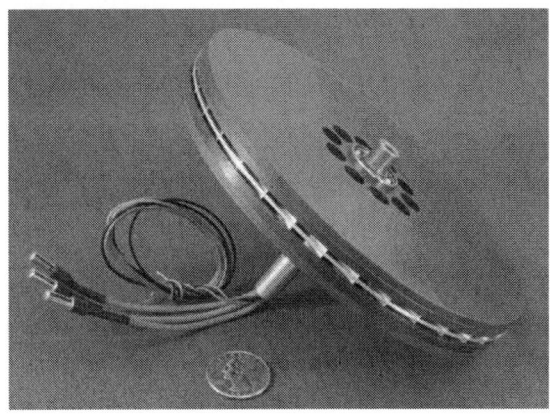

Fig. 1. Launchpoint Technologies disc type motor [11]

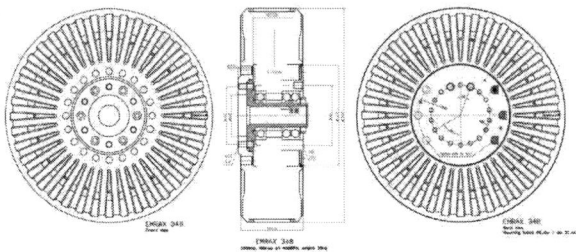

Fig. 2. Emrax disc type motor [12]

The YASA and Magnax also have a double rotor design with a yokeless and segmented armature stator [13, 14]. The wide interest of this designs is due to the simple technology of assembling the stator, and an efficient cooling system. Disc type motors YASA (Fig. 3) and Magnax (Fig. 4) are used for hybrid propulsion systems of aircraft. In these solutions, a disk type motors with a power density of up to 10 kW/kg at a power of up to 300 kW has been used to provide power generation functions on board, as well as an electric motor as part of the HPU system.

The presented disk type EMs are already widely used in the aviation industry. At the same time, they are commercial products with a serial release, and not separate research projects. As can be seen, in the aviation industry, disk type EMs with high specific characteristics are mainly used as a propeller drive and traction drive. At the same time, ironless disk type EMs have the greatest prospects due to their high power density and the possibility of creating effective air cooling.

Fig. 3. YASA disc type motor [13]

Fig. 4. Magnax disc type motor [14]

In systems where weight and dimensions are critical, EM with PMs is the best option for obtaining high power density using traditional materials. This makes it possible to build a low-sized machine, which allows the most rational use of the limited space in the installation place on aircraft. For example, for the propeller drive of the more electric aircraft, the propeller drive of the electric helicopter, the unmanned aerial vehicles propeller drives, the aircraft landing gear drives, there are restrictions on the axial length of the EM. The small axial length of disk type motors allows them to be integrated directly without the risk of increasing the dimensions of the aircraft.

III. STUDY OF EM WITH AXIAL MAGNETIC FLUX WITH PM

In the literature, the issues of forming the appearance of disk type EMs with excitation from PMs are not sufficiently covered. Therefore, there is a need to study electromagnetic processes in disk type EMs in various designs and determine the most effective versions that provide the maximum possible energy characteristics with minimum weight and dimensions.

Due to geometric features, disk type machines can be most fully calculated by computer simulation methods only in a three-dimensional formulation. Despite the high accuracy obtained with 3D finite element analysis, such simulations are too time consuming.

In connection with the above-mentioned difficulties of three-dimensional calculation, it is quite common to simulate a disk type EM in a two-dimensional (2D) formulation. The 2D simulation is performed on the average diameter of the stator and rotor, which is equivalent to a permanent magnet linear machine model. Modern computer simulation software packages, such as Ansys Electromagnetic Desktop, make it possible to perform disk type EM calculations in 2D without a significant reduction in accuracy compared to 3D.

In EMs with conventional slotted stators, iron losses often make up a significant portion of the total losses. In some PM machines, the stator teeth are shortened or eliminated altogether to reduce spatial harmonics and iron losses. More conductors can be placed in the space saved. This design is called slotless, since the winding is located in the air gap. Air gap windings are often made of licendrate (a bundle of very small strands of conductors twisted together) to reduce high frequency eddy current losses in the copper. Lower losses and higher electrical load allow higher power density.

Another design of EM with PMs is the Halbach magnetic array of PMs on the rotor. The Halbach array is a configuration in which some magnets are magnetized not only radially but also tangentially. This configuration increases the overall MMF in the magnetic flux circuit. The Halbach array can be used to keep the magnetic flux at the desired value, even with an increase in the equivalent air gap. With magnets magnetized in different directions, the iron yoke in the rotor can also be eliminated for saving weight. Designs of disk type EM with PMs and axial magnetic flux have advantages such as better utilization of the core material and suitability for designs with a large number of poles.

To determine the version with the smallest mass, five designs of disk type EM designs has been considered (Fig. 5). Electromagnetic calculations of designs of disk type EM has been carried out by the finite element method in the Ansys Electronics Desktop software package. Disk type EMs has been designed for an output power of 100 kW, a rated speed of 3200 rpm. The overall dimensions, such as the outer and inner diameters of the considered disk type EMs, remained the same.

Table I shows the results of electromagnetic calculations of disk type EMs structures at the generator mode. Simulation parameters: to speed up numerical calculations, the assumption was made that the disk type EMs are represented as equivalent linear EMs with PM. The attached disk type EMs operate in the generator mode with an active load.

One-sided designs of disk type EMs (Fig. 5) has been presented in options *a* and *b* are distinguished by the magnetic system of the rotor. In design *a*, the rotor consists of a standard alternating pole magnet system, and design *b* consists of a Halbach magnetic assembly. The height of permanent magnets in design *b* is 42.5% higher than in design *a*. This is due to the need to obtain high values of magnetic flux in the air gap.

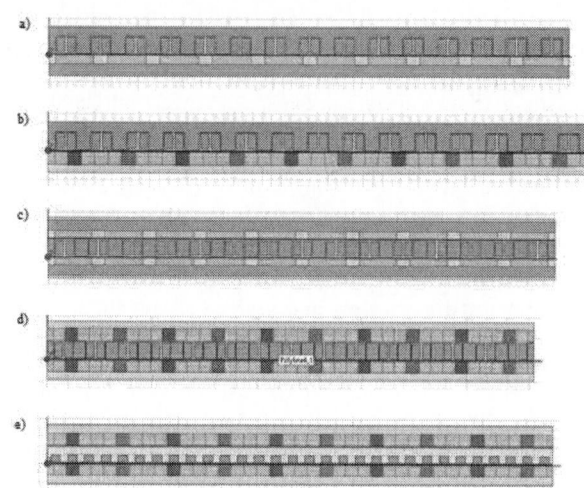

Fig. 5. Disk type designs: a) with one rotor and one stator; b) with one rotor with Halbach array and one stator; c) with two rotors and a yokeless and segmented stator core; d) with two rotors with Halbach assemblies and a yokeless and segmented stator core stator core; e) with two rotors with Halbach assemblies and without a stator core

TABLE I.   SUMMARY OF CONSIDERED GENERATORS

Design No.	No. 1	No. 2	No. 3	No. 4	No. 5
Output power, kW	100				
Frequency, Hz	270				
Line voltage, V	460				
Phase current, A	146	146	147	147	147
Number of poles/stator slots	10/15				10/30
Winding type	Concentrated				Distributed
Winding factor	0.866	0.866	0.866	0.866	1
Current density, A/mm²	7.3	7.4	7.4	7.4	7.5
Electrical load, A/m	48210	48990	48990	48990	47700
Air gap magnetic flux density, T	0.7	0.7	0.7	0.7	0.6
General losses, W	2170	2210	1700	1700	1660
Efficiency, %	97	97	97	97	98
Stator core material	FeCo Steel (49K2FA), sheet thickness 0.2 mm				–
Permanent magnet material	SmCo, residual induction 1.051 T, coercive force 740 kA/m				
Weight of the active part, kg	74	72	63.3	61.4	48.1
Rotor outer diameter, mm	450				
Rotor inner diameter, mm	290				
Stator outer diameter, mm	450				

Design No.	No. 1	No. 2	No. 3	No. 4	No. 5
Stator inner diameter, mm			290		
Axial length, mm	107	90.5	122	99.8	94

In design options *c*, *d* and *e*, double rotors of disk type EMs make it possible to obtain output characteristics similar to single-sided designs *a* and *b* with a lower weight of active parts, but by increasing the height of the PMs, which is necessary to obtain high values of magnetic flux in the air gap.

In addition, in designs with magnetic Halbach assemblies, inductances along the longitudinal and transverse axes are reduced compared to designs with a steel cores. Which is due to a decrease in the magnetic conductivity of the rotor magnetic circuit due to the absence of the rotor core.

Also use of Halbach magnetic assemblies makes it possible to reduce the total weight of the active part of disk type EMs by an average of 2.7 to 3%. In this case, such a slight decrease in weight is due to the fact that the mass density of the permanent magnet material made of *SmCo* is 8300 kg/m^3, and the mass density of the rotor yoke steel material is 7800 kg/m^3. In the Halbach magnetic assembly, the PMs occupy the entire volume of the disk type EM rotor and have an increased height compared to standard magnetic systems. These factors cause such a small decrease in the active weight of EM. Based on this, the use of the PM NdFeB material will make it possible to more noticeably reduce the active weight of the disk type EM due to the fact that the mass density of this material is 7600 kg/m^3.

From the results presented in Table I, the design options for disk type EMs produce a power of 100 kW with the same overall dimensions of the active parts. At the same time, structure *e* has the smallest weight of active parts of 48.1 kg, which is 35% less than the mass of structure *a* by 33%, structure *b* − 33%, structure *c* − 24%, structure *d* − 21%.

Conclusion

Numerical modeling of various designs of disk type EMs has been carried out and the most efficient options have been determined to ensure the maximum possible energy characteristics.

In the presented simulation of disk type EMs, the models has been considered in steady-state operation at 3200 rpm and output power of 100 kW with active load in generator mode. In order to maintain the equality of electrical and magnetic loads in the presented models with ironless rotors, the PM height has been increased, as can be seen from the results and conclusions. The generator output load was also selected so that the phase currents were the same in all models.

It has been established that the mass of active parts of the ironless disk type EM with the Halbach assembly has the smallest weight of active parts in comparison with other disk type EM designs. This allows us to conclude that ironless disk type EMs with magnetic Halbach assembly have the highest power density for the same dimensions and electromagnetic loads.

References

[1] C. Friedrich and P.A. Robertson, "Hybrid-Electric Propulsion for Aircraft", Journal of Aircraft, vol. 52, no. 1, pp. 176-189, 2015.

[2] A. M. El-Refaie, "Motors/generators for traction/propulsion applications: A review," in IEEE Vehicular Technology Magazine, vol. 8, no. 1, pp. 90-99, March 2013, doi: 10.1109/MVT.2012.2218438.

[3] S. Wu, L. Song and S. Cui, "Study on Improving the Performance of Permanent Magnet Wheel Motor for the Electric Vehicle Application," in *IEEE Transactions on Magnetics*, vol. 43, no. 1, pp. 438-442, Jan. 2007, doi: 10.1109/TMAG.2006.887705.

[4] W. Fei, P. C. K. Luk and K. Jinupun, "A new axial flux permanent magnet Segmented-Armature-Torus machine for in-wheel direct drive applications," 2008 IEEE Power Electronics Specialists Conference, 2008, pp. 2197-2202, doi: 10.1109/PESC.2008.4592268.

[5] P. B. Reddy, A. M. El-Refaie, K. Huh, J. K. Tangudu and T. M. Jahns, "Comparison of Interior and Surface PM Machines Equipped With Fractional-Slot Concentrated Windings for Hybrid Traction Applications," in IEEE Transactions on Energy Conversion, vol. 27, no. 3, pp. 593-602, Sept. 2012, doi: 10.1109/TEC.2012.2195316.

[6] W. Fei and P. C. K. Luk, "Design and performance analysis of a high-speed air-cored axial-flux permanent-magnet generator with circular magnets and coils," *2009 IEEE International Electric Machines and Drives Conference*, 2009, pp. 1617-1624, doi: 10.1109/IEMDC.2009.5075420.

[7] F. R. Ismagilov, A. A. Zherebtsov, V. E. Vavilov and I. F. Sayakhov, "Design and experimental investigation of BLDC motor for aircraft electromechanical actuator," in International Review of Aerospace Engineering, vol. 13, no. 1, pp. 10-15, 2020.

[8] F. R. Ismagilov, V. E. Vavilov and I. F. Sayakhov, "Analysis of Performance of Disc-Type High-Speed Generators Design with PMs," 2018 International Conference on Industrial Engineering, Applications and Manufacturing (ICIEAM), 2018, pp. 1-6, doi: 10.1109/ICIEAM.2018.8728891.

[9] F. R. Ismagilov, V. E. Vavilov, I. F. Sayakhov and N. Uzhegov "Review of the application of composite materials in electrical machines International," in International Review of Electrical Engineering, vol. 15, no. 1, pp. 31-40, 2020

[10] I. Sayakhov and G. Zinatullina, "Testing of High-Torque Electric Motor For Gearless Electromechanical Actuator," 2019 International Conference on Electrotechnical Complexes and Systems (ICOECS), 2019, pp. 1-4, doi: 10.1109/ICOECS46375.2019.8949987.

[11] LaunchPoint. Electric Vehicle Propulsion. Accessed: Jan. 24, 2022. [Online]. Available: https://www.launchpnt.com/portfolio/transportation/electric-vehicle-propulsion.

[12] EMRAX electric motors / generators. Accessed: Jan. 24, 2022. [Online]. Available: https://emrax.com/e-motors/emrax-348/

[13] YASA P400 R Series. Accessed: Jan. 24, 2022. [Online]. Available: https://www.yasa.com/products/yasa-p400/.

[14] Traxial – Yokeless Axial Flux Motor Technology. Accessed: Jan. 24, 2022. [Online]. Available: https://www.traxial.com/technology/.

AUTHOR INDEX

Aarniovuori, Lassi 55
Alassaf, Omar 79, 91
Aleksandr, Lukin 79
Ali, Yousef 122
Aliamkin, Dmitry 116
Andreevich, Nakataev Anton 61
Anuchin, Alecksey 3, 10, 116, 122
Asad, Bilal 20, 49, 85
Autsou, Siarhei 49, 85
Bayda, Sergey 73
Bolam, Robert Cameron 10
Bondarenko, Daria 27
Chervinsky, Mikhail 5
Dadoyan, Razmik 128
Das, Moumita 2
Demidova, Galina 79, 91
Dianov, Anton 67
Do, Duy Hiep 122
Florentsev, Stanislav 73
Ghahfarokhi, Payam Shams 32
Husev, Oleksandr 4
Hyunh, Van Khang 49
Ismagilov, Flur 128
Iurii, Plotnikov 102
Jassim, Haider M. 42
Kallaste, Ants 20, 32, 85
Kartashova, Margarita V 36
Kazemirova, Yulia 116
Khalghani, Shahab 55
Korunets, Alina 6
Kovyazin, Alexey 116
Kozlov, Gleb 91
Krasovsky, Alexander 96
Kudelina, Karolina 49, 85
Kulik, Egor 122
Lashkevich, Maxim 116
Lukichev, Dmitry 79
Lukin, Aleksandr 91
Moiseenkova, Elizaveta D. 36
Naseer, Muhammad Usman 20
Nurieva, Albina 128
Osipov, Oleg 27
Poliakov, Nikolai A. 36
Poliakov, Nikolai 91
Polyuschenkov, Igor 14
Pyrhönen, Juha 55
Rassõlkin, Anton 20, 49, 79, 85
Rassudov, Lev 6

Roque, Jhon Paul C. 10
Sarap, Martin 32
Sayakhov, Ildus 128
Szabó, Loránd 108
Tecle, Samuel Isaac 61
Tiismus, Hans 32
Vagapov, Yuriy 10
Vaimann, Toomas 20, 32, 49, 79, 85
Vasic, Miroslav 1
Vavilov, Vyacheslav 128
Vladimir, Polyakov 102
Volkhontsev, Andrey 91
Voronin, Igor 27
Voronin, Pavel 27
Vostorgina, Elena 96
Zharkov, Alexandr 116
Zherebtsov, Alexey 128
Zhurov, Igor 73
Ziuzev, Anatolii 61
Ziuzev, Anatoliy 42

IEEE
445 Hoes Lane
Piscataway, NJ 08854-4141

ISBN 978-1-6654-6787-2